Building Construction Technology
SI Metric

Building Construction Technology

SI Metric

Kenneth F. Cannon
Vocational Instructor
Department of Indian Affairs and Northern Development
Ottawa, Ontario

Frederick G. Hatley
Carpentry and Drafting Instructor
Carlton Comprehensive High School
Prince Albert, Saskatchewan

McGraw-Hill Ryerson Limited

Toronto   Montréal   New York   St. Louis   San Francisco
Auckland   Bogotá   Guatemala   Hamburg   Johannesburg
Lisbon   London   Madrid   Mexico   New Delhi   Panama
Paris   San Juan   São Paulo   Singapore   Sydney   Tokyo

Building Construction Technology

Copyright © McGraw-Hill Ryerson Limited, 1982

ISBN 0-07-548036-0

23456789   D   1098765432

Printed and bound in Canada

Canadian Cataloguing in Publication Data

Cannon, Kenneth F., date
   Building construction technology

Includes index.
ISBN 0-07-548036-0

1. House construction.   I. Hatley, Frederick G.
II. Title.

TH4811.C36   690'.8   C81-094831-1

Cover photos by Warren MacDonald

TABLE OF CONTENTS

ACKNOWLEDGEMENTS viii

INTRODUCTION: xi
    Purpose of the book xi
    Metrication xiii

CHAPTER 1: HAND AND POWER TOOLS 1
  HAND TOOLS 1
    Measuring tools 1
    Hand saws 6
    Hammers 8
    Wood chisels 10
    Hand planes 11
    Sharpening the plane iron and wood chisel 13
    Boring and drilling tools 14
    Fastening tools 16
  POWER TOOLS 17
    Stationary electric power tools 18
    Portable electric hand tools 22
    Powder actuated tools 26
    Air actuated tools 28

CHAPTER 2: HOUSE STYLES AND FABRICATION 31
    House styles 31
    Low-energy houses 33
    Solar houses 36
    Fabrication of houses 40

CHAPTER 3: BLUEPRINT READING 44
    Types of blueprints 44
    Types of working drawings 45
    Language of blueprints 49
    Building code 53

CHAPTER 4: SITE PREPARATION AND EXCAVATION 58
    Site planning 58
    Lot or property layout 59
    Instrument layout 61
    Theory of leveling 64
    Elevation work 65
    Erection of batter boards 67
    Soil 68
    Water table 69
    Excavation 69
    Determining the number of cubic metres
      of earth in an excavation 70

CHAPTER 5: CEMENT AND CONCRETE 73
    Portland cement 73
    Manufacture of Portland cement 73
    Types of Portland cement 75
    Concrete 75
    Steps in obtaining quality concrete 76
    Proportioning the concrete mix 77
    Mixing the ingredients 80
    Placing concrete 81
    Finishing concrete 82
    Curing 85
    Hot and cold weather concreting 86
    Plain and reinforced concrete 86
    Controlled cracking in concrete 87

CHAPTER 6: FOOTINGS AND FOUNDATIONS 90
    Types of footings 90
    Foundation walls 97
    Foundation wall forms 101

Masonry foundations	112
Preserved wood foundations	114
Drainage and backfill	121

CHAPTER 7: FLOOR FRAMING — **127**

Beams and girders	127
Columns	128
Floor joists	129
Floor joist installation	129
Subfloor	136

CHAPTER 8: WALL AND CEILING FRAMING — **140**

Wall framing	140
Framing members	140
Openings in exterior walls	147
Assembly of wall framing members	149
Ceiling framing	152
Framing for low energy houses	153
Wall sheathing	155

CHAPTER 9: ROOF STYLES AND TERMINOLOGY — **162**

Types	162
Roof terms and components	165
Roof slope	166
Ratio and proportion	166
Roof triangle	167

CHAPTER 10: ROOF FRAMING — **168**

Framing square	168
Common rafter	168
Laying out and cutting a common rafter	170
Ridge board	171
Collar ties, dwarf walls, and struts	171
Eave projections and gable ends	172
Hip rafter	174
Laying out a hip rafter	176
Hip jack rafter	178
Laying out a hip jack rafter	180
Laying out a hip jack pattern	180
Ridge board on a hip roof	181
Material requirements for a hip roof	182
The intersecting roof	182
Main and minor roof	183

Valley rafter	184
Laying out a supporting valley rafter	185
Laying out a shortened valley rafter	186
Jack rafters	186
Valley jack rafter	186
Hip-valley cripple jack rafter	188
Valley cripple jack rafter	188
Laying out a valley jack rafter	189
Laying out a hip-valley cripple jack rafter	189
Laying out a valley cripple jack rafter	190
Ridge boards on intersecting roofs	190
Truss rafters	191
Roof sheathing	193

CHAPTER 11: CORNICE CONSTRUCTION AND ROOF COVERINGS — **196**

Types of cornices	196
Roofing materials	201
Installation of asphalt shingles	202
Installation of low slope asphalt shingles	203
Wood shingles and shakes	204
Installation of wood shingles and shakes	205
Built-up roofs	207

CHAPTER 12: EXTERIOR DOORS AND WINDOWS — **211**

Exterior doors	211
Exterior door frames	211
Hanging an exterior door	214
Installation of locksets	215
Types of doors	216
Windows	218
Window types	219
Installation of windows	221

CHAPTER 13: EXTERIOR WALL FINISH — **224**

Types of exterior wall finish	224
Application of bevel siding using mitred corners	228

CHAPTER 14: INSULATION AND VAPOUR BARRIER — **231**

Heat loss	231
Insulation	233
Types of insulation	233
Application of insulation	237

Exposed floors 242
Exterior walls 243
Ceiling and roofs 244
Vapour barrier 247
Application of vapour barrier 249
Air-to-air heat exchangers 250

CHAPTER 15: INTERIOR WALL AND CEILING FINISH **254**
Types of interior finish 254
Hollow wall fasteners 254
Panelling 256
Application of panelling 256
Ceiling finish 259
Application of ceiling tile 259

CHAPTER 16: INTERIOR DOORS AND TRIM **262**
Types of doors 262
Installation of door jambs 264
Techniques for hanging doors 266
Trimming doors and windows 267

CHAPTER 17: STAIR CONSTRUCTION **271**
Terms used in stair construction 272
Types of stairs 273
Types of stair stringers 275
Basic rules in stair construction 276

Floor framing 277
Methods of fastening stairs in place 278
Calculating the actual rise in a flight of stairs 278
Determining stairwell opening length 279
Determining the maximum run of a stair with a fixed opening 279
Split entries 280
Determining actual headroom 281
Building a set of stairs 282

CHAPTER 18: CABINETRY **286**
Kitchen planning 286
Types of cupboards 288
Drawers 290
Standard measurements of kitchen cupboards 292
Custom built kitchen cupboards 292
Pre-built units 293

SUPPLEMENT TO CHAPTER 10
Roof Framing 295

GLOSSARY **299**
ANSWERS TO QUESTIONS **305**
INDEX **309**

ACKNOWLEDGEMENTS

The authors of *Building Construction Technology* wish to express their sincere thanks to the following agencies, companies and manufacturers for their assistance in preparing this book.

Aerofoam Chemicals, Division of Border Fertilizer (1972) Limited

Automated Building Components Incorporated

Caterpillar Americas Company, Illinois, U.S.A.

Cellufibre Industries Limited

Concept Construction Limited, president-Mr. K. Funk

Council of Forest Industries of British Columbia

Dow Chemical of Canada Limited

Enercon Consultants Limited

Faber-Castell Canada Limited

Fiberglas Canada Limited

Fiodrain Limited

Frederickson Metric Systems Limited

Goldblatt Tool Company, Kansas, U.S.A.

Government of Saskatchewan, Office of Energy Conservation, Saskatchewan Mineral Resources

Grace Company of Canada

Inland Cement Industries Limited

JCB Excavators Limited

Kennedy Sky-lites Incorporated, Florida, U.S.A.

Kramer Tractor Limited

Lockwood Limited

Loewen Millwork

Merit Industries Ltd.

Moose Jaw Sash and Door Company (1963) Limited

National Concrete Producers Association

National Research Council of Canada, Division of Building Research
—Miss M. Gerard, Editor, National Office
—Mr. H. Orr, Prairie Regional Office
—Mr. R. Dumont, Prairie Regional Office

Nelson Lumber Company Limited

Peachtree Doors Canada Limited

Portland Cement Association, Illinois, U.S.A.

Ramset Limited

Rockwell International of Canada

Sandvik Canada Corporation, Disston Division

Senco Fastening Systems, Senco Products Inc.

Skil Canada Limited, subsidiary of Emerson Electric Company

Smith-Roles Limited

Stanley Works Limited

Sun Centre Systems Limited, president-Mr. V. Ellis

University of Saskatchewan, Saskatoon Campus
—Professor R. Besant, professor of Mechanical Engineering
—Dr. K. Nasser, professor of Civil Engineering

Westcon Construction Products Limited

Special recognition must be given to the following individuals for their assistance and the superb quality of work which they prepared for us. A sincere thank you to Mr. Terry Burt who prepared the numerous line drawings and illustrations which helped to make our book what it is. Secondly, we wish to thank Mrs. Jeanette Nutting for her typing skills and for arranging typists for us. And to Mr. A. C. Watson for his assistance in taking a number of the photographs appearing throughout the book. To those people too numerous to mention, who throughout the years have given us their knowledge and expertise—we owe you a great deal.

As well, we would also like to express our appreciation to our individual families who so understandingly stood by us during the assembling of this book.

—Ken F. Cannon and Fred G. Hatley

ABOUT THE AUTHORS

KENNETH F. CANNON obtained his apprenticeship training and interprovincial journeyman certificate in carpentry in the province of Saskatchewan. He worked as a general contractor in his home town of Marquis for a number of years before joining the staff of the Saskatchewan Technical Institute in Moose Jaw as a carpenter instructor for apprenticeship training. During that time he obtained a Vocational Teacher's Certificate from the University of Saskatchewan.

In 1968 he was employed by the Department of Indian and Northern Affairs as a carpenter instructor in a pre-vocational school in Churchill, Manitoba. He taught natives there, mainly Inuit from the Eastern Arctic.

In 1974 he was transferred to Ottawa to become an instructor counsellor, providing on-the-job apprenticeship training in carpentry to Indian and Inuit apprentices. Mr. Cannon and an apprentice crew of 5 to 6 travel extensively across northern Canada, building houses and other structures that provide suitable training for the apprentices. During the two months of each year that these apprentices are attending classes in Fort Smith, N.W.T., he frequently accompanies them and teaches classes either there or in the vocational school at Whitehorse, Yukon.

FREDERICK G. HATLEY received his early schooling at Riverhurst, Saskatchewan. After completing high school, he was employed by Horosko Construction Company Limited. During this time he served his apprenticeship and obtained his provincial journeyman and interprovincial certificates. He continued his employment for a number of years, serving as a carpenter and later as a site foreman.

In 1966 he joined the staff of the Saskatchewan Technical Institute in Moose Jaw as a carpentry instructor for apprenticeship training. In the fall of 1968 he moved to the provincial high school system, teaching one year at Biggar before moving to the Technical High School in Prince Albert in the fall of 1969. He has remained with the school, which, after a major addition in 1975 became known as the Carlton Comprehensive High School.

He received his Vocational Teacher's Certificate in 1970 from the University of Saskatchewan, Saskatoon Campus. From 1973 to 1975 he served as a member on the provincial curriculum committee for the Department of Education, setting up and introducing the present Saskatchewan Division IV curriculum in carpentry. Over the years he has been a member of the Saskatchewan Industrial Teachers' Association, a special subject council of the Saskatchewan Teachers' Federation, serving two years on its provincial executive.

Currently teaching grades X to XII, he is instructing in the areas of construction, drafting, and woodworking.

INTRODUCTION

THE PURPOSE OF THE BOOK

The fundamental purpose of *Building Construction Technology* is to make available to senior high school students, apprentice carpenters, journeymen carpenters, prospective homeowners and builders of homes, the basic principles involved in residential house construction. In preparing this book, such things as design data, terminology and factual information have been kept as simple as possible. Particular attention has been paid to the very real energy crisis which affects our lives today. From this crisis there has evolved a need for energy-efficient homes, especially regarding heating and cooling. With this in mind, it is the hope of the authors that the reader will gain an appreciation for, and recognition of, sound building practices for today's needs.

Wherever we look, residential construction surrounds us; in the form of single-dwellings, duplex homes, apartments and multiple or row housing. The work force required to complete today's homes is made up of executives, staff officers, supervisors, tradespeople and research teams. The individual operating a computer is as much a part of the industry today as is the architect, engineer, carpenter, labourer or payroll clerk.

The bulk of the labour required, however, comes from skilled and semi-skilled workers. Of these, the work of the carpenter must be co-ordinated with the other trades to ensure the progress of the building—from the driving of the first stake into the ground to the owner's final inspection before moving into a new home.

The arrangement of the chapters in the book allows the reader to systematically follow the operations that would be performed in the construction of a home. Because of the importance of energy today, no one chapter has been set aside to cover energy-efficient changes to the home. Many ideas and changes which are now in limited practice in various parts of Canada, and generally have not been outlined in any previous textbook, can be found throughout the book. The authors have endeavoured to show the reader that with very minor changes in the design and construction techniques of a building it is both possible and above all economically feasible to build an energy-efficient home.

The use and care of hand and power tools common to the trade are covered in Chapter One. This topic is so large, it would be impossible to cover all the tools that could be used. Therefore, only those tools commonly used have been chosen. Specialty tools such as the transit-level and concrete tools are also covered.

In keeping with developments that have taken place in the areas of solar and low energy consuming homes, the chapters on house styles and blueprint reading cover the essential information required by the homeowner in the planning and drawing stages of this type of home. In this section, the reader will find plans for a low energy home that has been built by one of western Canada's leading contractors in this field.

Another important topic is the role of the Canada Mortgage and Housing Corporation. Included is its involvement in financing, preparation and checking of house plans to ensure conformance to the National Building Code of Canada as well as its inspection of homes along with local building codes.

The sequential layout of the home on the building site includes the staking, elevation work and assembly of batter boards before the excavation work can begin. These topics are thoroughly discussed in the chapter on site preparation. Types of soils that may be encountered are discussed, as well as preventative measures that can be taken if poor soil conditions exist. Concrete, one of the most widely used building materials, is covered from the manufacture of cement through to the placing and curing of the concrete mix. Included in the section on foundations is the use and application of preserved wood foundations in the housing industry.

The latest techniques are described in the chapters on framing. Floor and wall framing is covered from the plate and header layout through to the assembly and erection of the component parts. Emphasis is continually placed on the low energy system using the minimum 38×140 mm stud and/or the double wall assembly. Chapters on roof construction include a study of the layout and assembly of all the individual rafters. By using the ratio-proportion method along with the metric square, roof assembly is greatly simplified. The design, style and use of the truss rafter conclude a totally new and up-to-date look at the framing package of a home.

A complete chapter has been devoted to insulation and vapour barriers—types, application and restrictions. The new approach to the application of the vapour barrier will be of interest to the energy-conscious reader. In looking at the energy-efficient home, control of air infiltration is discussed and the principle of the air-to-air heat exchanger explained.

Roof materials and their application, soffit construction and the use of flashings—all essential parts of the home—are described in the text. Exterior and interior finishes are discussed along with the application of many of the finishes provided. The use of energy-effective exterior doors and windows, along with their installation help to complete a study of the energy efficient home. Thus, this section of the book is as important to the concept of energy efficiency as are all the other sections of the book.

Stair construction, always a challenge, is simplified both through the use of metrication and the construction approach used. The assembly and installation of interior doors and door and window trim all have their place in the construction of the home.

Prior to the energy crisis, having a more comfortable home would have meant a larger furnace or auxiliary heating systems. The crisis of the mid 1970s, when Canada began importing oil to maintain our level of daily consumption, has changed that. Consumers today are realizing that prices for fossil fuels and electricity are not going to decrease. We are beginning to accept the fact that we cannot afford to continue our present wasteful consumption of energy.

Conservation is one way to reduce the consumption of energy. To this end, we are seeing changes in the design of our homes. This is unlike any other change we have experienced in the last fifty years in the housing industry. Ideas are being explored in the area of conservation—with or without help from Federal/Provincial sources. The implementation of these ideas centres around three aspects of the home;

(a) the most appropriate and economical design

(b) construction and operational changes

(c) recognizing the peculiarities of each particular region of our country.

The results of these steps are, to say the least, very encouraging. Reductions of 50 to 80% in energy requirements for heating are not uncommon in maintaining an average 100 m² home.

To achieve the energy conservation mentioned, some of the more notable changes found in the book and which are being implemented in today's energy-efficient homes include the following:

(a) Preplanning of house design so that major window areas face south, along with the re-establishment of air-lock entries or porches.

(b) Use of active or passive solar heating systems to carry or supplement the load within the home during the heating season.

(c) Use of a minimum of 38×140 mm studding, with the double wall assembly as an alternative.

(d) Insulating of the basement wall and/or grade foundation.

(e) Recommendation of wall and ceiling insulation to RSI factors well above the present minimum.

(f) Better application methods for the installation of the vapour barrier to form an airtight seal have been developed and easily implemented.

(g) Elimination of ceiling outlets in areas where controlled wall outlets exist.

(h) Removing the attic access inside the home and placing it on the outside of the home if possible.

(i) Reducing the headroom to reduce the number of cubic metres of air that must be heated to maintain a comfortable temperature.

(j) Weather-stripping of all exterior doors and windows to control infiltration of air into the home.

(k) Use of double glazed window units

TABLE 1   METRIC (SI) PREFIXES COMMONLY USED IN CONSTRUCTION AND RELATED INDUSTRIES

Prefix	SI Symbol	Multiplication Factor
mega	M	$10^6 = 1\ 000\ 000$
kilo	k	$10^3 = 1\ 000$
hecto	h	$10^2 = 100$
deca	da	$10^1 = 10$
* base unit to which prefix is added		
deci	d	$10^{-1} = 0.1$
centi	c	$10^{-2} = 0.01$
milli	m	$10^{-3} = 0.001$
micro	μ	$10^{-6} = 0.000\ 001$

Note—this is only a partial list of the SI prefixes used in the metric system.

(minimum) with triple glazed or two double glazed window units where moisture problems could exist or a higher RSI value is desired.

(l) The use of air-to-air heat exchanger to remove foul air from inside the home while supplying fresh air from outside.

METRICATION

Canada's decision to go to the metric system has given us a unique opportunity to rethink and better plan construction and related industries as a whole. Metric measurement, since its beginnings around the time of the French Revolution, has been continually developed and standardized over the years. Consolidated into its present form, the **System International** or **SI**, represents the system that approximately 90% of the world's population is either using or implementing.

The metric system has as its greatest advantage its simplicity. In the entire system there are only seven base units. Of these, only three will be widely used in the construction industry—the *metre* (for length), the *litre* (for liquid volume) and the *gram* (for mass). All other parts of the base unit required, whether larger or smaller, are simply obtained by multiplying or dividing the base unit by ten (10) and adding the appropriate prefix before the base (Table-1).

With the three basic units and their prefixes established, their use is relatively easy. Some prefixes are going to be used more than others (Table 2). Linear dimensions given in either metres (m) or millimetres (mm) have been established at the level of government and industry. The common unit for area will be the square metre (m²), while the unit for volume will be the cubic metre (m³). In volume measure, the litre (liquid measure) is equal to a cube 100 mm on all sides. Under standard conditions one litre of water will

TABLE 2 METRIC UNITS COMMONLY USED IN THE BUILDING INDUSTRY

Metric Unit	Symbol	Common Examples of Application
Length		
metre	m	1. Width and length of a building 2. Linear measurements
millimetre	mm	1. Thickness, width and length of lumber 2. Thickness, width and length of panel material 3. All layout work and measurements in respect to constructing parts of the home
micrometre	μm	1. Specifying the thickness of air-vapour barrier material (i.e. polyethylene)
Area		
square metre	m²	1. Area of the building site 2. Specifying the area of the home 3. Exterior and interior wall and roof claddings
hectare	ha	1. Area of a group of building sites together. *NOTE*—One hectare (ha) = 10 000 m²
Volume		
cubic metre	m³	1. Large volumes of dimensional lumber and panel materials 2. Volume of excavated materials 3. Volume of gravel, concrete and fill materials.
litre	L	1. Amount of water used in the cement-water ratio of a concrete mix
Mass		
kilogram	kg	1. Quanitative unit of bagged or bulk cement
Engineering Units		
Force		
newton	N	1. Used in designing live loads for wood beams, floor joist and rafters. Expressed as kilonewtons per square metre (kN/m²) at specified centres
Pressure		
pascal	Pa	1. Used when determining the compressive strengths of materials (i.e. soil tests or concrete tests) and is expressed as megapascals (MPa)

have a mass of 1000 g or 1 kg; expressed another way one millilitre (ml) of water has a mass of one gram (g).

In differentiating between mass and weight, **mass** is the quantity of matter which an object contains and the basic unit is the kilogram (kg). This is the only base unit which contains a prefix. **Weight** is the gravitational force of the object and is expressed in newtons (N). A newton (unit of force) is based on the kilogram (unit of mass), the metre (unit of length) and the second (unit of time). In terms of the other units it is expressed as kg • m/s². Because the gravity of the earth varies, the weight of any given mass will vary slightly depending upon where the mass is located. **Pressure**, the result of a given weight (newtons) over a specific area (square metre), is called the pascal (Pa). Table 2 shows the basic units and prefixes that will be used by the industry along with some common examples.

The commitment to metric (SI) has allowed us time to look at the total building process. With the ever-increasing trend to prefabricated component parts in building, it has become increasingly clear that the manufacturer, designer and builder must all use a basic standard throughout. To achieve this, a basic module of 100 mm has been accepted. Multiples of the module have then been set throughout to further simplify the process. Since manufactured parts will be standardized to the basic module, designers and builders will require less time and money to ensure that parts fit.

Wherever government and industrial guidelines have been established and hard conversion measurements set, they have been used in the book. For example, in the plywood industry, the length and width of a standard sheet has been set at 1200 × 2400 mm to accommodate the new stud, joist, and truss spacings of 300, 400 and 600 mm on centre. Thicknesses of metric plywood will vary from sheathing to sanded grades. This is because sanded grades will now be produced from the same veneers as the sheathing grades (Table 3).

In the case of the dimensional lumber industry, where hard conversion sizes have yet to be completed, the existing sizes will be used and expressed in metric terms. Soft converted for now, they will reflect the actual size of the material. If and when it becomes necessary to change the existing sizes to hard conversion, it would not affect anything other than the dimensions of the material. An example here is a 38 × 89 mm piece of material (soft conversion) eventually becoming 35 × 85 mm (hard conversion). Table 4 indicates the standard sizes now in use, soft converted to the metric system.

We know the technology is there, from the metric system to the research done on energy-efficient housing, to the individuals working with this research. It is now up to the designer, builder and owner to take this technology and make it work to our advantage. If, in some small way, the ideas and methods of construction compiled and presented here provide better construction techniques which help to lower fuel bills for individual homeowners and the nation as a whole, then this book has been a success. It should certainly be a welcome addition for anyone interested in, or concerned about, energy-efficient house designs, energy-saving materials and methods of applications. Who among us can afford to be unconcerned indefinitely?

TABLE 3   THICKNESS OF SOFTWOOD PLYWOOD

Sheathing Grade	Sanded Grade
7.5 mm	6 mm
9.5 mm	8 mm
12.5 mm	11 mm
15.5 mm	14 mm
18.5 mm	17 mm
20.5 mm	19 mm

NOTE: New thicknesses of 8, 11 and 14 mm in sanded grades have been established with the metric system.

TABLE 4   METRIC SIZES OF SURFACED BOARD AND DIMENSIONAL LUMBER (SOFT CONVERSION)

	Thickness		Width	Length
Boards		Dimensional Lumber		
19 mm		38 mm	38 mm	2.44 m
19 mm		38 mm	89 mm	3.05 m
19 mm		38 mm	140 mm	3.66 m
19 mm		38 mm	184 mm	4.27 m
19 mm		38 mm	235 mm	4.88 m
19 mm		38 mm	286 mm	5.49 m
19 mm		38 mm	337 mm	6.10 m

REVIEW QUESTIONS

INTRODUCTION

Answer all questions on a separate sheet

A. Write full answers

1. Plan a field trip to a construction site to see what measures are being used in the conservation of energy. Using the ideas presented in the introduction as a starting point, be prepared to talk to the workers on the site and obtain their views. Prepare a written report on conservation in housing from your observations and findings on the trip. Reference material found in your library will contain helpful ideas.

2. What are the three base units of the metric system used in the construction field?

3. Of the base units referred to in Question no. 2, which one will be used the most? In what form(s) will it be commonly expressed?

4. A basic modular unit has now been accepted. What effect will this have on the construction industry?

5. Explain the difference between ''hard'' and ''soft'' conversion of materials.

6. What may result as materials which are now ''soft converted'' become ''hard converted'' in the future?

1 HAND AND POWER TOOLS

The carpenter must master many hand and power tools in acquiring the various skills necessary to the trade. The ability to use these tools carefully and correctly is evidenced by the quality of the finished product and the time required to produce it.

The key factor in using any tool is the ability to choose and use the correct one for each job. When combined with quality work, good tools used properly not only last longer, but work better. Quality tools naturally cost more initially than ordinary tools, but they pay for themselves many times over in performance and in the quality of the finished product. **Remember:** Good tools are a lifelong investment.

To cover the complete range of hand and power tools in use today would require a complete book devoted to this subject. The purpose then is not to cover *all* tools but rather those which are common to the trade. Included is a brief description of some of the more important tools used by the carpenter, along with the care, serviceability and safety of each.

HAND TOOLS

MEASURING AND LAYOUT TOOLS

The **capability** of the students to measure accurately and lay out work correctly is dependent on their **familiarity** with different tools and their **ability** to read them correctly. As indicated in the introduction, in the building construction field the two principle units of measurement are the millimetre (mm) and the metre (m). Because of manufacturing and cost restrictions in the manufacture of some tools it is not uncommon to find them marked off in units of two millimetres, five millimetres and even ten millimetre or one centimetre units.

Measuring Tapes

Of the two measuring tapes shown in Figure 1-1 (a), (b), the **pocket tape** is the most widely used measuring device today. The size of a pocket tape is dependent on the blade width, which varies from 13 to 25 mm, and on its length, which ranges from 2 to 7.5 m. Equipped with a belt clip to keep it readily available, most pocket tapes are spring loaded to retract the blade on completion of the measurement. They may be purchased with a locking device to hold the extended blade at a predetermined position and may come equipped with a power return button in place of the spring loaded retracting mechanism.

(a)

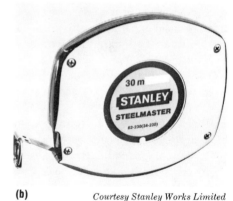
(b)   *Courtesy Stanley Works Limited*

Figure 1-1 Pocket (a) and steel (b) measuring tapes

Measurements beyond the range of the pocket tape may be obtained with a **steel tape**. Used extensively for the major layout work, they are available in lengths

1

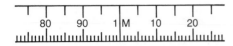

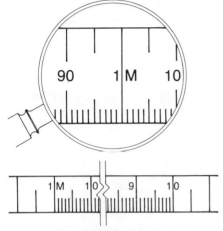

Figure 1-1 (c)   Magnified section of a measuring tape

of 15, 20 and 30 m with a standard blade width of 9.5 mm. Graduated in metres, centimetres and millimetres, (Fig. 1-1 (c)) it is equipped with a folding end clip for attaching to the material when a measurement is taken and folds back out of the way when not in use. Because of its length, a hand crank is usually provided rather than a recoil spring for retracting the blade.

Observe these **precautionary measures** when using a pocket and/or steel tape:

(a) Keep dust, dirt and moisture from collecting within the case on the blade. This possibility can be reduced by occasionally wiping the blade with a slightly oiled rag.

(b) When withdrawing and /or recoiling the blade do not bend the blade back or twist it as kinking, and eventually, breaking of the blade may occur.

(c) When recoiling a pocket tape, do not grasp the blade by the edge as it can be sharp and produce a very painful cut.

(d) Never risk breaking the blade of a measuring tape by leaving it with the blade exposed where it may get stepped on or have something dropped on it.

A number of the layout tools can also be used as measuring tools. They are, however, restricted by the length of their blades and by the graduation intervals. Some tools are marked off at every millimetre while others are marked in 2 mm intervals.

Try Square and Combination Square

The **try square** (Fig. 1-2) and combination square (Fig. 1-3) are widely used layout tools. The try square is used for *testing* and/or *squaring* the *surfaces, ends* or *edges* of material. It has either a wood or metal handle and the blade length varies from 150 to 300 mm.

The **combination square** performs the same operations as the try square but has a number of added advantages. Among these is the *45°* as well as the *90° layout;* a level bubble enclosed in the handle allows for *plumbing* and *levelling* while the sliding or movable handle enables it to be used as a *scribing* tool.

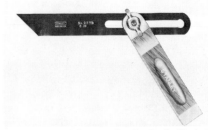

Figure 1-4   Adjustable T-bevel

Adjustable T-Bevel

An adjustable T-bevel (Fig. 1-4) is used when setting and/or obtaining any angle other than a 45° or 90°. It has a wood, steel or plastic handle combined with a blade 200 mm long and is extremely handy for angular work. A thumbscrew on the handle which locks the blade at the desired angle allows the marking and/or transferring of the angle. A slotted handle permits storage and protection for the blade when not in use.

Framing Square

The framing square is an accurately manufactured tool, capable of performing the same squaring procedures as the tools

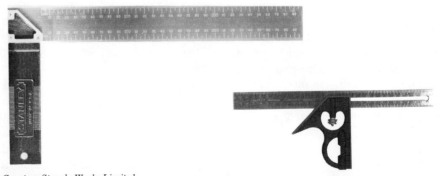

Figure 1-2   Try square

Figure 1-3   Combination square

previously mentioned. It can also be used for joist, stud and rafter layout, to obtain the length of rafters, for stair layout, polygon and brace layout, making it a very versatile tool.

It consists of two parts, the *body* (50 × 600 mm long) and the *tongue* (40 × 400 mm long). The arms meet at a 90° angle at the heel, or corner, of the square. It is graduated every mm or every second mm depending on the manufacturer of the square. The metric framing square is marketed by many manufacturers, and is made of various metals—from aluminum to polished steel. The authors have chosen to use the Frederickson Framing Square (as shown on this page) and the Stanley Metric Square (as shown on page 295).

On the *face* side of the body of the square is the *rafter* table. On the face side of the tongue is the *brace* and *formulae* table, indicating lengths of common braces and the procedures used to perform the intricate cuts of the hip rafter, jack

rafter and roof sheathing (Fig. 1-5 (a)).

Located on the *back* of the body of the square is the *octagon* table, used for framing an octagonal roof system. Also on the back, located along the tongue of the square is the *polygon* table (Fig. 1-5 (b)).

The **rafter table**, Figure 1-6, is used to determine the line length per unit of run. The unit of run is based on 250 mm for all rafters with the exclusion of the hip rafter, which has a unit of run of 353.6 or 354 mm. Knowing the run and the slope of the roof, you are able to calculate the total line length of any given rafter. Chapter Ten, (Roof Framing) gives complete instructions on how to use the tables and calculate the line lengths of all rafters.

Figure 1-7 (a) shows a portion of the **brace table**. This table gives the length of braces commonly used. Consisting of three numbers, the first two represent the projections of the right angle triangle, while the last figure represents the diago-

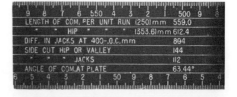

Figure 1-6   Rafter table taken from the Frederickson square

Figure 1-7 (a)   Brace table

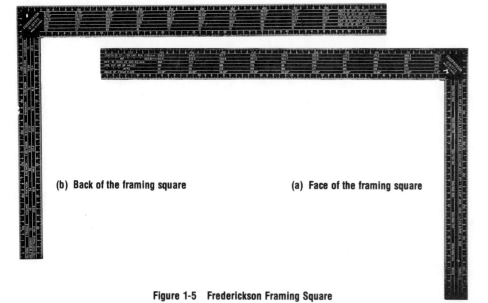

(b) Back of the framing square   **(a) Face of the framing square**

Figure 1-5   Frederickson Framing Square

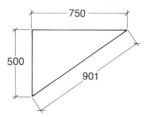

Figure 1-7 (b)   Measurement from the brace table

nal measurement between the two points (Fig. 1-7 (b)). The length of any 45° angle brace is easily found by using any two equal numbers, as shown in Figure 1-7 (c).

The **octagon table** on the back of the square gives the necessary information

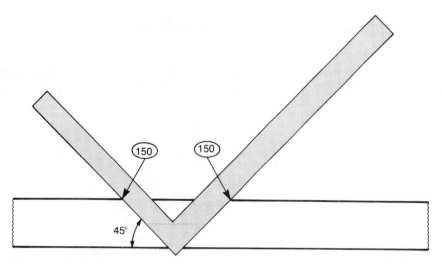

Figure 1-7 (c)   Finding a 45° angle

for laying out an octagon roof. Used in the same manner as the rafter table on the face side of the square, calculations are required for two rafters. The hip rafter on this roof has unit of run of 271 mm while the jack rafters remain the same as a regular roof, or 250 mm per unit of run (Fig. 1-8).

Figure 1-8   Octagon table

On the back of the square along the tongue is the **polygon table**. Used to lay out any polygon from the triangle (3 sided) to the dodecagon (12 sided), it gives the corresponding figure, used

along with the base number (250 mm). Also given is the cut angle required. The angle required is marked along that portion of the square used to measure off the corresponding number relative to the number of sides of the polygon (see Fig. 1-9).

Used extensively in stair construction (Chapter 17), the framing square can save countless hours in layout time. Because of its great importance, and since we were unable to give it the required space, the authors suggest that students refer to other books which are devoted entirely to the framing square and its uses.

Hand Level, Straight Edge and Plumb Bob

In addition to measuring and layout, some tools are also used for *testing*. The common tests or checks required are those for materials and/or sections of the structure to be level, plumb and straight. To ''level'' means to check or adjust an object in reference to the horizontal. To ''plumb'' refers to a vertical surface which is at right angles to the horizontal or level surface.

The **hand level** may vary from a compact pocket level 127 mm long, to a large hand level 1829 mm in length. Based on the principle that an air bubble will rise to the highest point in a curved glass, the vials are placed in both the plumb and vertical positions on the hand level. In some cases vials may also be set at 45°. The body of the level can be made of wood, metal or in some cases plastic. The vials are either fixed in the body or are

Courtesy Stanley Works Limited

Figure 1-10   Hand level—Line level

Figure 1-9   Polygon table showing figures used and the required layout.

Figure 1-11   Straight edge

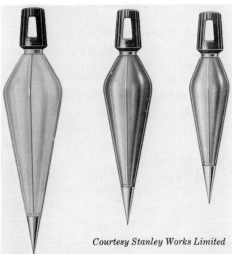

Courtesy Stanley Works Limited

Figure 1-12   Plumb bob

mounted in such a way that they can be adjusted and/or replaced if broken (Fig. 1-10).

Line levels are designed with a single horizontal vial. With special hooks for attaching it to a building line, it can be used for rough leveling purposes. When using such a level, care must be taken to ensure that the line is taut and that the level is placed in the centre of the span. This ensures that any deflection of the building line from the mass of the level will be equal on both sides of the level.

A **straight edge** is generally made by the carpenter from a seasoned, straight-grained piece of dimensional stock or plywood. Ranging in length from 1200 to 2400 mm, it is used for checking the straightness of materials and/or sections of the structure. It may be used for drawing straight lines between two points. It can be used in conjunction with the hand level to check or set the level and/or plumb line of the building or building parts. Figure 1-11 shows a common design for a straight edge used by the carpenter/builder.

Plumb bobs are either precision turned or cast weight, tapered to a point, and suspended by a builder's line, steel wire or cable. They are extremely accurate, and are used both to transfer points and for general work, when an object must be directly below a known point or 90° to the horizontal (refer to Chapter 4–Site Preparation and Excavation). Normally ranging in size from 142 to 342 g, they can be any size depending on the job requirements. A plumb bob with a mass of

113.4 kg and suspended by a steel cable was used to plumb the main shaft of the famous CN Tower in Toronto, the tallest free standing structure in the world today.

Marking and Scribing Tools

The simplest device used for marking is the **carpenter's pencil**. It is rectangular in shape and has a large, soft lead which is well protected by the wooden case surrounding it. Used for rough work, it leaves an easily identifiable mark. For finer work a standard pencil with a medium to hard lead works well. When precision is required, a pocket knife and/or utility knife is recommended for scribing or laying out the work.

The **builder's line** is not to be confused with a chalk line. A builder's line is usually nylon and is used to establish a line between two or more known points (refer to Chapter 4—Site Preparation and Excavation, Figs. 4-24 to 4-26). As shown, it is used when the carpenter wants to indicate the position of an ob-

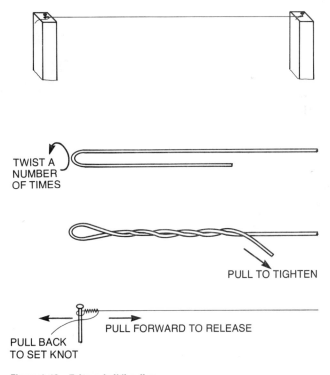

TWIST A NUMBER OF TIMES

PULL TO TIGHTEN

PULL BACK TO SET KNOT   PULL FORWARD TO RELEASE

Figure 1-13   Tying a building line

ject. Suspended between two points, a simple knot holds the line in place (Fig. 1-13).

Hooking one end over the partially driven in nail, the carpenter then makes a loop at the opposite end where it will be attached. Then, twisting the loop a number of times, it is placed over the nail at the other end. After pulling the line taut, the loose end is then pulled back over the twisted area, causing the line to fold over itself. To release the line a simple tug opens up the twists and releases the line. This eliminates the need for a formal knot (Fig. 1-13).

A **chalk line**, either self-chalking or one requiring chalk applied by hand is used as a layout and marking tool. In a self-chalking reel, the line is pulled out and powdered chalk from inside the case sticks to the line. The line is stretched between two points, tightened and snapped, and the chalk is transferred to the material (Fig. 1-14). Some chalk line reels are designed in such a way that they can be used as plumb bobs.

Gauges

Gauges are available for a number of purposes. The **marking gauge** (Fig. 1-15) is designed for general layout work where a line is to be drawn or scribed parallel to a given surface. The sliding head is adjusted up to 150 mm. A protruding metal scribe scores the wood, plastic or wallboard as the marking gauge is drawn along the material.

A **butt gauge** is used to mark the position and thickness of the hinge or butt on the door or door jamb. It can also be used to mark the location of lock and striker plates when installing the door lock.

Butt markers are specifically designed for marking the outline of the hinge butt prior to chiseling out the hinge area. They are available in 76, 89 and 102 mm sizes. Examples of all three gauges are shown in Figure 1-15.

Like measuring tapes, other measure-ment tools require proper care. Rust, a common enemy of all tools, can be avoided with an occasional wiping of the tool with an oily rag. This will normally prevent or remove rust. Be sure you have a proper storage facility for your tools, whether it be a tool box, tool crib, or board mounted to the wall of the shop.

HANDSAWS

Handsaws are used by the carpenter/builder wherever the squaring, shaping and fitting of individual components are required. Handsaws are divided into two main groups; (a) Crosscut saws and (b) Ripsaws. The most commonly used handsaws in the trade are the *crosscut* or *panel,* the *rip,* the *backsaw* and the *coping* saw.

The length of the saw blade or number of points for every 25 mm of length identifies the handsaw as to its use. Handsaws

Courtesy Stanley Works Limited

Figure 1-14   Self-chalking line

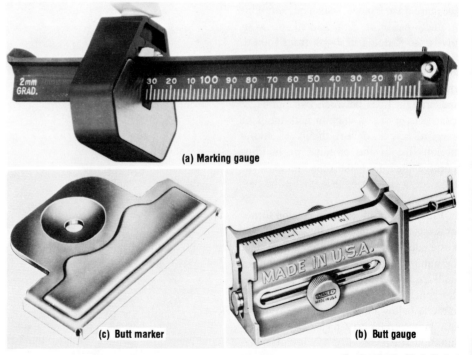

(a) Marking gauge

(c) Butt marker

(b) Butt gauge

Courtesy Stanley Works Limited

Figure 1-15   Gauges

are available in either the skewback or straightback style. The standard saw blade is 660 mm long. Panel saws are smaller, at 600 mm or less. The number of points designate the size of each individual saw tooth. For example, a saw with a small number of points (8) has relatively large teeth and will cut quickly but leaves a rough edge on the board. When handsaws have 10 points or more they produce smoother cuts but are slower than the coarser toothed handsaw.

On all handsaws, the upper half of each tooth is bent or *set* alternately left and right to assure blade clearance. The width of the groove or *kerf* cut out by sawing is determined by the set. The set width is an important factor in determining which saw to use. Wet wood requires a saw with a coarse tooth and a wide set, while dry, seasoned material can be cut easily with a finer toothed, narrow set blade.

Crosscut Saw

As its name implies, the crosscut saw is designed to cut *across* the grain of the wood. The teeth are designed like a series of *knives* which cut on both the forward and back stroke. The number of points on a crosscut saw will vary from eight for a standard saw to twelve for a very fine panel saw. A typical crosscut saw is illustrated in Figure 1-16 along with the shape and cutting action of the teeth.

Before starting to cut with any saw be sure the measurement is correct. Decide what pieces you need and what will be waste. Always cut on the **waste** side of the line, **never** on the line.

When beginning to cut with a crosscut saw, place the heel of the saw in position and draw it toward you a few times until a kerf has been formed. Maintaining an angle of approximately 45° between the saw and the board, begin cutting, using the full length of the saw blade. Continue cutting using full even strokes but do not

Courtesy Sandvik Canada

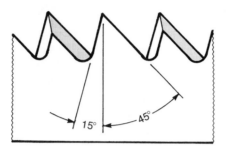

TOOTH SHAPE

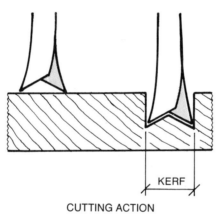

CUTTING ACTION

Figure 1-16   Crosscut handsaw

force the saw. As you complete your cut, support the waste material and take shorter, lighter strokes to avoid splintering the material. Never twist the saw to break off waste material.

Ripsaw

A ripsaw is designed to cut *with* the *grain* of the wood. The teeth differ from those of a crosscut saw in size, shape and cutting action. A ripsaw usually has five and one-half points for every 25 mm of length. It has teeth that resemble a small

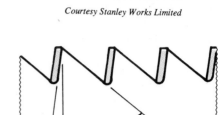

Courtesy Stanley Works Limited

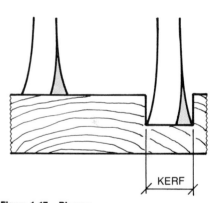

KERF

Figure 1-17   Ripsaw

chisel and each cuts in the same way. Each tooth chips out a small portion of the wood fiber. Cutting is achieved on the forward stroke only (Fig. 1-17).

When cutting with a ripsaw, lift the saw slightly so as not to take a full bite and carefully work the saw forward and back until a kerf has been started. An angle of 60° is maintained when using the ripsaw. Cutting is accomplished in the same manner as with the crosscut saw, using the full length of the blade. **Avoid** forcing the saw.

Backsaw

The backsaw is essentially a crosscut saw designed for precision work. Ranging from 300 to 650 mm long and from 11 to 14 points for every 25 mm in length, it is used almost entirely for cabinet and finish work. A metal reinforcing strip along the back stiffens the blade and prevents flexing while cutting. This produces a smooth, accurate cut. When starting to make a cut with the backsaw, use the same procedure as with a crosscut saw. However, the angle between the teeth and the work is decreased until you are cutting with the blade on a horizontal plane.

A backsaw in conjunction with a mitre box (either wood or metal) is used almost *exclusively* for cutting mouldings and trim work. It may also be used for square or angular cuts. The table on the metal mitre box is designed to facilitate 90° cuts and most angular cuts up to 45° to the right or left. Figure 1-18 illustrates a backsaw and metal mitre box saw assembly.

Coping Saw

The coping saw is used to form or *cope* the end of a piece of moulding at the point where it fits into a similar piece. It is also used for cutting irregular shapes when required. Consisting of a frame, handle and a thin fine-toothed blade, it is primarily a finishing saw. The blade has the advantage that it can be set in any position the worker requires, depending on the work involved. However, due to the narrow thin blade it must **not** be forced because it will break easily. Figure 1-19 shows a coping saw and a typical coped joint.

Good quality saws deserve good treatment. By following a few of the simple procedures listed here you will receive better service from them.

(a) Guard against moisture; *rust* will eventually pit the surface of the metal, causing the saw to drag. Over a period of time rust will ruin your saw.

(b) Always place the *cutting edge* so that no other tool will knock against the teeth and damage them.

(c) After using the saw, lay it down carefully—*do not drop it!*

(d) Check over used material and/or repair work before cutting to see that there are no nails in the path of the saw.

HAMMERS

One of the first tools purchased and probably one of the most misused tools is the hammer. Designed specifically for the *driving* and *pulling* of nails, its abuses range from driving hardened nails (concrete nails), to breaking up concrete, to attempting to remove hardened nails and/or fasteners. Hammers should be selected for one purpose and used for that purpose only. Figure 1-20 illustrates two common claw hammers and their parts.

Nail hammers are available in two styles, the *curved claw* and the *straight* or *ripping claw*. The striking face of the

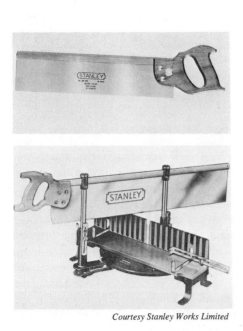

Courtesy Stanley Works Limited

Figure 1-18   Backsaw and mitre box saw

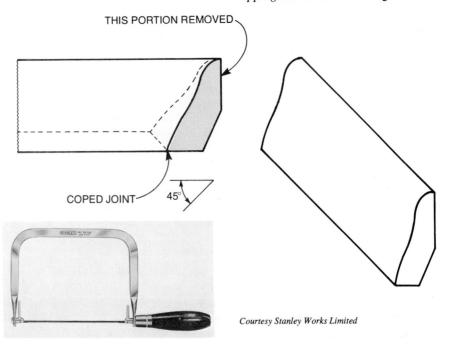

THIS PORTION REMOVED

COPED JOINT   45°

Courtesy Stanley Works Limited

Figure 1-19   Coping saw and a typical coped joint

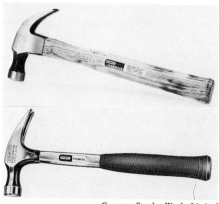

Courtesy Stanley Works Limited

Figure 1-20 The curved and straight claw hammer

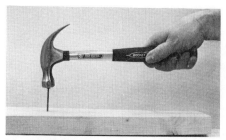

(a) Driving the nail

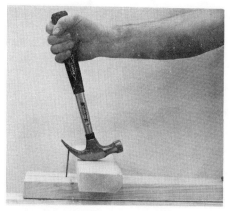

(d) Pulling a nail

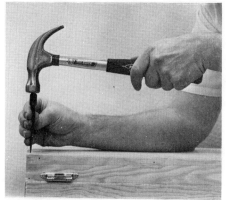

(b) Setting a finishing nail

Courtesy Stanley Works Limited

(c) Nail set

Figure 1-21 Correct use of a nail hammer

hammer comes in three different styles: the belled face, the flat face and the checkered face. Each has its advantage, the belled face being the most commonly used. Handles are available in wood, steel and fiberglass. The last two are usually covered with a rubber, vinyl, or leather grip.

Hammers are *sized* according to the mass of the hammer head and range from 198 to 794 g; hammers 369 g and down are referred to as *finishing* hammers while those 454 to 567 g are *general purpose* hammers. Hammers heavier than 567 g are called *framing* hammers.

When driving nails, grasp the hammer by the *end* of the handle to get the maximum swing and power out of the hammer. Choking up on the handle reduces the power of your stroke and increases the chance of missing the nail. The force of the blow is delivered through the wrist, elbow and shoulder with each or all being used at any given time. When starting and/or driving small nails only the wrist and elbow motion is required. Large nails require the co-ordination of all three parts of the body. Always *strike* the nail squarely. A glancing blow will result in bent nails and marred wood (Fig. 1-21 (a)). In rough work (forming or framing), the nail is driven flush with the sur-

face of the wood. When using finishing nails leave the nail 3 mm *above* the wood. Finish driving the nail (b) with a nail-set (c), setting the nail approximately 2 mm below the surface of the wood.

To pull a nail, set the claws of the hammer firmly into the nail shank. Withdraw the nail until the handle is in a nearly vertical position. If the nail is not removed at this point, place a small block of wood under the hammer head and continue pulling the nail out. This not only increases your leverage on the nail, but also prevents damage to the surface of the wood and the possibility of breaking a handle (Fig. 1-21 (d)).

Sledge-hammers are used for driving stakes, setting forms for concrete, straightening walls and other heavy work. They come in a variety of sizes and styles. The *hand drilling* hammer may have a mass of 0.9, 1.4 and 1.8 kg while full sized *sledge*-hammers range from 5.4 kg or more (Fig. 1-22).

Like hammers, **hatchets** are designed for a particular purpose. Those commonly used are the *half* hatchet, which combines a hammer head with a single or double edged cutting blade, and the *shingling* or *lathing* hatchet.

Figure 1-22   Hand drilling (top) and sledge-hammers (bottom)

The half hatchet is used for rough hewing—sharpening stakes, forming wedges, etc. It can also be used in place of a heavier hammer if required. The shingle and/or lath hatchet is used for lighter work. As their names imply, they are used for shingling roofs and for applying gyproc and gyproc lath to interior walls.

For all types of striking tools:

(a) The proper *selection* of the tool is essential.
(b) Damaged heads should be *discarded,* and cracked handles should be *replaced.*
(c) **Never** strike one hammer head against another.
(d) Always hit the surface of the object *squarely,*–avoid glancing blows.
(e) Exercise *caution* when using hatchets. Be aware of the cutting edge when swinging them.

WOOD CHISELS

The wood chisel is made from a steel blade, heat treated to hold a sharp edge. Available in a variety of lengths, widths and thicknesses, each type of chisel is suited for a specific application. Wood chisels are divided into four groups: (a) **paring** (150 mm or longer) (b) **firmer** (100 mm) (c) **butt** (63 to 75 mm) and (d) **mortice** (50 mm). Sized accord-

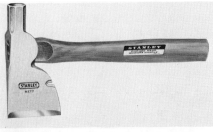

Figure 1-23   Half hatchet (top) and lath hatchet (bottom)

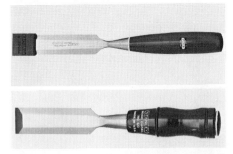

Courtesy Stanley Works Limited

Figure 1-24   Firmer (top) and butt chisel (bottom)

ing to the *width* of the blade, wood chisels range in width from 6 mm to 50 mm (Fig. 1-24).

The wood chisel can be used for cutting, fitting or trimming, on both rough and finishing work. When removing large amounts of material (rough cutting), the *bevel* of the chisel is placed *down.* The problem of splitting and of having the chisel go too deep is avoided this way. With the bevel down, it will have the ten-

dency to work back out to the surface of the wood. If additional pressure is required to make the chisel cut, strike the head of the chisel with a hammer or mallet.

When finish cutting, the *bevel* of the chisel is placed *face up* and the pressure needed to finish the work is applied by hand in a paring motion. Turn the chisel slightly sideways when working across or with the grain of the wood and slide it through the material to produce a smoother cut. The chisel in this position cuts like a knife and slices through the wood fibers.

As in sawing, the chisel is used with or *across* the grain of the wood. It is important to always chisel with the grain of the wood (Fig. 1-25 (a)). Going against the grain usually results in a rough cut or at worst a split or damaged board. When cutting across the grain, work from *both* sides toward the middle, removing the inner section last. When large areas are to be removed with a chisel, such as a hinge gain on a door or door jamb, it is good procedure to make a series of cuts first and then remove the wood chips. If the first series of cuts are not deep enough, simply repeat the process until the proper depth is attained.

Precautionary measures to use when working with wood chisels include:

(a) Keep both hands *back* of the chisel blade.
(b) Keep the chisel *sharp* (see section on hand planes for sharpening of chisels and planes).
(c) **Avoid** dropping the chisel.
(d) Use one hand to *control* the chisel, the other to apply the *pressure* required to make the cut.
(e) Always chisel *away* from yourself.
(f) Always work *with* the grain of the wood.
(g) With rough cuts, hold the bevel *down,* and with finishing cuts, keep the bevel *up.*

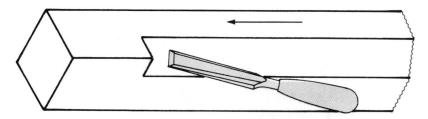

(a) Cutting with the grain of the wood

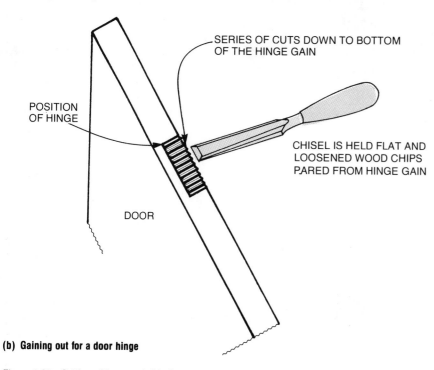

SERIES OF CUTS DOWN TO BOTTOM
OF THE HINGE GAIN

POSITION
OF HINGE

CHISEL IS HELD FLAT AND
LOOSENED WOOD CHIPS
PARED FROM HINGE GAIN

DOOR

(b) Gaining out for a door hinge

Figure 1-25   Cutting with a wood chisel

(h) Make a *series* of cuts prior to chiselling out the bulk of the wood in a large gain.

(i) Chisel from the edge to the *end* or from the edge to the *middle* to produce a neater and smoother cut.

HAND PLANES

Originating from a wood chisel mounted on a block of wood at an angle, the metal hand plane is used for *smoothing* the rough surfaces of material after they have been rough cut to size. Categorized according to their *use* and/or *length,* types of hand planes include: **bench, block,** and **specialty** planes. Figure 1-26 shows the main parts of the hand plane.

Bench Plane

All bench planes use the same design and cutting principle. Differences in the width of the cutting blade or iron, length and mass are the determining factors in choosing the correct hand plane. It should be noted that *all* bench planes have the plane iron installed with the bevel *down* and have a plane iron cap installed *above* the plane iron to break and curl the shaving

(Fig. 1-27). It is generally accepted that the *longer* the plane bottom the straighter the surface and/or edge of the board.

The **smooth** plane is designed for *final* trimming and smoothing of edges and the surfaces of a board. Being the shortest, it is therefore the lightest, and has a blade width of 44 to 51 mm and a length of 235 mm.

The **jack** plane is the all purpose hand plane. Used for preparing sawn material before final finishing, it will do the work of all other bench planes discussed here. The most popular size available has a 51 mm wide blade and is 356 mm long (Fig. 1-28).

A **fore** plane *combines* the operations of the jack plane and the jointer plane. Being longer (457 mm) in length it can be used more effectively on larger work where the edges and surfaces need smoothing and straightening.

The **jointer** plane, largest of the bench planes (559-600 mm long) is used on longer edges where straightness is important. As implied by the name, it produces long straight edges for fitting or joining together the edges of stock where accuracy is of prime importance.

Block Plane

The smallest of all the hand planes (152-178 mm), the block plane is designed to be used on *end* grain and other small jobs. Used as a one-handed plane, the plane iron is set at a very flat angle (Fig. 1-28, 1-29). It comes *without* the plane iron cap common to the bench plane. The plane iron, because of its angle and absence of a cap is placed in the plane with the *bevel up*. An adjustable mouth allows for different types of planing. Generally, cross or end-grain requires only a small opening, while planing with the grain requires a larger mouth opening.

Bench plane:

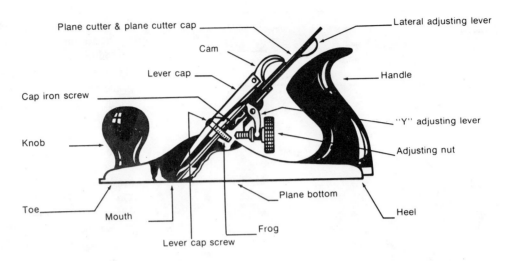

Plane cutter & plane cutter cap

Cam

Lever cap

Cap iron screw

Knob

Toe

Mouth

Lever cap screw

Frog

Plane bottom

Lateral adjusting lever

Handle

"Y" adjusting lever

Adjusting nut

Heel

Block plane:

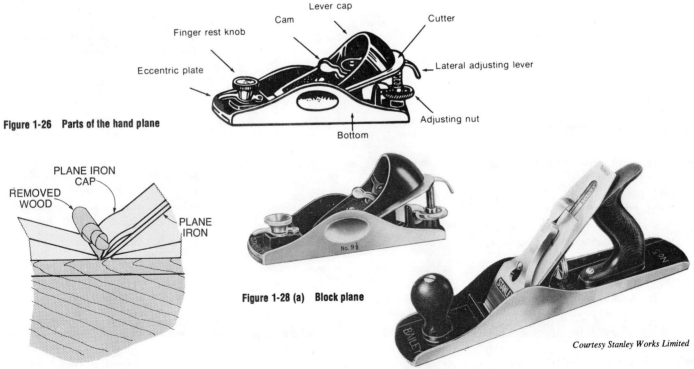

Lever cap

Cam

Cutter

Finger rest knob

Eccentric plate

Lateral adjusting lever

Adjusting nut

Bottom

Figure 1-26   Parts of the hand plane

PLANE IRON CAP

REMOVED WOOD

PLANE IRON

Figure 1-27   Plane iron and plane iron cap

No. 9½

Figure 1-28 (a)   Block plane

Courtesy Stanley Works Limited

Figure 1-28 (b)   Jack plane

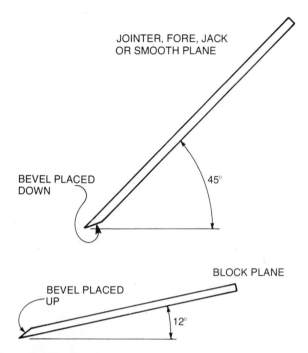

JOINTER, FORE, JACK OR SMOOTH PLANE

BEVEL PLACED DOWN

45°

BLOCK PLANE

BEVEL PLACED UP

12°

Figure 1-29   Cutting angle of the plane irons

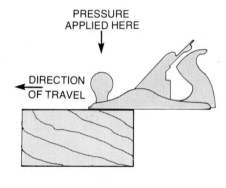

PRESSURE APPLIED HERE

DIRECTION OF TRAVEL

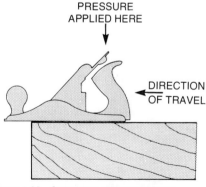

PRESSURE APPLIED HERE

DIRECTION OF TRAVEL

Figure 1-30   Correct use of the hand plane

Specialty Planes

Specialty planes are those planes designed for a particular purpose. Included in this group of planes are the **rabbet, router** and **trimming** plane. Although used quite extensively in the past, they are now being replaced with power operated equipment which can do the work much faster.

When using any hand plane a few basic rules should be observed:

(a) Keep the plane iron *sharp*. A dull or poorly sharpened plane iron will not produce the desired results.
(b) Always plane *with* the grain of the wood if possible (Fig. 1-30).
(c) To plane a straight line, keep the pressure on the *front* of the plane at the beginning of the cut and on the handle at the *end* of the cut.
(d) If working on end grain, work from the *edge* toward the *centre*, to avoid splitting the wood.
(e) Always lay the plane on its *side* between uses to prevent blade damage.
(f) Retract the blade when *storing* the hand plane.

Sharpening the Plane Iron and Wood Chisel

Two operations are usually employed when sharpening the plane iron and wood chisel: **grinding** and **whetting**.

Grinding is necessary in the sharpening process to restore the correct *bevel* before the final whetting of the plane iron or chisel. Grinding may not be required unless the blade has had a number of whettings and the bevel has gone or the cutting edge has been nicked.

The tool-rest on the power grinder should be set to achieve a *hollow ground level* of 25° to 30°. While grinding the plane iron or chisel, move it from side to side, using the full width of the grindstone (Fig. 1-31). Frequent dipping in water will keep the blade cool and prevent *burning* or *softening* of the tempered blade. Check the ground edge with a square to make sure the cutting edge is at right angles to the side of the blade.

After the grinding is complete, the beveled edge is whetted. This produces a small *second* bevel and a sharp cutting edge. An **oilstone**, the kind with a coarse grit on one side and fine grit on the other, is preferred. Using a small amount of oil as a *lubricant*, place the blade on the coarse side of oilstone with the bevel down. Raise the blade to an angle of approximately 30° to 35°. Stroke the blade back and forth a number of times until a second bevel appears (Fig. 1-31). To maintain a flat surface on the oilstone, always use the entire width.

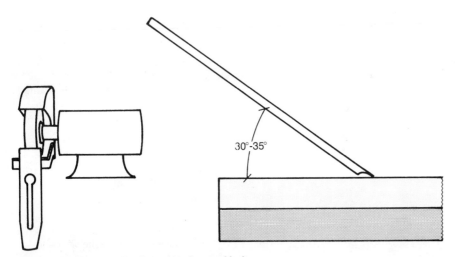

Figure 1-31 Grinding and whetting a plane iron or chisel

In preparing to finish the whetting procedure, wipe off the oilstone and turn it over. On the fine side of the oilstone, using the same angle as before, continue rubbing the blade back and forth a few times until a *wire edge* is produced. To remove the wire edge, place it **flat** on the oilstone, bevel side up and work it back and forth until it is removed. Be sure **not** to raise the blade during this operation as this will produce a double bevel on the blade and ruin the edge you have been striving for.

BORING AND DRILLING TOOLS

The **bit brace** (ratchet brace), and portable electric **hand drill** are commonly used for holding and turning a variety of bits and drills. This section will deal with those drilling tools specifically used by hand. The electric drill and bits or drills common to it can be found in the power tool section.

Bit Brace

The **bit brace** is used to hold and turn bits which have a square tapered *shank* or *tang* (Fig. 1-33). Hold the head of the brace in position with one hand or against your body. Grasp the handle and turn it

through the *sweep* or diameter of swing. Common sizes include 203, 254, and 305 mm bit braces. Figure 1-32 shows a bit brace and its parts.

The ratchet device is controlled by the cam ring and can be set so that the bit may be turned to the left, right or both, depending on the requirements. This allows for the drilling of holes close to corners or awkward places where a full sweep of the handle is not possible. The V-grooved, two-jaw chuck holds the tang of the bit solidly within the shell of the chuck assembly.

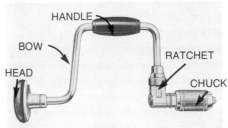

Figure 1-32 Bit brace

Wood Auger Bits

Auger bits vary from 6 to 25 mm in diameter and from 178 to 250 mm in length. The principle parts include the *shank, twist* and the *head*. The **shank** terminates with the square end or tang, and fits into the jaws of the bit brace. Stamped on the shank and/or the tang is the *diameter size* of the bit. Figure 1-33 shows a typical auger bit and its parts.

The **twist** on the auger bit is designed to *remove* shavings from the drilled hole to prevent binding of the bit in the hole while drilling. Available in three different designs; the *single* twist, the *single twist solid centre* and the *double* twist, each has advantages and disadvantages. The auger bit shown in Figure 1-33 is a single twist solid centre bit.

The **head** of the auger bit contains the *screw, spur(s)* and *cutting lip(s)*. The screw centres the bit and draws it into the wood fibre. Available in coarse, medium or fine thread, the first two are used for general work where speed is a consideration. If quality is desired a fine thread will produce a smoother cut hole.

The spurs on the auger bit score the outer circumference of the hole and help to prevent chipping around the hole. The cutting lips cut and lift the chips of wood from the area being drilled. The chips are passed up through the throat to the twist which then removes them from the hole.

Courtesy Stanley Works Limited

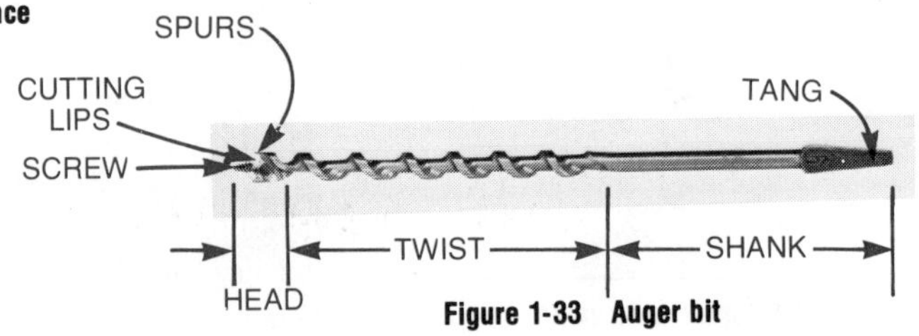

Figure 1-33 Auger bit

Expansive Bit

An **expansive bit** is adjustable and is used for boring holes larger than 25 mm in diameter. It has a movable cutter and spur, and is capable of drilling holes up to 125 mm in diameter (Fig. 1-34).

Although some manufacturers provide a dial setting to adjust the cutter, it is a good policy to *check* the size before cutting the hole. This can be accomplished either by measuring the radius (from the spur to the screw) or by starting the bit into a scrap piece of wood. Measuring the diameter as indicated by the spur mark will confirm your setting. Always be sure the locking screw is tightened to avoid movement once the actual cutting has begun.

Courtesy Stanley Works Limited

Figure 1-34   Expansive bit

Automatic Drill

The **automatic drill** (Fig. 1-35) is used to drill small *pilot* holes for nails and/or wood screws. Drill points range in size from 1.6 to 4.4 mm and are contained in the handle of the drill for easy access. Specially designed, they lock into the chuck of the drill. The tool is held with one hand leaving the other free to hold the work in place. Drilling is accomplished by pushing *down* on the handle. When pressure is released, a reversing spring turns the drill counter-clockwise, clearing the wood chips.

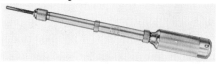

Courtesy Stanley Works Limited

Figure 1-35   Automatic drill

There are many accessories which can be purchased to aid in drilling special types of holes. The **countersink bit** is used for enlarging the top of the hole for countersunk or flat head wood screws so that they will not protrude above the surface of the material. The **bit gauge** (either commercially made or one made for the job), controls the depth of the drill bit. A **dowelling jig** is used to establish the exact location of dowel pins in a dowelled joint.

Forstner bits are used for drilling holes partway through material. An ordinary auger bit would go through or damage the back of the work due to the screw point. **Bit extensions** are used to increase the reach or depth of drill bits. Figure 1-36 shows a number of these accessories.

Note: For other types of drill bits and accessories see page 23 under Portable Electric Hand Drill.

In caring for and using the bit brace, drilling tools and special accessories the following should be observed:

(a) Lightly oil the head and chuck mechanism of the bit brace to *ensure* its smooth operation.

(b) As with other tools, rusting of the bit brace and bits can be avoided by storing them in a dry place and with an occasional wiping with an oily rag.

(c) Bits which become bent can be straightened by placing a hardwood block on the bent portion and lightly tapping with a hammer on the block.

(d) Bits should be stored in a pouch or case which *protects* the cutting head and parts from knocks or damage.

(e) When sharpening auger bits, use an auger bit file, maintaining the original angles. **Never** file on the outside of an auger bit as this will change the diameter of the bit.

(f) When boring a hole through a board, use a backup board to *prevent* damage to the back of the material. The alternative to this is to *reverse* the bit when the screw begins to come through the opposite side of the

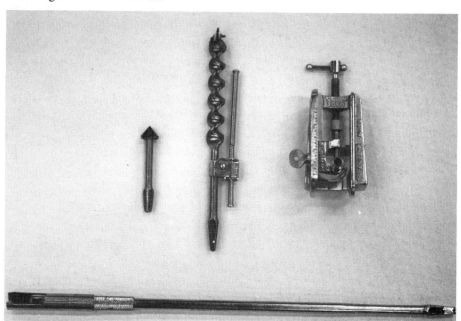

Figure 1-36   Special accessories for drilling

board. This prevents the wood around the hole from splintering as the bit comes through.

(g) Use *light* pressure when using the automatic drill. Too much pressure results in bent or broken bits.

FASTENING TOOLS

Fastening tools are used to temporarily hold parts together while laying out work, fitting parts, gluing or assisting in the permanent securing of those parts. Various clamps and screwdrivers are used for a number of jobs in the construction of the home. Generally, their most extensive use occurs during the exterior-interior finishing and cabinet work as the house nears completion.

Clamps

Clamps are manufactured in a wide range of styles. The **C-clamp** is probably the most widely used. It ranges in size from 25 to 300 mm and comes with either a sliding pin or wing nut style handle for tightening. When using the C-clamp on finished work, *clamping blocks* should be used at the pressure points (the head and fixed jaw) to avoid marring the wood. With precision work, care must be taken after gluing to ensure the pieces are not pulled out of position during the final tightening (Fig. 1-37).

The **bar** or **pipe clamp** is a variation of the C-clamp. Consisting of a flat bar or pipe frame, it has a *fixed* jaw at one end with a *sliding* or *adjustable* jaw at the other. A bar clamp can range in size from 150 mm to 2 m in length, while a pipe clamp is limited by the length of pipe being used. Jaws on a bar clamp are fixed in position. The jaws on a pipe clamp may, however, be turned, allowing two shorter clamps to be put together for clamping extra long work. Like the C-clamp, clamping blocks should be used on finished work to prevent marring of the wood (Fig. 1-38).

Figure 1-37   C-clamp

A **parallel** or **hand screw** clamp is one of few clamps made that do not require clamping blocks under the jaws. The blocks are eliminated by the wide hardwood jaws which spread the pressure over a wider area. Using a right and left-handed screw, the jaws may be set either *parallel* or *tapered,* depending on the shape of the work being clamped (Fig. 1-39).

The **band** (web) clamp is specially designed for holding *irregular* shapes together while applying even pressure. Using either a screw tightener or ratchet device (Fig. 1-40), pressure is applied by drawing the band together. Clamp bands are made of steel, canvas or nylon.

One of the more recent clamps to be put on the market is the **spring clamp**.

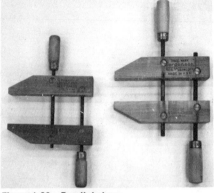

Figure 1-39   Parallel clamp

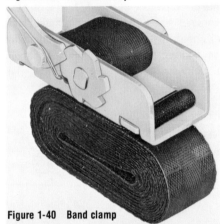

Figure 1-40   Band clamp

Courtesy Stanley Works Limited

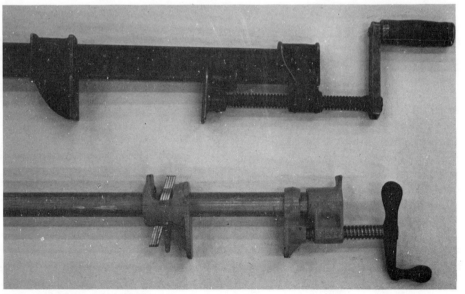

Figure 1-38   Bar and pipe clamps

Generally associated with light-duty work, some spring clamps require two hands to open the jaws. The jaws may be formed to hold flat or round objects and are plastic coated to prevent marring of wood surfaces (Fig. 1-41).

Courtesy Stanley Works Limited

Figure 1-41   Spring clamp

Screwdrivers

Screwdrivers are available in a variety of shapes and sizes but have only one use— the *driving* and *extracting* of threaded fastening devices such as wood screws, machine screws and self-threading screws. **Always** use the *size* and *type* of screwdriver to match the head of the screw (Fig. 1-42). Used correctly, the screwdriver will perform the job for which it was designed–driving screws. In spite of that, the most common mistake made is using a screwdriver that does not fit or match the screw.

The **spiral ratchet** screwdriver saves both time and work when a large number of screws need driving. Like the automatic drill it is operated by pushing down on the handle. A shifter mechanism on the shaft allows for a right or left hand drive. Setting the shifter determines whether the screw will be driven in or extracted (Fig. 1-43).

With the range of screwdrivers available today, it is possible to buy any length, tip size and style required.

Proper *use* and *care* of fastening tools are reflected in the work when it is com-

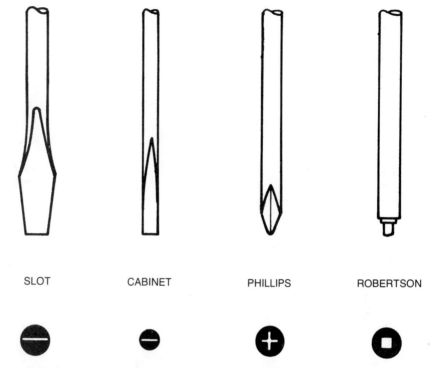

SLOT   CABINET   PHILLIPS   ROBERTSON

Figure 1-42   Commonly used screwdrivers

Courtesy Stanley Works Limited

Figure 1-43   Spiral ratchet screwdriver

pleted. In selecting the proper clamp(s) for a job, check the following:
(a) How *wide* is the work requiring the clamping?
(b) Is a *deep* or *shallow* clamp required?
(c) Will it be necessary to use *clamping blocks* to prevent marking the work?
(d) What *pressure* is required to clamp the work together; is a light or heavy duty clamp required for the job?
(e) After evaluating the situation, choosing the correct clamp will not only do the job but will also save time. When the clamps are no longer needed be sure to *return* them to their respective storage areas.

When using screwdrivers, consideration should be given to the following:
(a) Use the proper screwdriver(s) for the job. One that is the *wrong size* or has a *poor tip* will often result in damage–to the screw head, the work or to the person using it.
(b) **Never** use screwdrivers for prying, chiseling or scraping.
(c) Store screwdrivers in a pouch or rack where *selection* of the right one is easy.

POWER TOOLS

Today the carpenter has within arm's reach a multitude of power tools. This has led to an increase in production and reduces the number of hours spent on the many intricate operations required to construct a building. The amount of energy spent to increase production, save time and still retain the quality of the product is only possible when power tools are used in a safe and correct way.

Power tools may be grouped into three general categories: *stationary, portable,* and *powder* or *air actuated tools.* **Stationary electric tools** include those machines which come equipped with a bench or floor stand. Although they can be moved from one location to another, once they are set up it is common practice to leave them in one place for the duration of the job. Materials requiring machining are brought to the machines for cutting, shaping, etc., and are then moved to an assembly area. On the other hand, **portable electric tools** are brought to where the material is or to where assembly takes place. Being the most convenient and widely used, the portable power tools arrive at the building site with the carpenter and do not leave until the home is completed. **Powder** and **air tools** are spe-cialty tools and are designed for one specific use. Tools of this nature are driven by either explosive cartridges or compressed air.

In this overview of power tools it is important to realize that these tools are designed to increase production while reducing the work load of the person using them. However, along with that comes a certain degree of skill necessary to operate the tool correctly. **Safety** can never be over stressed nor can the proper use and care of the power equipment. It is recommended that manufacturer's instructions, operator's manuals, safety charts, and textbooks devoted to the setting up and operation of these tools be used to complement and expand on the information given here.

STATIONARY ELECTRIC POWER TOOLS

Table Saw or Bench Saw

Of the stationary power tools used by the carpenter, the table saw is probably the most versatile and most widely used. There are two designs–*tilting table* or *tilting arbor* (most common). The capacity of a table saw depends on the largest diameter blade it will take. Figure 1-44 shows a typical table saw and its component parts.

Noted for its wide use, the table saw performs these basic cuts; crosscutting, bevel crosscutting, mitres, bevel mitres, ripping and bevel ripping with ease. With experience, the right accessories and in some cases jigs, it is easy to produce many of the bevels, mitres, dadoes and mouldings used today.

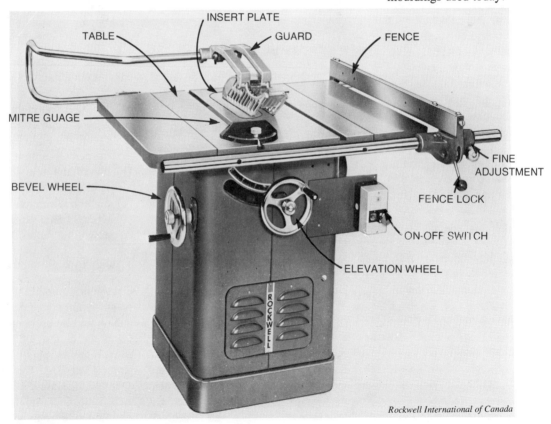

Rockwell International of Canada

Figure 1-44   Tilting arbor table saw

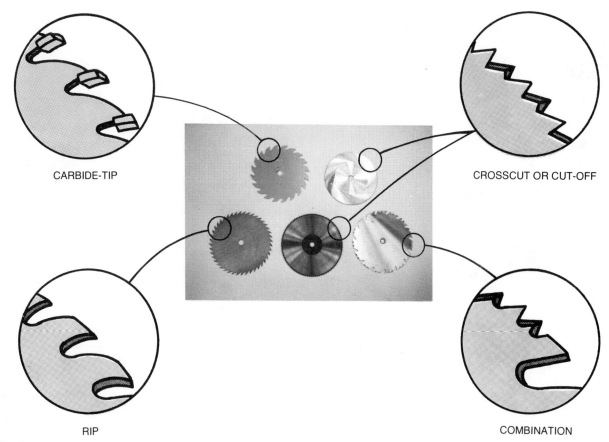

CARBIDE-TIP

CROSSCUT OR CUT-OFF

RIP

COMBINATION

Figure 1-45   Types of circular saw blades

Circular Saw Blades

Circular saw blades for all power saws (stationary and portable) are essentially the same and can be interchangeable. The only restriction is the size of blade and the diameter and/or style of arbor hole the power saw will accommodate.

Types of Circular Saw Blades

Types of circular saw blades vary depending on the material to be cut, the type of cut and the smoothness of the cut. In a **spring set** blade, the individual teeth are bent alternately to the left and right to provide the required kerf width, preventing the blade from binding. Available in *rip, crosscut* and *combination* style (Fig. 1-45), they are recommended for cutting

lumber (both hardwood and soft wood) and flooring, and rough-cutting of plywood.

The **hollow ground** blade does not require setting because it is tapered from the outer rim towards the centre. The taper can extend inwards to the arbor hole or only part way, like that used on the thin rim plywood blade. Used where extremely smooth saw cuts are required, they are most useful in exterior-interior finishing of the home.

Carbide-tipped blades may be used for all cutting operations. Carbide teeth slightly wider than the blade material are welded into place, providing a blade which will not dull easily and therefore requires little maintenance. However, due to the hardness of the carbide tips, they

will chip easily and care must be taken not to drop or bang them against a hard surface. Available in rip, crosscut, or combination style for all types of wood, they can also be used to cut asbestos board, laminates, sidings and some roofing materials (Fig. 1-45).

A **dado head** assembly consists of a special set of saw blades used on either the table saw or radial arm saw to produce dado and/or rabbet joints. Consisting of two outer blades called the *cutters,* and a number of other blades of various thicknesses called the *chippers,* they can be used to produce joints from 1.6 to 20.6 mm wide (Fig. 1-46 (a)).

Moulding head assemblies may also be purchased for use on the table saw or radial arm saw. Available with cutters of

Figure 1-46 (a)   Dado head assembly

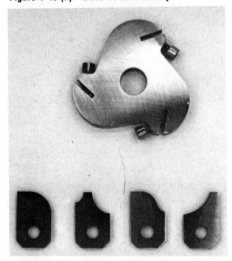

Figure 1-46 (b)   Moulding head assembly

different shapes, they may be used to produce many of the standard mouldings used today (Fig. 1-46 (b)).

Radial Arm Saw

The radial arm saw (Fig. 1-47) is used in many cases to either replace or complement the table saw. A motor unit and blade mounted above the table is positioned in a yoke assembly, which moves back and forth on an overarm extending across the table. Rotating a full 360° on the yoke assembly, coupled with the ability of the arm to swing to the right or left

Courtesy Rockwell International of Canada

Figure 1-47   Radial arm saw

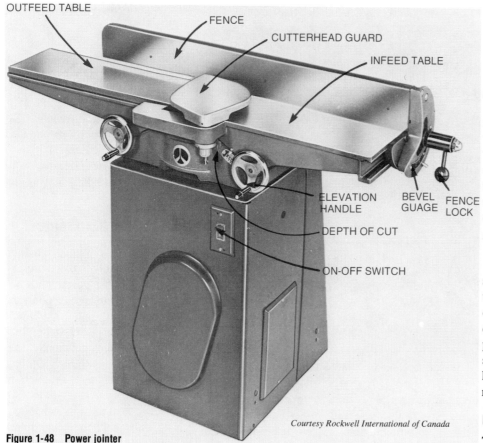

OUTFEED TABLE

FENCE

CUTTERHEAD GUARD

INFEED TABLE

ELEVATION HANDLE

BEVEL GUAGE

FENCE LOCK

DEPTH OF CUT

ON-OFF SWITCH

Courtesy Rockwell International of Canada

Figure 1-48   Power jointer

and the capability for tilting the blade gives it all the features of a table saw.

Basic cuts common to the table saw can be performed on the radial arm saw and in many cases the operation is faster and safer. This is especially true when the saw is being used as a cutoff saw for pattern cutting. All types of saw blades, dado and moulding heads may be used on the radial arm saw.

Power Jointer

A power jointer is designed to plane the surfaces, edges and ends of a board in the same manner as the hand plane. In many cases it can be used to plane tapers, chamfers, bevels and rabbet joints. The capacity of a power jointer is determined

by the cutting width of the knives which are mounted in the cutterhead on the machine (Fig. 1-48).

A cutterhead usually contains three knives and is mounted between the split bed or table. The *front table* is adjusted up or down depending on the thickness of the cut required. The *rear table* is also adjustable, but adjustment occurs **only** when a new or re-sharpened set of knives is installed (Fig. 1-49 (a)). If the rear table is not level with the cutting edge of the knives, the jointer will not function as designed. When the rear table is higher than the knives, the result will be an uneven cut. Having the table lower than the cutting edge of the knives will result in a gouge at the end of the board (Fig. 1-49 (b) and (c)).

Jointer knives are made of high speed steel or may be carbide tipped. Carbide-tipped knives have the same advantage as carbide saw blades, in that the cutting edge remains sharp for an extended period of time. This not only produces smoother-jointed parts, but reduces the honing and knife maintenance required of regular jointer knives.

Bench Grinder

The bench grinder is a common tool to all workshops because of the constant need to sharpen and re-sharpen both hand and power tools. The extended motor shaft on most of the models becomes the spindle,

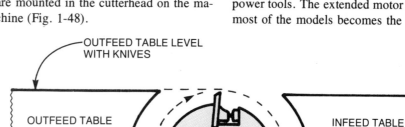

OUTFEED TABLE LEVEL WITH KNIVES

OUTFEED TABLE

INFEED TABLE

KNIFE

CUTTERHEAD

Figure 1-49 (a)   Proper setting of the jointer tables

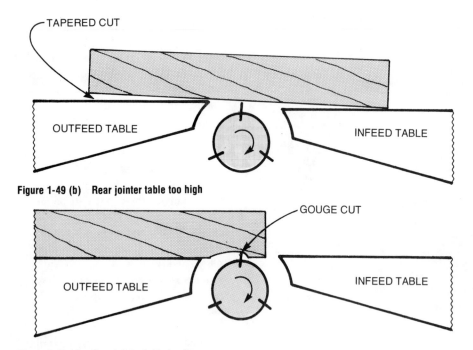

Figure 1-49 (b)   Rear jointer table too high

Figure 1-49 (c)   Rear jointer table too low

to which the grindstone(s) is attached. Grinder size is determined by the largest abrasive wheel the grinder will receive (Fig. 1-50). Guards enclosing the wheels should be as full as possible to protect the operator. Tool rests are adjustable, and as stone wear occurs, tilt to the required angle for precision grinding or dressing of the different tools. To obtain the most use from the bench grinder, it should be fastened to a workbench or mounted on a floor pedestal.

Safety of Stationary Equipment

Attitude is the most important factor in developing safe work habits around all

Courtesy Rockwell International of Canada

Figure 1-50   Typical bench grinder

power equipment. Simply learning how a tool operates is not sufficient. A responsible person knows how to operate the equipment in a safe and correct manner; thus protecting not only himself but those around him. Points to keep in mind are:

1. **Proper** clothing is a must. Loose fitting or torn clothing, dangling cuffs on shirts, rings or other objects have no place around power equipment.
2. Be totally familiar with the operation of the machine. **Consult** operator's manuals, etc., or personnel who have the necessary knowledge about the machine.
3. Keep the work area around the machine **free** of debris.
4. Keep cutting tools sharp. **Remember:** a dull tool is a dangerous tool.
5. Be sure to make *all* necessary adjustments before starting up the machine.
6. Use the *safety devices* at your disposal when operating the machine.

Make sure guards, etc. are functioning properly.
7. Know what you are going to do before turning the switch on.
8. Do not let activities of persons around you distract you while operating the machine.
9. **Avoid** force-feeding a machine. Listen to the sound it produces and feed the material in accordingly.
10. **Always** shut off the machine, waiting until it stops before leaving the area.

PORTABLE ELECTRIC HAND TOOLS

Portable electric tools have evolved from a number of hand tools. Like stationary equipment, all portable power tools are designed to speed up production and lessen the workload at hand. Therefore, it is imperative that the operator be completely familiar with the operation of each portable hand tool.

Grounding (connecting a neutral conductor in the circuit to the ground) of portable electric hand tools is very important. It is not enough to simply use an extension cord and plug that are in good condition; they must also be grounded to prevent a serious electrical shock to the user. Only approved, 3-wire extension cords, properly grounded from the source should be used. Today, most manufacturers utilize the *double insulated* system on a number of their tools. This provides two separate insulation systems between the operator and the electrical current for greater safety. It would be necessary for both systems to fail for the operator to receive a shock using tools so designed.

Hand Drill

Of all the portable tools, the hand drill is probably the most common. Specified by the largest drill bit the chuck will accommodate, common sizes are 6, 10 and 13 mm. The standard 3-jawed chuck can be either keyed or keyless. From the basic model to those which come equipped with variable speeds and full reversible motors, the range of styles seems endless. Figure 1-51 shows a 10 mm pistol grip, variable speed, reversible hand drill.

Like the electric drill, drill accessories are many and varied. Depending on the material being drilled, bits can include high speed steel twist bits, spade bits, carbolide bits, hole saws, circle cutters, and countersinks. Figure 1-52 shows a number of the accessories used with the hand drill.

A specialized drill, the **rotary hammer**, uses a hammering stroke, combined with a rotating motion to drill through hard materials such as brick, stone and concrete. Along with a number of carbide-tipped percussion drills, demolition tips are also available for remodelling or repair work (Fig. 1-53 (a)).

Courtesy Rockwell International of Canada

Figure 1-51   Electric hand drill

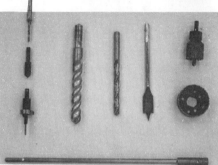

Figure 1-52   Hand drill accessories

Figure 1-53 (a)   Rotary hammer drill

Courtesy Rockwell International of Canada

Figure 1-53 (b)   Power screwdriver

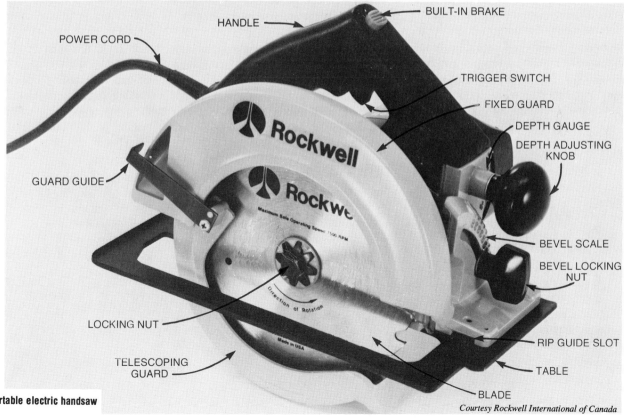

POWER CORD

HANDLE

BUILT-IN BRAKE

TRIGGER SWITCH

FIXED GUARD

DEPTH GAUGE

DEPTH ADJUSTING KNOB

GUARD GUIDE

BEVEL SCALE

BEVEL LOCKING NUT

LOCKING NUT

RIP GUIDE SLOT

TELESCOPING GUARD

TABLE

BLADE

Figure 1-54   Portable electric handsaw

Courtesy Rockwell International of Canada

The **power screwdriver** has reduced the painstaking labour of driving and extracting screws. Shown in Figure 1-53 (b), it comes equipped with interchangeable tips for a full range of wood, machine or self-tapping screws. One area where the power screwdriver is used extensively today is in the application of drywall. Fully reversible, with a built-in slip clutch to prevent over-driving of the screws, the screwdriver has a wide range of uses and possibilities.

Portable Handsaw

The portable electric handsaw is now a common tool in the carpenter's tool chest. The size of a power handsaw depends on the largest diameter blade it uses. Common sizes are 184 to 235 mm. Figure 1-54 shows a typical power handsaw and its parts.

This saw performs the same basic cuts as the table saw or radial arm saw, and all types of blades can be used in it with the exclusion of dado and moulding heads. Because the table on the handsaw is positioned *above* the work, the teeth on the blade point *up* instead of down as is the

Courtesy Skil Canada Limited

Figure 1-56 (a)   Portable electric jigsaw

case on stationary saws. Used for both crosscutting and ripping, the depth of cut is fully adjustable, depending on the material being sawn. Bevel adjustments allow for angular cuts from 90° to 45°. A telescoping guard returns to cover the exposed blade once the cut has been completed.

A new power saw recently released on the market is the **electric mitre saw.** Designed to complement the hand mitre box (Fig. 1-18), it is fast earning its place as a cutoff saw for framing as well as finishing work. The common size available is 254 mm. Calibrated for angular cuts from 90° to 45° both right and left, it functions as a plunge saw when hinged from behind the table fence (Fig. 1-55).

Portable Jigsaw

The portable jigsaw (sabre saw) is used for straight, bevelled or scroll cutting. Using a narrow, reciprocal-action blade, it is easily manoeuvered through most materials. Models include single-speed,

two-speed and variable speed saws, which allow for the specific needs of the job (Fig. 1-56 (a)). Blades are available for cutting wood, metal, or plastic. Therefore, it is important to select the correct blade for the material being cut. One of its most important features "plunge cutting" is its ability to cut out holes for sink installation or wall and floor openings, without first having to drill a hole for the blade.

With its operation similar to that of a jigsaw, the **reciprocating saw** (Fig. 1-56 (b)) is used for a variety of cutting jobs. With a larger, longer blade combined with a longer blade stroke it has a greater capacity than that of the jigsaw.

Portable Electric Power Plane

The portable electric power plane is designed primarily to eliminate jointing operations that cannot be handled safely on the power jointer or where it is not practical to use the jointer (Fig. 1-57 (a)). A fence may be mounted onto the side of the tool, allowing for better control and precision-planing while machining edges or bevels. Adjustment for the *depth of cut* is controlled by the front portion of the plane bottom. With many models this may be adjusted if required as planing progresses.

Designed to complement the power plane, the **block plane** is used where smaller planing jobs are required. It is however, capable of planing materials up to 45 mm wide. It is particularly useful during exterior-interior finishing and cup-

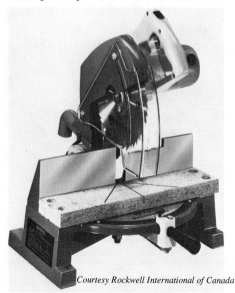

Courtesy Rockwell International of Canada

Figure 1-55   Portable electric mitre saw

Figure 1-56 (b)   Electric reciprocating saw   *Courtesy Rockwell International of Canada*

Courtesy Rockwell International of Canada

Figure 1-57 (a)   Portable electric power plane

board construction (Fig. 1-57 (b)). High speed steel and carbide-tipped cutters are available for both planes. Use a constant pressure on the plane while it is in operation. **Remember** to keep the pressure forward when beginning to make the cut, moving it to the back of the plane as you finish the cut.

Courtesy Rockwell International Canada Ltd.

Figure 1-57 (b)   Portable electric block plane

Portable Electric Sanders

When sanding and finishing surfaces prior to painting, varnishing or removing old finishes, portable sanders such as the *belt* sander and/or *finishing* sanders may be used. Sanders are rated according to the size of the sanding belt or by the size of the pad on the sander. Abrasive belts and sandpaper come in a variety of grits and must be chosen according to the job requirements.

When using a **belt sander** (Fig. 1-58) care must be taken to ensure that the belt is installed correctly and that it is tracking

properly. Adjustments must be made before starting to sand. Always be sure the work is securely fastened and that you have complete control of the machine as considerable force is exerted by the belt sander when operating. The weight of the machine alone provides all the necessary force required to make the sandpaper grit cut. Used at a sharp angle to the grain, belt sanders should *never* be used across the grain as gouging may occur.

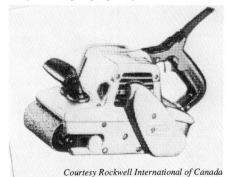

Courtesy Rockwell International of Canada

Figure 1-58   Belt sander

Finishing sanders are used where fine sanding is required. They move in an oval (orbiting) or forward and backward (oscillating) motion. Both types may be used with or across the grain to produce a smooth surface. Much lighter than the belt sander, they are quite often used to touch up joints in cabinet work, trim, etc. where the belt sander would be too big and cumbersome to use (Fig. 1-59).

Portable Electric Router

The electric router is a highly versatile power tool because of the numerous operations that can be performed. The basic router consists of a high speed motor (20 000 to 27 000 r/min) and a base which regulates the depth of cut while it holds the motor in position. With the wide selection of bits available, you can perform freehand routing, dadoing, scrollwork, morticing, edging, etc. With its numerous accessories, the router has

(a) Pad sander

Courtesy Rockwell International of Canada

(b) Block sander

Courtesy Rockwell International of Canada

Figure 1-59   Finishing sanders

many uses in the finishing work on a house. Routing out for door hinges, stair treads, cupboard doors, drawers, and laminate trimming is done with the portable router.

Courtesy Skil Canada

Figure 1-60   Portable electric router

Safety in Using Portable Equipment

When using any portable equipment, points to keep in mind include:

1. Properly *grounded* extension cords are a must. **Remember:** tools equipped with an off-on switch should be turned "off" before plugging them into the extension cord.
2. Keep the extension cord *back* out of the way of the tool and of the work being done.
3. **Clear** the work area of debris, unnecessary material and equipment.
4. **Support** all work adequately. Portable tools must be held, so be sure your work is fastened properly.
5. Become familiar with the tool before attempting to use it. **Consult** operator's manuals, etc. or personnel who have a good working knowledge of the machine.
6. Keep all cutting edges *sharp;* be sure guards are functioning properly.
7. Make necessary adjustments *before* plugging in the tool.
8. Wear tight fitting clothing so that it will not become tangled in the moving parts of the machine.
9. Use necessary safety protection when required.
10. **Always** unplug the power tool when you are finished using it.

POWDER ACTUATED TOOLS

When it is necessary to fasten materials to concrete or steel, the powder actuated tool is often used. This eliminates the need to drill, sleeve and then bolt material into place. Using the energy from a charge of powder, these tools drive a special heat-treated steel fastener into the concrete or steel (base material) in a single operation. Attachments to the base material can be wood, steel, aluminium or eyepins for anchoring.

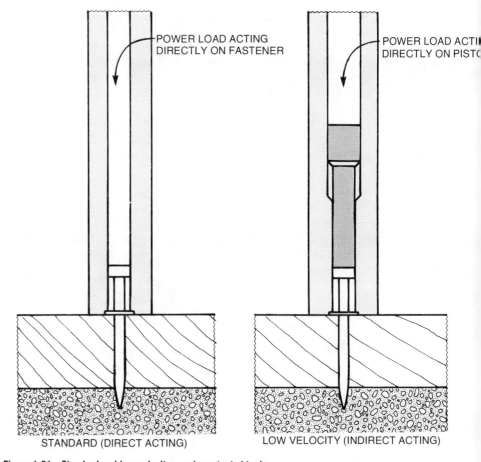

Figure 1-61   Standard and low velocity powder actuated tools

Two basic types of tools are available, one being high velocity, the other low velocity. The **standard velocity** tool uses the energy of the power load directly on the fastener to drive it into the base material. To decide which fastener to use and which powder load is needed to drive the fastener, it is recommended that trial shots be taken to determine what strength of cartridge to use. Always start with the lowest cartridge strength and work up until proper pin penetration has been achieved.

A **low velocity** tool uses the energy of the powder load to drive the piston within the tool. The piston in turn drives the fastener into the base material. Since the travelling distance of the piston is restricted by the barrel of the tool, the velocity at which the fastener is released is greatly reduced. This feature makes the low velocity tool considerably safer to use when compared to a standard velocity tool. This is because the fastener can never exceed 91.4 m/s; therefore, even if the fastener should leave the tool in uncontrolled free flight, it can cause no harm. This is because the energy of the cartridge is transmitted to both the heavy piston and then to the fastener. As the piston is limited in its travel within the tool, the excess energy of the piston is absorbed by the buffer in the tool.

Tools which qualify as low velocity must bear the C.S.A. label, stating that it is a low velocity tool. This assures the operator that the tool was tested by the C.S.A. laboratory, that it has all the required safety features, and that the velocity of the fastener is within the required limits.

Another feature of the low velocity tool is the ability to redrive fasteners that are not properly set. When this occurs another cartridge is inserted into the tool, repositioned over the fastener and driven again. .**Note:** This should never be attempted with a standard velocity tool. Figure 1-61 indicates the difference between the standard and low velocity tools.

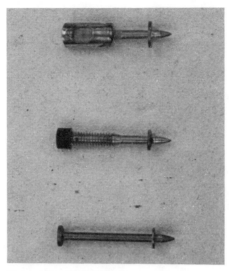

Figure 1-62   Basic fasteners

Courtesy Ramset Limited

Figure 1-63   Aligning a powder actuated tool for firing

Cartridges for the two tools are rated in terms of the energy that they release. They are designated by a colour code: **Grey, Brown, Green, Yellow, Red** and **Purple**. Grey is the lightest and purple the most powerful load. The low velocity tool covers the first four levels (grey to yellow) only, while the standard velocity tool covers the full range. If there is any doubt as to which cartridge to use for a fastening job, always start with the lightest load and work up from that point until proper fastening is achieved.

Fasteners, which are used for securing materials to the base material (concrete or steel), are manufactured specifically for forced entry fastening (Fig. 1-62). Basic fastener types include **headed fasteners**, which are used to fasten directly to the base material where a permanent connection is required. If the material being fastened is prone to pulling, metal washers (supplied by the manufacturer) can be used to increase the size of the head. This will prevent the material being fastened from pulling over the head of the pin. **Threaded studs** are used to bolt an object

to the base material on either a temporary or permanent basis after the threaded stud is set. **Eyepins** are used where materials are suspended or hung from the base material.

Base material should always be tested before attempting to drive in the fastener. When using a fastener as a centre punch check the base material. If the point is blunted and the material undamaged, *do not* attempt to fasten to it as the material will be too hard. If the material cracks or breaks up, it is too brittle. Too rapid a penetration indicates that the material is too soft for proper fastening.

When using the standard or low velocity tool, **safety** is of prime importance to everyone. Every tool has standard built-in features to prevent accidental discharging of the tool too early. On the standard velocity tool, the guard must be in position; removing it renders the tool inoperable (low velocity tools do not require such a guard). Both tools must be held at right angles to the work while attempting to discharge them. Compression of the tool

engages the trigger mechanism, allowing it to be fired (Fig. 1-63).

Other safety rules designed to protect the operator:

1. **Never** use a powder actuated tool without having its operation thoroughly explained by a specialist.
2. When using a standard velocity tool, keep the shield in the centre position if at all possible.
3. **Never** try to place a fastener through a pre-drilled hole or in a spot where a fastener was previously placed.
4. Always **inspect** the base material. Will a fastener penetrate it properly?
5. Check the barrel for obstructions. Make sure it is clear before reloading.
6. Always use **safety glasses** when operating a powder actuated tool.
7. **Never** carry cartridges or fasteners in your pocket.
8. Do not load the tool until you intend to use it.
9. Remove the spent cartridge after the fastener has been installed.

10. **Never** use a powder actuated tool if you are unsure of the results it will yield.

Note: In a number of provinces, it is against the law to use a powder actuated tool without holding a valid operator's certificate certifying that you are qualified to operate one.

AIR ACTUATED TOOLS

Air actuated (pneumatic) tools depend on compressed air to operate them, to drive the fastener (nail or staple) into the connecting member. Used in every stage of assembly from the foundation through the finishing work, pneumatic nailers and staplers can be found in manufacturing plants and on-site construction. They have become popular because they speed up the production of the article, and have eased the labour involved in assembly. In many cases, they also provide a more effective connection between pieces of material. Figure 1-64 shows a typical air actuated nailer-stapler.

Nails and staples are available in *standard wire, stainless steel* (corrosion resistant) and some manufacturers have a *hardened nail* available for fastening into concrete and concrete block. Nail lengths range from a 25 mm finishing nail to a 90 mm framing nail. Staples vary from 5 to 90 mm depending upon their use. Available in rolls or sticks, they are held together with a plastic coating, enabling them to be loaded into the magazine on the tool quickly and easily.

Depending on the size and capacity of the nail or staple the tool will handle, and on the material being fastened, an in-line *regulator* is required to control the air pressure to the tool. Along with the regulator, a filter (to take out any water in the line) and a lubricator are required if the

(a) Nailer

Courtesy Senco Products, Inc.

(b) Stapler

Figure 1-64   Air actuated tools

tool is to function properly. Depending on the make, the magazine which holds the fasteners will be located under, in front of or behind the trigger mechanism. The fasteners in the magazine are fed into the driving chamber as required.

Manufacturers use a *dual release* system to prevent accidental firing of the nail or staple and to increase the safety of the tool. Squeezing the trigger without touching the tool to the surface will not activate the driving mechanism. Likewise, touching it down on the surface of the material without squeezing the trigger will not drive the fastener in. **Both** operations **must** be done *simultaneously*.

Courtesy Senco Products, Inc.

Figure 1-65   Typical applications of air-actuated tools

REVIEW QUESTIONS

1—HAND AND POWER TOOLS

Answer all questions on a separate sheet.

A. HAND TOOLS

A. (Select the correct answer)

1. The smallest division on the pocket tape is
   (a) 1 cm
   (b) 10 mm
   (c) 1 mm
   (d) 2 mm
2. The tool commonly used to determine the squareness of a piece of stock is a
   (a) Framing square
   (b) Try square
   (c) Combination square
   (d) T-square
3. A T-bevel is used to
   (a) Lay out curves
   (b) Test and mark angles
   (c) Square the board
   (d) Scribe lines
4. Tables found on the Frederickson framing square are
   (a) Angle, octagon, brace
   (b) Brace, rafter, hexagon
   (c) Hexagon, brace, rafter
   (d) Rafter, polygon, brace
5. When installing door hinges, you may require the aid of a
   (a) Butt marker
   (b) Marking gauge
   (c) Combination square
   (d) Try square
6. The handsaw which has teeth shaped like a row of chisels is a
   (a) Ripsaw
   (b) Panel saw
   (c) Backsaw
   (d) Crosscut saw

7. When referring to the handsaw, the word "kerf" means
   (a) The type of cutting tooth
   (b) The width of the groove made by the saw
   (c) The bending of the saw teeth
   (d) Changing the size of the tooth
8. To relieve strain on the hammer handle and increase leverage when pulling nails,
   (a) Grasp the handle near the hammer head
   (b) Place a block of wood under the hammer head
   (c) Twist the hammer handle to one side
   (d) Use both hands when pulling the nail
9. One of the following hand planes is normally used with only one hand:
   (a) Smooth plane
   (b) Jack plane
   (c) Block plane
   (d) Rabbet plane
10. One of the following clamps requires the use of clamping blocks:
    (a) C-clamp
    (b) Parallel clamp
    (c) Web clamp
    (d) Spring clamp

B. Replace the Xs with the correct word on your answer sheet

11. To prevent splitting the wood, always work XXXXXXX the grain of the wood when paring with a chisel.
12. A XXXXXXX screwdriver has a square shaped tip.
13. The bevel on the plane iron or wood chisel should be ground to an angle of approximately XXX degrees.
14. The size of the bit brace is determined by the XXXXXXX of the brace or diameter of swing.
15. A XXXXXXXX is used to drive nails below the surface of the wood.

C. Mark as either TRUE or FALSE

16. An accurate method of establishing plumb lines is to use a plumb bob.
17. The backsaw is used for fine work.
18. When driving a nail, grasp the hammer handle well in from the end of the handle.
19. The size of a chisel is determined by its length.

D. Write a full answer

20. List four precautions to take when using, caring for and maintaining hand tools.

E. Select the correct answer

B. POWER TOOLS

1. The capacity of a circular table saw is determined by
   (a) The size of the table on the saw
   (b) The diameter of the saw blade
   (c) The radius of the saw blade
   (d) The number of teeth on the blade
2. For most cutting purposes, the blade best suited for the job is the
   (a) Crosscut blade
   (b) Rip blade
   (c) Spring-set blade
   (d) Combination blade
3. A power jointer may be used to
   (a) Mortice
   (b) Rout
   (c) Rabbet
   (d) Shape
4. Setting the rear table of the power jointer occurs
   (a) When a new set of knives is installed
   (b) Before starting up the jointer
   (c) Each time the jointer is used
   (d) When making a tapered cut

F. Replace the Xs with the correct word(s)

5. The size of the portable electric hand drill is determined by the size of its XXXXXXX.

6. The depth of cut on the portable hand plane is adjusted by moving the XXXXXXX up or down.

7. Portable electric sanders are rated according to XXXXXXXX.

8. Powder actuated tools include two basic types, namely XXXXXXXX and XXXXXXXX.

9. Air actuated tools use a dual system to prevent accidental firing. To use the machine both operations must be XXXXXX.

10. A specialized drill uses a XXXXXX motion while drilling to aid in drilling materials such as concrete and masonry.

11. Power tools can be divided into two classes: XXXXX and XXXXXXX.

12. The teeth on the portable electric handsaw point up because the table is positioned XXXXXXX the material being cut.

13. The numerous XXXXXXX make the portable electric router a versatile tool.

14. Electrical shock, a potential hazard with all portable tools, is eliminated through using a XXXXXX cord. Manufacturers of tools today have virtually eliminated this hazard by introducing the XXXXXX system on most tools.

G. Give a full answer

15. List six precautions to be observed when using, caring for and maintaining stationary and portable power tools.

2 HOUSE STYLES AND FABRICATION

Approximately 80% of all houses constructed in Canada today are of the wood-frame type. Until the mid 1970s, standards important to the design of a home related primarily to the needs of the inhabitant. Varying degrees of attention were paid to the interior and exterior appearance. Little, if any attention was paid to designing the home with energy-efficiency as a primary consideration.

A factor which affects the efficiency in the design stage of the home is utilizing the building site to the fullest. Positioning the home with respect to the sun and prevailing winds can be used to advantage, or if ignored, can mean poor energy usage. Other energy saving features which may be designed into a home include: the building form (style), building volume, orientation of the window area, materials used and, to a somewhat lesser degree, the colour of the exterior of the home. Even with all the energy-efficient features which can be designed into the home, and even with the utmost care taken in assembling the structure, energy conservation depends as much upon the final operation of the home as it does upon its initial design.

Assembled under a broad range of house styles, homes can range from the energy-efficient home to low energy homes designed to use solar energy as a heating alternative. The decision presently rests in the hands of the designer, owner, and builder. However, this decision may not be ours to make in the future as some energy supplies dwindle.

HOUSE STYLES

In the designing stage of the home, the prospective owners must choose the style of home they feel will meet their needs. There are essentially **four basic designs** and within each design exist a number of variations. When correctly designed, each one may reflect the person's needs as well as the minimum energy requirements for that particular design.

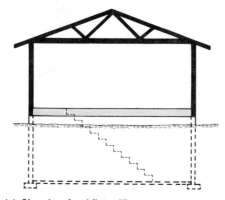

(a) Sloped roof and flat ceiling

Figure 2-1 Single-storey house

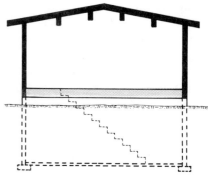

(b) Sloped roof with sloping ceiling and exposed beams

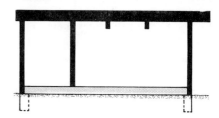

(c) Flat roof with exposed beams and/or bearing walls

Single-Storey Home

The single-storey bungalow, designed with or without a basement area is the most popular. It is consequently built in greater numbers in Canada than the other basic house styles. There are also more design variations in the one-storey home than in any other.

Because it minimizes the traffic flow between floors, the single-storey home adapts itself well to all age groups. With the living area on the main floor, the basement area is usually developed for recreation, storage and utility needs. Once the foundation and floor frame is constructed, the wall sections are framed-up using the *platform framing* technique (Chapter 8—Wall Framing). When the exterior-interior walls are assembled and erected, the roof is constructed. Roof designs vary from the common sloped roof, which has a flat ceiling area (using a truss rafter) to homes using either a sloped roof and ceiling (cathedral) or a flat roof and ceiling with exposed or concealed beams to carry the roof load (Fig. 2-1).

Due to its simplicity, the single-storey home is an excellent starting point for examining the structural components of a home and methods of assembly. Many of the structural components of one and a-half, two-storey and multi-level buildings are subjected to greater stresses but are really an extension of the single-storey structure.

One and a-half Storey Home

The one and a-half storey home is designed to make use of a portion of the home above the main living area. Built in a similar manner to the single-storey home, some have an extended exterior wall. The slope of the roof is generally much steeper, allowing for the living area on the second floor. *Dormers* built into the roof and extending outward allow for more usable floor space.

Lending itself well to narrower building lots and more heavily populated areas, the storey and a-half home makes good use of its available livable floor space. The second floor is usually designated as a bathroom and bedroom area, while the main floor makes up the balance of the livable area in the home (Fig. 2-2).

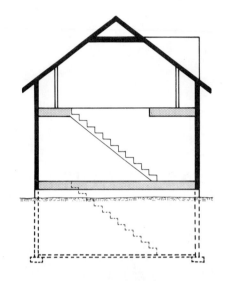

Figure 2-2   Storey and a-half house

Two-Storey Home

Traditionally, the two-storey home has represented prestige and elegance. Typically european in style, the two-storey home provides the maximum livable area from the smallest amount of displaced land area. This makes it quite economical from the point of view of land costs. Also, it usually results in a reduction in building costs and in comparison to the building surface area (walls and ceiling), it has the lowest area in relationship to its usable floor space. When following this style in designing an energy-efficient home, it is reasonable to assume that heating costs would be significantly less.

Framing a two-storey home can be done by using the platform framing technique, in which each floor level is framed independent of the others. The alternate framing method used is the *balloon framing* technique (Chapter 8—Wall Framing). Using this method, the exterior walls run the full length of both floors. The roof components are usually comprised of a truss rafter system similar to the single-storey home (Fig. 2-3). The in-

terior layout is similar to the storey and a-half home where the major living area is located on the main floor and the second floor level is used as the sleeping area.

Multi-level Homes

Multi-level homes are referred to as either **split-level** or **bi-level**. Both styles are comparatively new in comparison to other homes previously discussed.

In a **split-level** home, there are *three* major living areas. The highest level generally contains the bedrooms and bathroom; the main level contains the living-dining room and kitchen and the third level is used as a recreation, den, and/or bedroom area. The basement area is normally left for utility area and storage. This style of the house allows the designer to create many variations and combinations.

The four levels in the home are connected by a series of short-span stairs located at the point where the change in elevation occurs. The split-level home can be adapted for most building sites, although they lend themselves well to contoured lots. Roof styles vary; some build-

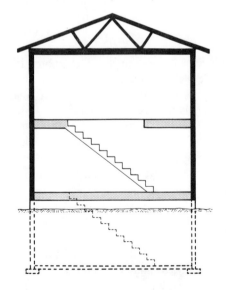

Figure 2-3   Two-storey house

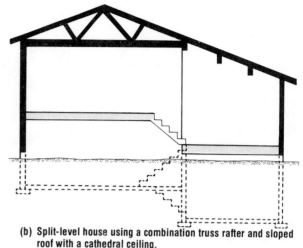

(a) Split-level house with adjoining roof slopes at right angles.

(b) Split-level house using a combination truss rafter and sloped roof with a cathedral ceiling.

Figure 2-4   Split-level house

ing structures use the truss rafter running at right angles to another over the opposing levels, while others use a combination of the truss rafter and the sloped or cathedral ceiling (Fig. 2-4).

Bi-level homes have *two* major living areas. It is similar in design to a single-storey home where the basement area is partially raised out of the ground and the interior completed, or to a partially sunken two-storey. It is the newest addition to the house styles of today. A *split-level* entrance is sometimes positioned approximately half way between the upper and lower floors, which allows for easy access to either floor. It can also be designed with entrances on either floor depending on the slope of the grade around the home. Properly oriented into the sloping site, it allows for unlimited flexibility in design.

Living facilities may be completely contained on the upper level with the lower floor left for recreation and utilities, while other designs allow for a split in the living accommodation between the two floors. Like the full two-storey home, the bi-level provides for the maximum amount of livable area for the least amount of land area consumed (Fig. 2-5).

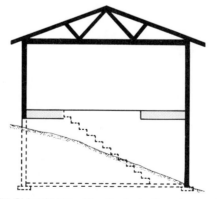

(a) Bi-level home with a sloped roof and ceiling, using a split entrance at the ground level.

(b) Bi-level home with a standard rafter or truss roof, using an entrance off the main floor.

Figure 2-5   Bi-level homes

LOW ENERGY HOMES

To accept the concept of low-energy homes, we have to first accept the fact that conservation is a source of energy. There are those who suggest it may be the largest source we have today as recycling industries spring up across the country. The question is, will it work for the housing industry? A low energy home built using any of the basic house styles, is one where the major design criteria is energy utilization with energy conservation as its main goal.

Before analysing the design of the home itself, land assembly regarding national and local restrictions must first be modified or changed. Nationally, the government must assist home builders in using more energy-conserving ideas in the future. Local governments have to become more concerned about utilizing the maximum number of street designs in developing subdivisions, which assures maximum exposure to the sun for the home builder. New building and landscaping bylaws may have to be passed or existing laws amended to provide assurances that those using heat gain areas will not have those areas obstructed by other homes, tall buildings, coniferous trees, etc., during the months of the year when sunlight is needed the most. As the nonrenewable sources of heat available today dwindle, people in the government planning sector will have to deal with these problems. Decisions made today will have a great bearing on whether or not home-owners will have access to the sun; the only known source of energy today which does not contribute to thermal pollution. For more information on factors concerning building orientation, see Chapter 4—Site Planning and Excavation.

Some of the proven energy-saving features in a low-energy home include: increased insulation values, correct design of roof overhang, maximum window orientation to a south-west exposure, airtight vapour barriers, air-to-air heat exchangers, minimal number of ceiling fixtures, air-lock entrances, exterior attic access, higher RSI values for door and window areas, and the use of thermal curtains or shutters. Many of these terms are uncommon in the vocabulary of the home builder yet, but they are sure to become widely used in the future. Incorporating these features into any one of the basic home styles does not create major changes which would detract from the home's appearance, serviceability or market value (Figs. 2-6, 2-7, 3-1).

Figure 2-6 shows a number of low energy homes funded through the joint co-operation of national, provincial, and civic government agencies and the home builders of Saskatoon, Saskatchewan. Two questions arise: do the homes look different from the homes being built in your area? and, although a low-energy home is developed for the prairie climate, will it not function as efficiently in other areas of Canada, that is, can it be adapted to your area? The one problem that can be reduced is the cost of heating fuels. In many cases, costs have been reduced to one-third or more when compared to a home of comparable size. As costs of present fuels increase this reduction will continue.

(a) Homes situated on the south side of the street

(b) Homes situated on the north side of the street

Figure 2-6   Group of low-energy homes

In Figures 2-7 and 3-1, perspective renderings of two low-energy homes are shown. Both homes use a solar-assisted system to further reduce the heating load. In Chapter 3 a number of the working drawings for Figure 3-1 are shown, which further explain its design.

Accommodating these features into a home begins with the foundation. It must be insulated! Whether it is applied on the inside or the outside of the wall will depend on the foundation wall material as well as the type of insulation used. In the upper exterior walls insulation far exceeding the minimum standards is used. To accommodate this extra thickness of insulation either the *double wall* system or a *minimum* 38 × 140 mm *stud wall* is erected. Figures 2-8 and 2-9 are examples of how the two systems are used.

With the **double-wall** system, the inner wall is the bearing wall, while the outer wall is simply suspended from it and is built to suit the insulation package. Framing material in contact with or below grade level must be pressure-treated. Exterior finishes are then applied to the exterior frame (Fig. 2-8).

Figure 2-8   Double-wall frame construction

Courtesy Enercon Consultants

Figure 2-7 Low energy home

The **single wall** system utilizes a thicker stud, often spaced at 600 mm on centre (single storey home). Batt insulation is used between the studs while the exterior is covered with a rigid insulation up to 50 mm thick. Headers and end floor joists are set in 50 mm on the sill or wall to allow for an additional layer of rigid insulation at this point (Fig. 2-9).

(a) Using a standard truss rafter (flat ceiling)

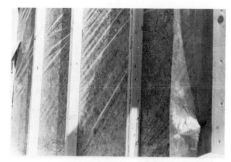

Figure 2-11   Vapour barrier installed on a double-wall system

Figure 2-9   Single wall system

Rafters (usually the truss rafter) in a low-energy home are constructed with a high heel so that the thickness of the insulation (RSI value 10.6 or better) can be carried out over the exterior wall (Fig. 2-10).

Creating an *air-tight* vapour barrier is the most vital operation in the construction of a low-energy home. In the double-wall system, the vapour barrier is placed on the **outside** of the inside wall (Fig. 2-11 and Chapter 14). This allows electrical and plumbing installations to take place without damaging the vapour barrier.

(b) Using a special rafter (sloped or cathedral ceiling).

Figure 2-10 Rafter construction

In the single wall system, the vapour barrier is installed in the same manner as in a conventional framed home. When it is necessary to join or cut through the vapour barrier for either system, all seams and edges are carefully caulked and stapled into place.

Because great care is taken when installing the vapour barrier in the low energy home, it becomes airtight. This makes it necessary to remove the stale air from inside the home and bring fresh air into it from the outside. An air management system or **air-to-air heat exchanger** is used to control the movement of the air. Essentially, a heat exchanger uses the warm stale air which it removes from the home to preheat the cold fresh air coming in. The heat from the stale air preheats the incoming fresh air, resulting in recovery rates of 70% to 80% of the heat being exhausted with this stale air.

Emphasis on artificial lighting in the home's interior results in area lights being used where they are most needed. Instead of ceiling mounts, wall hung fixtures are used on interior walls of the rooms. In some cases, suspended ceilings with flush mounted lights are used. The main concern is to keep fixture and wall plugs from penetrating areas where the vapour barrier is installed. The too common leaky attic hatch is hinged, insulated and weather-stripped. In homes with gable roofs, it is placed outside in the gable end with access from the outside only.

Air-lock entrances at the exterior doors reduce heat loss when entering or leaving the house. Major window areas face the south and west. Tests taken on prairie homes have indicated that up to 40% of normal heat loss through the window areas can be saved by this design feature alone. Window units include the double glazed thermo pane (minimum throughout the home), while triple glazed units or in some cases two sets of double glazed units are used. **Insulated curtains** or **thermal shutters** decrease the heat loss on large window areas during the night, and help to prevent heat build-up during the summer months. Extended soffits on the south and west sides control overheating by shading the window area in summer, but allowing the sun to enter in winter when it is needed the most.

Other energy-efficient features include automatic set-back thermostats, pre-heat water tanks which retain, and allow warming of the cold water coming into the home before it enters the system, and installation of smaller and more efficient furnaces. Whatever the device, the point to be made is that conservation can make a significant difference—it does work! If we are not aware of its benefits or do not consider it as a main component in building our home, we not only miss its advantages but we waste money as well. Once built into the home however, conservation must be maintained to be cost effective.

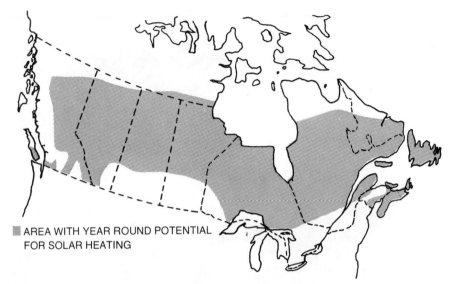

■ AREA WITH YEAR ROUND POTENTIAL FOR SOLAR HEATING

Figure 2-12   Potential for solar heating

SOLAR HOMES

The sun is our most important source of energy. Life on our planet simply would not exist without it. Trapping and utilizing some of this radiant heat, and using it to assist or even replace our present conventional heating systems is one way of conserving fossil fuels while lowering yearly heating bills.

Solar radiation has for years been helping to heat our buildings. Any structure which has glazed areas exposed to the sun has benefitted from solar radiation. Capturing that available heat was not a building priority until recently. As indicated in Figure 2-12, a large portion of Canada has the potential for year-round solar heating.

The shaded area in Figure 2-12 indicates the areas in which solar heat could be utilized on a permanent basis. For the area underlying the shaded section, the sun's energy could be used on a partial basis to heat the home during the winter season. Portions lying above the shaded area could take advantage of the summer sun to assist in their yearly heating requirements.

The *first* prerequisite for any solar home is a well-insulated, energy-efficient house. Using the basic principles outlined for low-energy homes, a significant difference in the heating load of the structure can be provided without any investment in solar components. However, a home designed to use the sun's energy as a heating source has the distinct advantage of being partially or even wholly self-sufficient.

Solar energy is, therefore, a viable source of heat for our homes. Capturing that heat, and storing, distributing and controlling it becomes the challenge.

ACTIVE SOLAR HEATING SYSTEM

The **active solar** heating system requires the assistance of another form of energy (usually electricity), to assist in the four stages of solar utilization. Using an external type of *collector* to trap the sun's energy, pumps circulate water (mixed in solution to prevent freezing) or air through the collector. Generally speaking, **liquid systems** require more attention and maintenance than air systems. Traditional problems with liquids are boiling, freez-

ing, leaky joints and bursting pipes. Air systems, on the other hand, do not suffer from these problems but require larger collector areas and a larger distribution network to attain the same degree of heating potential as that obtained in a liquid system.

During the day heat is collected, supplying the needs of the home. Once the daytime requirements are satisfied, the excess heat is moved into the *storage area* for later use. The way in which it is transported and the type of storage used usually depends on the type of collector system. A liquid collection system would likely use a liquid to transport the excess heat into the storage area. The storage area in turn would be liquid.

In collectors using **air** as the **heating medium,** air is normally used to transport the excess heat into the storage area. **Storage areas** for an air system can be *water, rock* or may be a recently developed *phase-change* material. One such system developed by Saskatchewan Minerals (a provincial crown corporation), uses a heat tray—(high-density polyethylene tray filled with a mixture of Glauber's salt; basically a mixture of hydrate of sodium sulphate and water).

When the mixture is subjected to heat over 31°C, a phase change occurs and heat is stored. When cooling occurs and the temperature of the tray drops below 31°C, the stored up heat is released for use within the home.

Water has the ability to accept heat rapidly but also loses it just as quickly. On the other hand, rock or coarse gravel is slow to heat up, but will also retain its heat for a longer period of time. The heat tray storage system being manufactured by Saskatchewan Minerals has the ability to store up to approximately four times as much heat as water, and approximately ten times as much as rock, per unit of volume. This makes it very attractive in terms of occupied area in the home for solar storage. Present estimates are that a container of heat trays measuring $1 \times 1 \times 2$ m high would be capable of storing 50% to 70% of the heating requirements of a 111 m² low-energy home. These heat tray systems are expected to be competitive with or cheaper than equal rock systems. They also offer the added advantage of space saving, architectural flexibility, and fewer maintenance problems.

Placed in an insulated storage area, the trapped heat remains there until required to maintain the house temperature. Generally referred to as *short-term* storage, it can retain heat for only a few days without being replenished. Although *long-term* storage systems have been developed, presently they are only viable on larger housing units such as row housing and apartments.

With any active solar heating system, a series of pumps, control valves and automatic sensors are required to control the collection, storage and distribution of the trapped heat. In most cases these are standard items available at most plumbing and heating outlets. Components for the solar collector and storage system how-ever, are not as readily available. As our solar technology advances, we see more products appearing on the market. Figure 2-13 shows a typical schematic drawing of an active solar heating system.

An active solar heating system with a liquid medium was used on a prototype home–"The Saskatchewan Conservation House"–built in 1975 in Regina, Saskatchewan. Geographically, its latitude is approximately half way between the 50th and 51st parallel. Financed through the government of Saskatchewan and the city of Regina, with research assistance from the University of Saskatchewan, the National Research Council and the Housing and Urban Development Association of Canada, many of the recent developments in low-energy housing have stemmed from its construction.

Shown in Figure 2-14, the 185.8 m² home uses a liquid collector made up of individual tubes mounted under an extended protective overhang. The liquid collector is coupled with a liquid storage tank (capacity 12 729 L) for storing excess heat at peak periods during the day.

Figure 2-14 **Saskatchewan conservation home (liquid medium)**

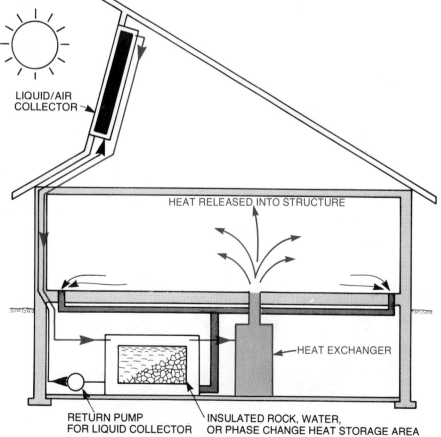

Figure 2-13 **Active solar heating system**

When required, heat is drawn off the storage tank through a heat exchanger set into an air duct. A fan circulates air past the heat exchanger and distributes the heated air through the home. Heat from the storage tank is also used to heat incoming city water normally heated by the conventional hot water tank.

An active solar heating system using an air medium has been constructed approximately 550 km north of Regina at LaRonge; geographically lying just north of the 55th parallel. Built by Mr. V. Ellis, president of Sun Centre Systems Ltd., construction of the home (176.5 m²) employed a different wall assembly than those previously mentioned.

Constructed from a vertically placed octagonal log, which is split to form the inner and outer wall face, both portions of the log are aligned using a tongue and groove system. Insulation and an air-tight vapour barrier were placed between the halves of the log to complete the wall structure.

A flat plate absorption collector (31.2 m²), designed by Mr. Ellis, was installed on the upper south facing wall (Fig. 2-15). Heat from the sun is stored in a reservoir containing 45.88 m³ of rock to be utilized later if not required immediately. Fans built into the air duct system draw the air from the collectors into the reservoir and then back into the house. Providing about 70% of the home's heating requirements, the balance is picked up from a back-up furnace system.

PASSIVE SOLAR HEATING SYSTEMS

Like an active solar home, the **passive solar house** relies heavily on the energy-efficiency of the structure and the individuals occupying it. Energy-efficient features incorporated into the home are designed to maximize the collection of heat during the day and minimize the loss of heat during the night.

The words *passive solar home* refer to the ability of a structure to collect and utilize the sun's energy without artificially distributing the heat within the structure by any other means. It is not a new concept—the caveman used it in ancient times by choosing caves facing South to live in. The house is designed so that the mass of the building acts as both the collector and storage system. Passive solar heating then is not a separate system as is the active system, but functions as a part of the house unit.

Advantages of a passive system versus an active system include lower initial costs when constructing the home and little or no ongoing maintenance. However, to function properly and efficiently it requires a firm commitment from the occupants to become a part of this total system. This commitment includes the acceptance and operation of the energy conserving features built into the home.

Types of Passive Solar Systems

Although there exist today five basic types of passive solar heating systems, the Canadian climatic conditions restrict our usage to three of these.

The **direct gain system** is the most simple and inexpensive passive solar home. Existing homes with windows facing south fall into this category as well. In most cases however, the heat gain in existing homes has not been designed to be part of the home and is therefore, in most cases, uncontrolled.

Window areas are the *solar collectors* on a direct gain home. Double or triple glazed, they are fitted with external or internal shutters or curtains to retain the heat absorbed by the mass of the house during the day. The *storage mass* of the home can be the walls of the house—exterior (with insulation placed on the outside), interior or variations of both. Examples would include wall and floor areas constructed of concrete, brick, ceramic, etc. **Remember** carpets work against passive systems.

Distribution of stored heat will depend upon the materials used in the storage area, its location in relationship to the windowed area, the volume of mass and its ability to store heat. **Time-lag,** a term used when referring to a storage medium's ability to first absorb heat and then release it, becomes a critical factor. Materials with a short time-lag tend to heat up quickly and usually release it just as quickly, causing extreme temperature changes. Choose materials that absorb as much heat as possible over a given time

Figure 2-15   Active solar home (air distribution medium)   *Courtesy Sun Centre Systems Ltd.*

span and release heat at a controlled rate to get optimum results.

Summer control is also necessary to prevent overheating of the building. Design features which help to control this problem include proper overhang, blinds, curtains, etc. to shade or block out the summer sun. Adequate ventilation, also a necessity, helps to maintain summer temperatures.

A schematic drawing of the direct gain system is shown in Figure 2-16. The perspective drawing (Fig. 2-7) is a hybrid of this system built into a low-energy home. This one and one-half storey home built in Regina by Enercon Building Corporation has a living area of 163.9 m², and utilizes a southern glazed solar gain area. *Containerized water* was used as its heat storage medium for the first year of operation. This has since been replaced by a *phase-change* storage system and has achieved comparable results. Early reports show that compared to conventional homes of comparable design and size built in the area, heating requirements of the *Pasqua House* are about 25% that of other homes.

With the **indirect passive** solar heating system a thermal storage wall is placed directly behind the sun-oriented window area. This concept, developed by French architect **Felix Trombe** usually bears his name—*trombe wall system*. Although usually constructed of concrete, other materials such as *brick, stone, masonry block* filled with sand can be used. Containerized water (referred to as a drum wall) and phase-change materials also have a place in this system.

As in the direct gain, the window area is the collector. The dark storage wall (dark colours draw heat rather than reflect it), placed directly behind the window area, is separate from the rest of the home. Heat build-up between the glass and wall is distributed to the home by convection currents (Chapter 14—Insulation and Vapour Barrier) through vents at the top of the wall. Bottom air vents in the wall complete the cycle. When the house demands are met, the vents can be dampered, thus allowing more absorption of heat into the wall.

Using a properly designed time lag, distribution of the heat build-up occurs by radiation through the wall during the night. A *thermal curtain* placed between the storage wall and the glass or shutters on the outside of the glass area prevent the loss of heat back through the glass during cold periods (Fig. 2-17). The perspective drawing and subsequent drawings in Chapter 3 are of a low energy home using the indirect solar gain design and incorporating a concrete trombe wall as the heat storage medium.

The third concept used in passive solar designs is referred to as the **sunspace or greenhouse** system. Essentially, an *isolated heat gain* area, the collection space must be attached to the house yet remain separate. The glazed area of the system, as in other passive designs is the collecting medium. Storage of heat over and above the immediate requirements of the home are similar to the other systems. Thermal storage mass can be in the form of walls, floors, containerized water or phase-change materials.

Because the system is segregated from the living area, the temperature of the house remains at a more constant level. If there is an excessive heat build-up, this will occur only within the greenhouse area. Directed into the thermal storage mass, the house can then draw on this stored heat as required (Fig. 2-18).

The attached greenhouse in Figure 2-19 (a), isolated from the living area by the concrete storage wall and patio doors, is shown during the construction stage of the home. To avoid overheating in the summer months, the glass panels adjacent to the roof are hinged to allow the build up of heat to escape. Further protection of the living area is achieved by using a thermal curtain adjacent to the storage wall which is lowered during summer days to keep out the heat. During the heating season the operation of the thermal curtain is reversed, thus conserving heat for use within the home. As shown in Figure 2-19 (b), the greenhouse becomes an integral part of the exterior of the home and blends in with the surrounding neighbourhood upon completion.

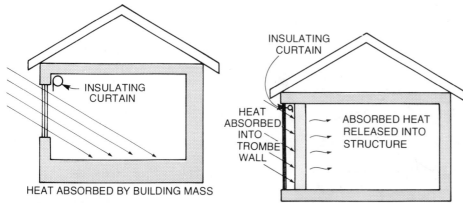

Figure 2-16   Direct gain system

Figure 2-17   Indirect gain system

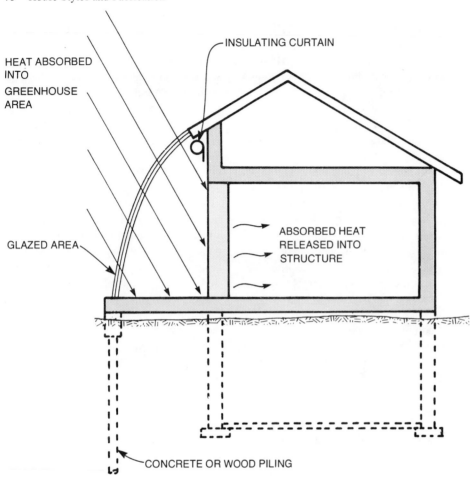

HEAT ABSORBED INTO GREENHOUSE AREA

INSULATING CURTAIN

GLAZED AREA

ABSORBED HEAT RELEASED INTO STRUCTURE

CONCRETE OR WOOD PILING

Figure 2-18   Isolated gain (greenhouse) system

FABRICATION OF HOUSES

Assembly of houses today takes the form of **site** fabrication, **factory** assembled or a **combination** of factory and assembled units. To *fabricate* means to make or put together by assembling any number of parts on the site. *Prefabricate* means the assembly of the component parts before they arrive on the site. Prefabrication allows for more control over size, and generally contributes to better working conditions and superior structures. It makes use of assembly-line techniques and lowers on-site labour costs which can then be passed on to consumers.

Total **on-site fabrication**, referred to as *stick framing*, is virtually non-existent. Builders continually make use of pre-cut lumber, pre-built roof trusses and window, door and cupboard units. Used throughout, they speed up assembly time and reduce costs. On-site fabrication usually begins once the foundation work is completed.

With the foundation and floor in place, the walls of the home can be laid out, assembled and erected. The method of assembly is often dependent on the design and framing technique used by the contractor. If using the *platform framing* method (Chapter 8—Wall Framing), the builder will likely frame-up all walls using ready cut **PET** (precision end trimmed) studs. Doors and windows in the meantime will be framed to exact standards; the size predetermined by millwork companies supplying the units. Figures 2-6, and 2-8 to 2-11 inclusive show low-energy homes built on-site at different stages of completion.

Rafters can be fabricated at the site, although prebuilt truss rafters speed up the construction of the roof. Builders who do not have the necessary shop space or equipment may purchase trusses from numerous outlets. Not only do they make a strong roof assembly, they ensure that the

(a) Construction stage of the greenhouse

(b) Completed attached greenhouse

Courtesy Concept Construction Ltd.

Figure 2-19   Low-energy home with an attached greenhouse

shell of the house is closed in sooner—important during inclement weather. Shingling of the roof area protects the interior, allowing work to proceed within that portion of the home with no danger of roof leakage. After door and window units are installed, exterior finishes are applied.

With interior work underway, the various sub-trades move in quickly to complete their various jobs. After installing their equipment in the necessary wall areas, insulating and vapour barrier application is completed. Drywall or other interior finishes are then installed, and taping and filling of the joints complete the operation.

By this time the carpenter can begin hanging the doors, and installing cupboards, vanities, (usually prebuilt) and door and window trim. The sub-trades can now complete their work, while painters stain, paint, and varnish both the interior and exterior. Linoleum and carpets are laid as required.

After any alterations, touch-ups and final inspection are completed, the house is ready for occupation and may be placed on the real estate market. In the past, the on-site built home required the expertise of a carpenter with help from a handful of other sub-trades people to erect a home from start to finish. Today, homebuilding requires a multitude of people from all walks of life and all parts of the country. Carefully coordinated, these people do not all see the completed home but each plays an important role in the construction of it.

PREFABRICATED HOUSES

Prefabricated houses can be partially or wholly manufactured at a factory or assembly plant. Prefabrication accounts for a large percentage of single family dwellings today. Fabrication of houses occurs in many forms including mobile, sectional, complete or packaged homes.

Mobile Houses

Mobile homes are mounted on a wheeled undercarriage and are completely assembled at the factory as a single unit or split, with the two halves joined together when it is set up on the site.

With the great mobility of many people in Canada, the mobile home industry has filled a large void in the housing industry. Used in all sectors, from those living in remote areas, where new development is taking place, to those who purchase land, set up their mobile home on a foundation and establish a permanent residence in the area of their choice.

Most mobile homes manufactured today come complete with all the facilities required by the average family. To take up occupancy, the family requires little more than to move in their personal effects once the home has been set up on the lot (Fig. 2-20).

Figure 2-20   Mobile home unit

Sectional Houses

A **sectionalized** house is a structure built in sections (usually two halves). Once completed, it is loaded up and moved to the site where final assembly takes place. They are almost entirely factory built, and arrive at the site with exterior and interior finishes applied. Wiring, plumbing and most fixtures are installed with hook-ups left below the floor section. These are then connected to their respective services once the home has been established on the foundation.

Once in place on the foundation, final on-site work completes the home. This work usually includes such things as joining the roof sections and the exterior finish on the outside of the home. Although the interior finish (usually gyproc) is applied at the factory, it is not crack-filled until the house is set on the foundation to avoid it cracking while being moved. With the drywalling complete, painting and floor covering is carried out. Because such a large amount of work is done before the house reaches its destination, the waiting period before occupancy is usually quite short.

Ready-to-Move Houses

The **ready-to-move** house (RTM) is a completely assembled home. It is moved from the fabrication plant to the site by a building mover, and the distance it travels is often limited due to the high cost involved in obtaining permits to transport a building of great size and height.

It is fully completed on the exterior, and requires only crack filling of the gyproc and painting and floor covering. Electrical, plumbing and heating services are installed in the structure and only require connection to their respective hookups once the home is placed on the foundation. Figure 2-21 shows an RTM home awaiting the building mover to move it out to its final destination.

Figure 2-21   Prefabricated ready-to-move (RTM) house

Packaged Houses

In prefabricating houses, the **packaged** home is the most commonly used. Referred to as a *factory-built* structure, the major components are assembled at the factory and then moved out to the site for assembling.

The packaged house industry is a specialized one, and begins with company brochures distributed to prospective home buyers. Over the years, companies have been able to provide the owner/builder with a wide variety of designs to choose from. Some companies offer up to one hundred or more designs, including in many cases the option to custom design and prefabricate a house to individual preferences.

Once the design has been chosen, blueprints are drawn up. Included with the blueprint of the house is a builder's guide (Fig. 2-22(a)), giving the builder all the information required to fabricate the home. Meanwhile, fabrication of the home is under way at the factory. Figure 2-22, (b-c-d) show assembly of different sections of the house.

Once the framing portion of the package is completed at the factory, it is transported to the building site. Automation and years of experience have streamlined this industry. All components that are required are numbered (keyed) to the blueprint. They are loaded on the truck in such a way that those sections required first are the last ones to be loaded; thus making it easier for the builder to find the parts as they are required upon arrival at the building site (Fig. 2-22(e)).

With the foundation in place, assembly can begin (Fig. 2-22(f)). Sub-trades required to install the mechanical portions of the house are normally hired locally. Once closed in and shingles are applied, the remainder of the package (insulation, interior finish, cupboards, etc.) arrives, allowing for the completion of the house and the arrival of its new owners.

(a) **Finalized home design allows for the plans to be completed**

(b) **Factory assembled floor joist**

(c) **Factory assembled roof trusses**

(d) **Factory assembled soffit**

(e) **House package arriving at the building site**

(f) **With the foundation and floor in place, the wall units can be erected.**

Photos Courtesy Nelson Lumber Co. Limited

Figure 2-22   Fabricated factory-built (Packaged) house

In the housing industry, it is possible to design and build houses to meet people's needs. The method of fabrication used is secondary to the design and quality work that must go into the house to ensure owners a maximum return for their investment. A delicate balance exists today between the home, its occupants, and the environment. The home must be designed to blend in with and take advantage of the surrounding area. Whether it is built as a low-energy home, an active or passive solar home, or a combination of these it must be delicately tuned to the environment in which it is placed. The occupants of such a house, who are prepared to use the environment to its fullest advantage, yet close off the environment when it becomes detrimental to the operation of the house and comfort of those within, will ultimately be living in an energy-efficient environment.

REVIEW QUESTIONS

2—HOUSE STYLES AND METHODS OF ASSEMBLY

Answer all questions on a separate sheet

A. Write full answers

1. Make a list of the factors which affect the efficiency of a home. How does the operation of the home affect these features?
2. List the four basic house styles used in the construction industry in Canada today.
3. Describe the differences between the double-wall and single-wall low energy home.
4. Why is it necessary to install an airtight vapour barrier in all homes?
5. What factors contribute to the use of the sun as a source of energy?
6. Explain how an active solar heating system functions.

B. Replace the Xs with the correct word(s)

7. The two transporting mediums used for collecting and distributing heat in the active solar heating system are XXXXXXX and XXXXXX.
8. Storage mediums used in the active system include: XXXXXXXX, XXXXXXX or XXXXXXX.
9. In the passive solar home energy efficient features XXXXXXX the collection of heat and XXXXXX the loss of the heat.
10. The ability of the storage medium to collect heat, hold it for a period of time and then release it is called XXXXXXXX.

C. Write a full answer

11. Describe the advantages and disadvantages of each of the passive solar heating systems.

D. Replace the Xs with the correct word(s)

12. The XXXXXXX passive solar heating system can be incorporated into an existing house with a minimum of expense.
13. On site fabrication of homes is referred to as XXXXXXX.
14. The letters PET mean XXXXXX.
15. XXXXXX of component parts of the home today account for a large percentage of the material and labour required in its construction.
16. A home built entirely off-site, then moved from the fabrication plant to the site is referred to as a XXXXXXX home.
17. Described as XXXXXX homes, components are factory assembled, then moved to the building site where final assembly takes place.

E. Write full answers

18. List from information supplied by the rest of your class:
    (a) The number of different styles of homes lived in
    (b) How many are of the low energy or energy-efficient variety
    (c) Those homes that are solar heated
    (d) Those homes that have an alternate heating source and the type of source.
19. Poll the housing industry in your area and list the part(s) of the house they can supply for you. Do they fabricate these component parts at their location or are they simply a distributor for the area?
20. If a prefabrication plant for homes exists in your area, arrange a tour to observe the plant operation and the assembly of the house parts. Prepare a report on the use of the assembly line technique in reducing labour costs.

3 BLUEPRINT READING

The carpenter's ability to read and understand a *set of working drawings* (blueprints) is just as important as knowing how to assemble the structure itself. Just as we learn how to read and write a language, the tradesperson must learn how to read and interpret the language of the architect or draftsperson. *Blueprint reading* then, is an individual's ability to interpret a drawing or series of drawings. This ability obviously cannot be gained overnight. Apprentices should take every opportunity to look at and become familiar with the blueprints for the structure they are working on. By doing this, they will soon be able to correlate between the drawings and the building. With time and experience, the apprentice will be able to transform the information from the drawings and anticipate problems within the structure before they occur.

This chapter will outline the devices used by architects to describe and present their ideas. It is not the purpose of the chapter to offer a course in drafting, but rather to offer an overview of the basic fundamentals of blueprint assembly. Once familiar with the views, lines, details and different symbols used in the drawings, even the most inexperienced person can begin to visualize the building portrayed by the architect.

TYPES OF BLUEPRINTS

Perspective Drawings

Drawings used by architects to portray their ideas to prospective home buyer(s) are known as *perspective renderings* (drawings). Sometimes called pictorial or presentation drawings, the perspective allows individuals to view the home from the exterior, which gives them a finished "picture" view of the home (hence the name pictorial), in its completed stage. These drawings are used in magazines, company brochures and other forms of advertisement. Quite often, trees, shrubbery and landscaping are used to accent the drawing.

To supplement the perspective view, the main floor(s) plans are often included to give an overall view of the arrangement and location of rooms within the parameters of the home. Little or no technical information is given on any view other than the square metres of floor area and possibly the overall dimensions of the structure. In Figure 3-1 a perspective drawing similar to those found in company brochures is shown.

Working Drawings

Once the concept for a building has been accepted, it is necessary to draft up a complete set of working or technical drawings so that the various tradespeople

Figure 3-1 Perspective drawing of a low energy passive solar home *Courtesy Concept Construction*

can carry out the operations required of them. These drawings include plan views, elevation views, sectional views, detailed views, mechanical and electrical views. No one view will give all the information necessary to carry out the concepts of the architect or draftsperson. Used as a ''set'', the working drawings give step by step details of the construction from the footings to the roof.

TYPES OF WORKING DRAWINGS

Site or Plot Plan

The site or plot plan shows information concerning the lot on which a house is to be built as well as where the house is located on the lot.

Looking at the area where the building is to be erected, lot lines or boundary lines enclose the parcel of land. Each line has a specific length (or radius if curved) and direction. Contour lines, both existing and new (as a result of grading) may be shown. The compass direction is also shown in relation to the lot lines. Figure 3-2 shows a plot plan which gives the exact location of the property on which the house is to be built.

The position of the house on the lot is located by dimensioning it from the front and side of the lot. This is termed *setback*. The amount of setback is locally controlled and therefore is different from city to city. Other proposed buildings such as garages are shown along with driveways, sidewalks and landscaping. Elevations (height of known points above or below a given point) are shown throughout the plan. The finished floor elevation of the house is indicated in relation to the ground level at the corners of the home. If sufficient differences occur or grading is required on the lot, grade elevations are shown on the contour lines, indicating both the existing and finished heights. (Fig. 3-3).

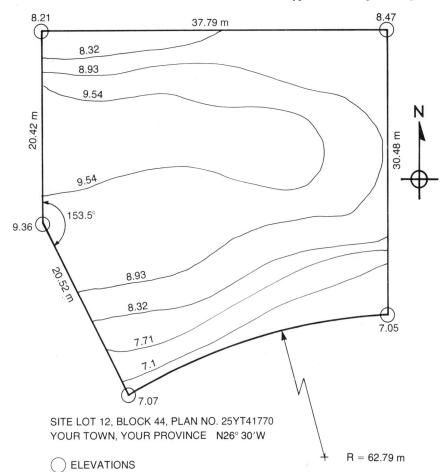

SITE LOT 12, BLOCK 44, PLAN NO. 25YT41770
YOUR TOWN, YOUR PROVINCE   N26° 30'W

◯ ELEVATIONS

Figure 3-2   Site plan showing location, lot and contour lines

On house plans specially designed for an individual or family by an architect or draftsperson, the above information is generally included within the set of working drawings. With house plans produced by Canada Mortgage and Housing Corporation and those produced by companies selling homes throughout a wide area, it is not customary to supply all the information required by the local governing agent issuing the building permit.

Plan Views

Plan views include all blueprints of a home which are in a horizontal plane to the structure. These are two-dimensional drawings showing *width* and *length*. Plan views generally include a foundation plan, main floor plan and successive floor plans if required. Some plans will also include a roof plan if needed in the construction of the roof components. A good example of this is found in pre-fabricated homes where rafters, roof sheathings etc. are pre-cut and a roof plan is necessary for assembly reasons.

Foundation Plan

The foundation, or basement plan of a house is drawn to include all construction requirements up to the first floor. Looking at this plan and starting at the floor joist you would expect to find the following information:

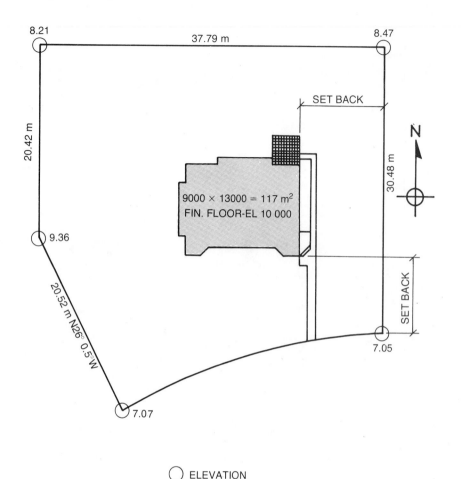

8.21

37.79 m

8.47

SET BACK

20.42 m

30.48 m

N

9000 × 13000 = 117 m²
FIN. FLOOR-EL 10 000

9.36

20.52 m N26° 0.5'W

SET BACK

7.05

7.07

○ ELEVATION

Figure 3-3   Proposed home on the site plan

Looking at the floor plan, information found on this plan includes the following:

a. Overall dimensions of the home, and all dimensions in relation to partitions, windows and doors.
b. Shape and arrangement of rooms.
c. Position of doors and windows.
d. Location of fixed utilities such as kitchen cupboards, bathroom utilities, closets, etc.
e. Necessary electrical and mechanical information.
f. Size and dimension of the second floor joist or ceiling joist.

Figure 3-5 shows a typical main floor plan including the information indicated above. Mechanical and electrical information may be supplied on the plan views. However, if there is overcrowding or confusion on the views it is not uncommon to have other views devoted entirely to this information.

Elevation Views

Elevation views are those blueprints which show the exterior of the home. They are also two-dimensional drawings which show length or width and height. There are generally four elevation views, known as the *front, rear, right* and *left* side views. Any one view will show the exterior as you face it in its finished form. Information found on the elevation views includes:

a. Type and size of window units.
b. Exterior finish of the home.
c. Grade level around foundation.
d. Type of shingles used.
e. Type and location of chimney and fireplace.
f. Elevation views may show any irregularities in the footing or foundation. An example of this would be the stepped system used in the footing

a. Overall dimensions of the foundation (this can be different from the floor and wall dimensions if a portion of the home is cantilevered).
b. Size and direction of floor joist, and position of bridging.
c. Size and location of the main beam supporting the joist and secondary beams.
d. Location of supporting posts/columns and independent footings under the beam.
e. Location of basement windows.
f. Foundation and footing sizes and shapes.
g. Stairwell and chimney requirements.

h. The location of partitions, doors and fixed equipment in the basement area. Mechanical and electrical installations may be shown as well.

Many of the items mentioned above can be found in Figure 3-4.

Floor Plan

The main floor plan(s) for a multi-level home continues on from where the foundation plan stops. From the first floor joist it continues up to the floor joist of the succeeding floor or to the ceiling joist, depending on the style of home, showing the interior construction of the home.

under the foundation wall of a multi-level home.

g. Venting system used in the attic area.

h. Type and position of exterior steps, landings and attached decks.

In Figure 3-6 the front elevation view of a typical bungalow-style home shows many of the items mentioned above.

Sectional Views

Sectional views are those portions of the blueprint which show a vertical cut through the home. This is used to show the interior construction of the various structural parts of the building which cannot be shown by elevations or plan views. By using sectionals, the architect is able to convey in detail the precise assembly of any portion of the home to the builder.

Sectionals normally used in a set of blueprints for a house are *cross-section* or *typical wall section* drawings. The loca-

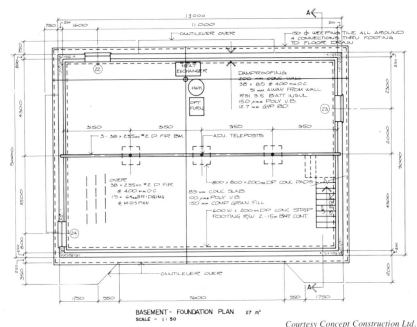

Courtesy Concept Construction Ltd.

Figure 3-4   Typical foundation plan of a bungalow-style home

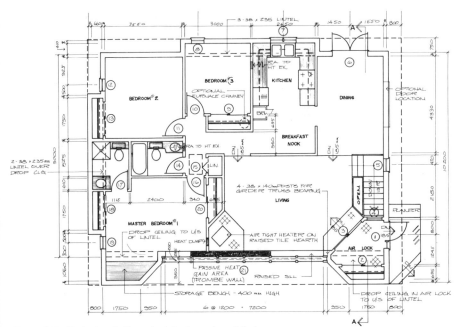

Figure 3-5   Typical main floor plan of a bungalow-style home   *Courtesy Concept Construction Ltd.*

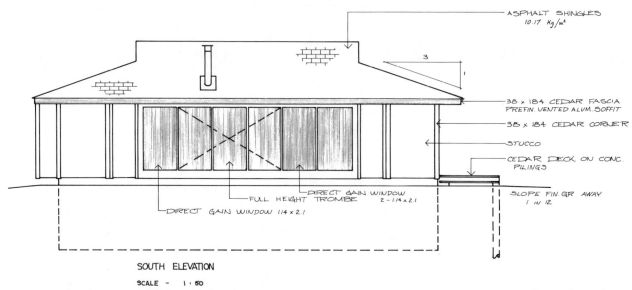

ASPHALT SHINGLES
10.17 Kg/m²

38 x 184 CEDAR FASCIA
PREFIN. VENTED ALUM. SOFFIT

38 x 184 CEDAR CORNER

STUCCO

CEDAR DECK ON CONC.
PILINGS

SLOPE FIN. GR. AWAY
1 IN 12

DIRECT GAIN WINDOW
2 - 1.14 x 2.1

FULL HEIGHT TROMBE

DIRECT GAIN WINDOW 1.14 x 2.1

SOUTH ELEVATION
SCALE - 1 : 50

Figure 3-6   Front elevation of a typical bungalow-style home

Courtesy Concept Construction Ltd.

tion of the section can be found on the floor plan and is generally noted as *cross-section A-A* or by some other appropriate numbering system. This system of identification allows the carpenter to see exactly where the cut occurs and clarifies for him or her the architect's intent. Figure 3-7 is a cross-sectional cut of the floor plan used previously in Fig. 3-5.

Detail Views

Detailed views are those which the architect uses to show any additional information about the home which cannot be included in plan, elevation or sectional views. Detail views may be cut either horizontally or vertically and quite often will show a cut both ways through an object. Illustrations of special details could include such things as structural connections, doorways, fireplaces, windows and stairways (Fig. 3-8).

Mechanical Details

As was indicated earlier in the chapter it is not uncommon to show electrical, plumbing, heating and air-conditioning information on the views discussed here.

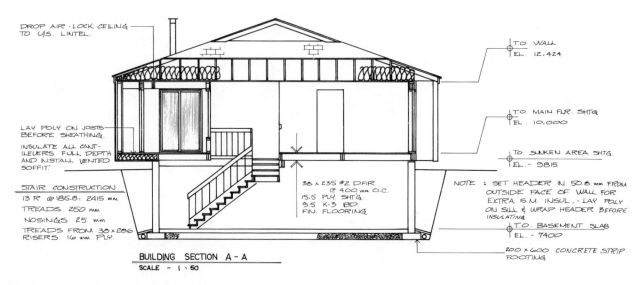

DROP AIR - LOCK CEILING
TO U/S. LINTEL.

T.O WALL
EL. 12.424

LAY POLY ON JOISTS
BEFORE SHEATHING.

INSULATE ALL CANT-
ILEVERS FULL DEPTH
AND INSTALL VENTED
SOFFIT.

STAIR CONSTRUCTION
13 R @185.8 = 2415 mm
TREADS 250 mm
NOSINGS 25 mm
TREADS FROM 38 x 286
RISERS 16 mm PLY.

T.O. MAIN FLR. SHTG
EL. 10.000

T.O. SUNKEN AREA SHTG.
EL. 9815

38 x 235 #2 D.FIR
@ 400 mm O.C.
15.5 PLY. SHTG.
9.5 K-3 BD.
FIN. FLOORING

NOTE : SET HEADER IN 50.8 mm FROM
OUTSIDE FACE OF WALL FOR
EXTRA S.M. INSUL. - LAY POLY
ON SILL & WRAP HEADER BEFORE
INSULATING

T.O. BASEMENT SLAB
EL. 7400

200 x 600 CONCRETE STRIP
FOOTING

BUILDING SECTION A-A
SCALE - 1 : 50

Figure 3-7   Cross-section of a typical bungalow-style home

Courtesy Concept Construction Ltd.

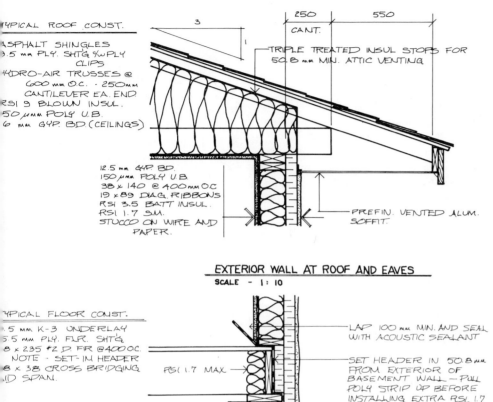

TYPICAL ROOF CONST.

ASPHALT SHINGLES
9.5 mm PLY. SHT'G %PLY
         CLIPS
HYDRO-AIR TRUSSES @
  600 mm O.C. - 250mm
  CANTILEVER EA. END
RSI 9 BLOWN INSUL.
150 mm POLY U.B.
6 mm GYP. BD. (CEILINGS)

3

250   550
CANT.

TRIPLE TREATED INSUL STOPS FOR
50.8 mm MIN. ATTIC VENTING

12.5 mm GYP. BD.
150 mm POLY U.B.
38 x 140 @ 400mm OC
19 x 89 DIAG. RIBBONS
RSI 3.5 BATT INSUL.
RSI 1.7 S.M.
STUCCO ON WIRE AND
PAPER.

PREFIN. VENTED ALUM.
SOFFIT.

EXTERIOR WALL AT ROOF AND EAVES
SCALE - 1:10

TYPICAL FLOOR CONST.

15.5 mm K-3 UNDERLAY
15.5 mm PLY. FLR. SHT'G
38 x 235 #2 D. FIR @ 400 OC.
NOTE - SET-IN HEADER
38 x 38 CROSS BRIDGING
MID SPAN.

RSI 1.7 MAX

LAP 100 mm MIN. AND SEAL
WITH ACOUSTIC SEALANT

SET HEADER IN 50.8 mm
FROM EXTERIOR OF
BASEMENT WALL — PULL
POLY STRIP UP BEFORE
INSTALLING EXTRA RSL 1.7
S.M. INSUL.

800 mm WIDE 150 mm POLY
STRIP LAID BETWEEN
DURING FRAMING

FLOOR HEADER AT EXTERIOR WALL
SCALE - 1:10

Figure 3-8   Detailed views taken from the bungalow-style home

Courtesy Concept Construction Ltd.

piling of repetitive information which would normally clutter up the drawings. As in sections and details a simple numbering or lettering system is all that is necessary for the carpenter to correlate between the schedule and the drawing (Fig. 3-9).

LANGUAGE OF BLUEPRINTS

To this point we have been discussing the different views architects employ to present their concepts. Such devices as different kinds of lines, symbols or conventions, abbreviations and methods of giving dimensions or working directions are used. Such devices are standardized throughout. It is therefore the carpenter's responsibility to know and be able to interpret these devices.

Principle Kinds of Lines Used in Blueprints

Main object lines are the fundamental lines used in any drawing. They are heavy unbroken lines, outlining the main walls on plan, section, elevation and other views. They clearly indicate to the reader the important parts of the construction.

Dimension lines are drawn mainly outside the object, although some may be re-

If the architect feels the addition of this material will clutter up the drawings and make them hard to read, they will be made upon separate sheets, showing all necessary views. This facilitates the mechanical part of the working drawings.

Schedules

Schedules are used by the architect-draftsperson to list information required on the working drawings. They may be used to indicate type, size, style, quantity, material used and location of recurring items such as rooms, room finishes, exterior finishes, doors and windows. The use of schedules facilitates the com-

DOOR AND WINDOW SCHEDULE								
NO.	SIZE	DESCRIPTION	ROUGH OPENING		NO.	SIZE	DESCRIPTION	ROUGH OPENING
1	810 x 2030	PEACHTREE 108	1275 x 2115		15	600 x 2030	MAH. BIFOLD	640 x 2100
2	1830 x 2030	MIRROR BIPASS	1880 x 2100		16	810 x 2030	ROT. MAH.	865 x 2100
3	1830 x 2030	VISTA GLIDE	1880 x 2100		17	762 x 2030	"	815 x 2100
4	1800 x 1600	SINGLE - FIXED	850 x 1690		18	1524 x 2030	BIPASS	1575 x 2100
5	600 x 1600	SASKO 2465 C	640 x 1690		19	600 x 1600	SASKO 2465C	640 x 1690
6	2 - 810 x 2030	PEACHTREE 108	1720 x 2115		20	1524 x 2030	BIPASS	1575 x 2100
7	1200 x 900	SASKO® 2436 FC	1250 x 960		21	SUN TRAP GLAZING WITH INSUL CURTAIN		
8	600 x 1600	SASKO® 2465 C	640 x 1690		22	1200 x 500	SASKO® 4820A	1250 x 550
9	1524 x 2030	BIPASS	1575 x 2100		23	1200 x 500	"	1250 x 550
10	810 x 2030	ROT. MAH.	865 x 2100		24	1200 x 500	"	1250 x 550
11	810 x 2030	"	865 x 2100					
12	600 x 1600	SASKO 2465 C	640 x 1690					
13	1524 x 2030	BIPASS	1575 x 2100					
14	760 x 2030	ROT. MAH.	815 x 2100					

Figure 3-9   Door and room schedule.

Courtesy Concept Construction Ltd.

quired inside to show the distance between two points. This line is narrow in comparison to the object line and is terminated by some type of symbol; a dot, slash or arrowhead. A break in the line is quite common and within the break the dimension between the two points is given. In conjunction with the dimension line are *extension lines*. These lines extend out from the object, but they **do not touch** it. The termination of the dimension line occurs at the point where it intersects the extension line. By observing where the extension lines project from the object, the dimension of that portion can be obtained. Any given dimension may not agree with the scaled distance between two points. Therefore it is important that the carpenter use the dimensions as shown.

Equipment lines on a drawing are those used to show the position of standard or permanent equipment in the building. These lines are narrow and continuous and are used to show equipment such as doors, windows, bathroom fixtures, kitchen cupboards, stoves, refrigerators, washing machines, clothes dryers, etc.

Break lines are generally used in sections and detailed views where larger scales are required and the architect chooses to leave out a portion or portions of the object. This enables the drawing to be condensed without leaving out important parts of the view. The break line is a heavy wavy line extending through the main object lines.

Invisible lines are those within the object which show parts of objects that are obscured by another portion of the structure. These lines are a series of short dashes showing the outline of the object. If more information is required about the hidden object, another view in the set of blueprints can be referred to.

Centre lines are used to locate the centres of objects such as doors, windows, basement walls, footings and parti-

tions throughout the house. The line is somewhat similar to a hidden line in that it uses a series of long and short dashes. It is common to have a dimension terminate at a centre line when the centre line is locating objects such as doors or windows.

Section lines are heavy thick lines used on plan views to indicate where a section or detail has been removed for clarity. As was mentioned earlier, the section is usually identified by a letter or numbering system. Section lines cut through the object lines and then turn at 90° to the cut line and terminate with heavy arrowheads which indicate the direction in which to view the sectioned portion.

Figure 3-10 gives examples of the different lines and their uses.

Symbols or Conventions

Symbols or *conventions* are symbolic representations used to show the tradesperson many of the different materials required in the construction of the home. All symbols for building materials, walls, floors, window and door openings, plumbing, heating and electrical fixtures have been standardized throughout to avoid misunderstanding or confusion. It is left to the tradesperson to become familiar with them and know their meanings. In Figures 3-11 to 3-15 some of the commonly used symbols in house blueprints are shown.

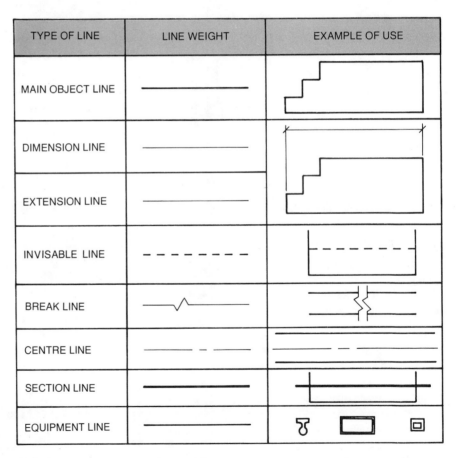

TYPE OF LINE	LINE WEIGHT	EXAMPLE OF USE
MAIN OBJECT LINE		
DIMENSION LINE		
EXTENSION LINE		
INVISABLE LINE		
BREAK LINE		
CENTRE LINE		
SECTION LINE		
EQUIPMENT LINE		

Figure 3-10   Principle lines used in blueprints.

Scales

For convenience in drawing and paper size the architect selects a scale which is in proportion to the size of the drawing and the dimensions of the object. The scale is always stated on the drawings, and is normally found in the title block unless there are individual drawings on a page requiring different scales. When this occurs the scale is noted under the individual drawing.

The simplicity of the metric system is reflected in the ease with which metric ratios can be used. Table 3-1 indicates the preferred scales to use when reducing an object's size. In some cases where greater clarification is needed, enlargements of the object may be necessary and would be noted as **Scale 2:1, 5:1**, etc.

Periodically it becomes necessary to obtain a measurement from the blueprint which is not stated. Like other materials, paper is affected by moisture, temperature and other conditions. Therefore, scaling from a blueprint can sometimes lead to inaccurate measurements. If possible, the dimension scaled should be checked against other dimensions on the blueprint and if errors are present these can be allowed for. If you are unable to do this, or if the dimension is critical, it is best to have it checked out before going ahead with the layout and construction of the project.

As materials are updated, changed or approved for use, discrepancies in the blueprints and changes in construction techniques make it necessary to update or revise the blueprints. A revised blueprint will be indicated by a note, usually within or near the title block, indicating the changes and the date when they take effect. In revising plans, dimensions are often changed without re-drawing the view. This is another reason for not scaling directly off the blueprint.

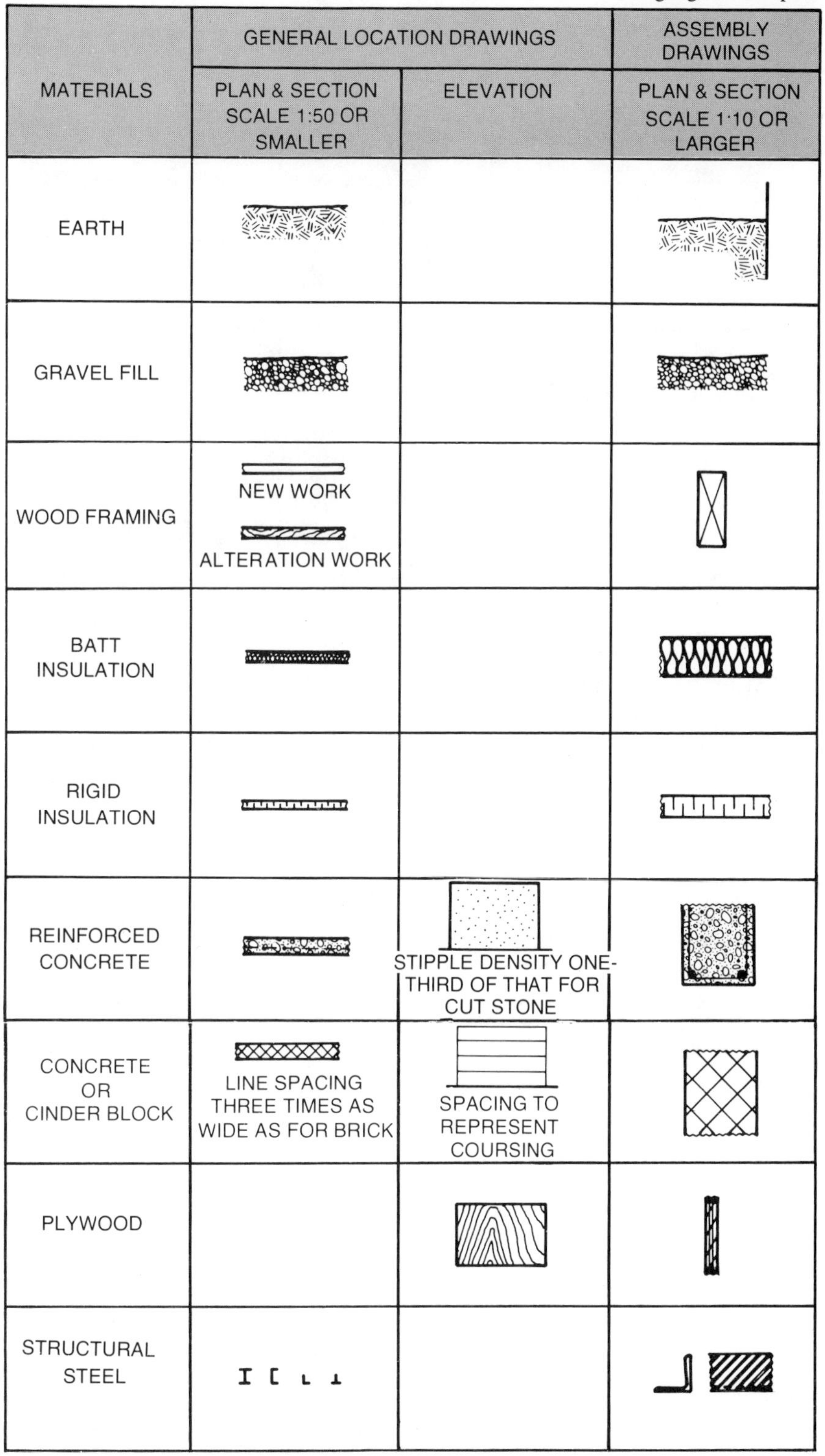

MATERIALS	GENERAL LOCATION DRAWINGS		ASSEMBLY DRAWINGS
	PLAN & SECTION SCALE 1:50 OR SMALLER	ELEVATION	PLAN & SECTION SCALE 1:10 OR LARGER
EARTH			
GRAVEL FILL			
WOOD FRAMING	NEW WORK / ALTERATION WORK		
BATT INSULATION			
RIGID INSULATION			
REINFORCED CONCRETE		STIPPLE DENSITY ONE-THIRD OF THAT FOR CUT STONE	
CONCRETE OR CINDER BLOCK	LINE SPACING THREE TIMES AS WIDE AS FOR BRICK	SPACING TO REPRESENT COURSING	
PLYWOOD			
STRUCTURAL STEEL			

Figure 3-11   Symbols used for building materials.

Courtesy National Research Council

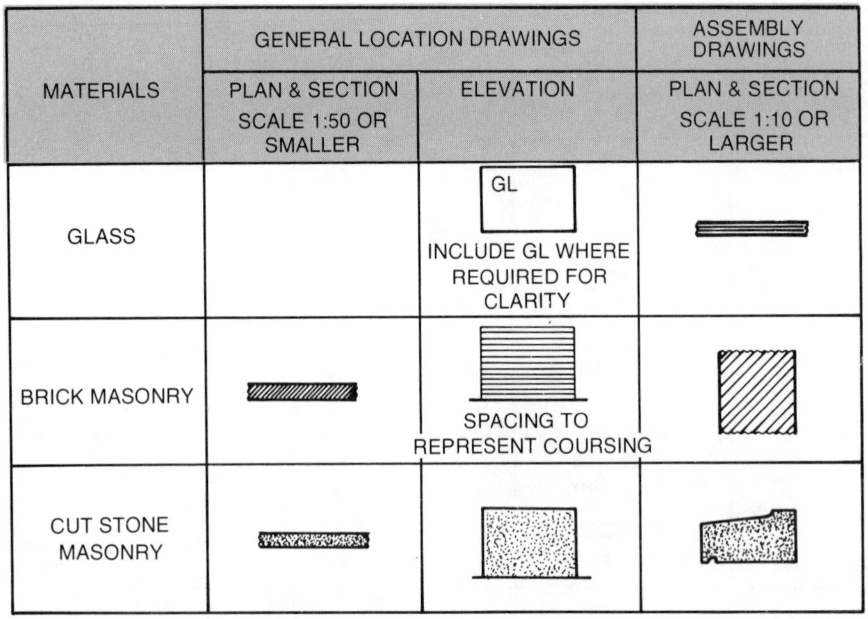

| MATERIALS | GENERAL LOCATION DRAWINGS | | ASSEMBLY DRAWINGS |
	PLAN & SECTION SCALE 1:50 OR SMALLER	ELEVATION	PLAN & SECTION SCALE 1:10 OR LARGER
GLASS		GL INCLUDE GL WHERE REQUIRED FOR CLARITY	
BRICK MASONRY		SPACING TO REPRESENT COURSING	
CUT STONE MASONRY			

Figure 3-11 (Cont'd)

TABLE 3-1   PREFERRED SCALES FOR BUILDING DRAWINGS.

Drawing	Recommended Scales	Use
Block Plan	1:2000 1:1000 1:500	To locate the site within the general district.
Site Plan	1:500	To locate building work, including services and site works, on the site.
Sketch Plans	1:200	To show the over-all design of the building.
General Location Drawings	1:100 1:50	To indicate the position of rooms and spaces, and to locate the position of components and assemblies.
Special Area Location Drawings	1:50 1:20	To show the detailed location of components and assemblies in complex areas.
Construction Details	1:20 1:10 1:5 1:1	To show the interface of two or more components or assemblies for construction purposes.
Range Drawings	1:100 1:50 1:20	To show in schedule form, the range of specific components and assemblies to be used in the project.
Component and Assembly Details	1:10 1:5 1:1	To show precise information of components and assemblies for workshop manufacture.

Courtesy National Research Council

Specifications

Specifications (Specs) are written instructions which accompany each set of blueprints. This additional information supplements the blueprint with descriptive instructions on such items as the kind and quality of material to be used, the method of construction, the accepted quality of work and special features of the job. In addition, many architects will include in the specs legal responsibilities, methods of purchasing materials and equipment, employment practices and insurance requirements for the building.

Although the architect or designer prepares the specifications for each individual building, it is not necessary to completely rewrite the specs for each project. It should be noted that many aspects of the specs fall under the heading of *common ground* thereby requiring revision of only those items affecting the building in question. Specifications serve three basic purposes for the owner, architect and builder: (a) They are legal documents outlining bidding procedures, owner-architect-builder responsibilities, bonding requirements of the contractor, and in the event of legal action taken by any party involved, they become evidence along with the blueprints. (b) They stipulate exactly what the owner is to receive and become a guide for contractor or sub-contractor when estimating costs and (c) They supply required technical data for fabrication and installation of materials to be used in the structural and finishing phases of the building.

Specifications are generally laid out according to the procedure required to erect the building and normally follow a format similar to the one shown in Table 3-2. This table uses as a reference the index found in the front of the specifications.

As blueprints and specs are legal documents between owner and the contractor,

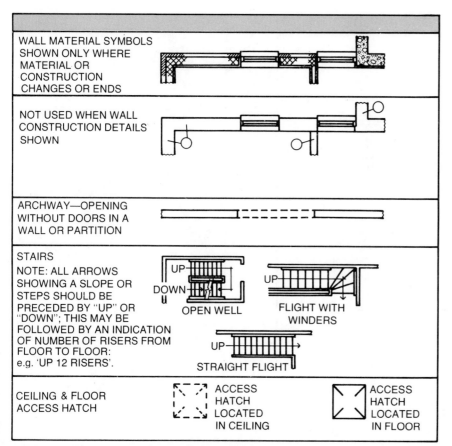

WALL MATERIAL SYMBOLS SHOWN ONLY WHERE MATERIAL OR CONSTRUCTION CHANGES OR ENDS	
NOT USED WHEN WALL CONSTRUCTION DETAILS SHOWN	
ARCHWAY—OPENING WITHOUT DOORS IN A WALL OR PARTITION	
STAIRS NOTE: ALL ARROWS SHOWING A SLOPE OR STEPS SHOULD BE PRECEDED BY "UP" OR "DOWN"; THIS MAY BE FOLLOWED BY AN INDICATION OF NUMBER OF RISERS FROM FLOOR TO FLOOR: e.g. 'UP 12 RISERS'.	UP DOWN OPEN WELL / UP FLIGHT WITH WINDERS / UP STRAIGHT FLIGHT
CEILING & FLOOR ACCESS HATCH	ACCESS HATCH LOCATED IN CEILING / ACCESS HATCH LOCATED IN FLOOR

Figure 3-12   Symbols used for walls, floors, and ceilings.   *Courtesy National Research Council*

TABLE 3-2 TYPICAL SPECIFICATION LAYOUT

**Specifications For
Your Building, Your Town, Province**

Section Title	Section Number	Pages
Tender form		1 to 3
Instruction to bidders		3 to 6
General and Supplementary General Conditions		6 to 9
Excavation, Grading and Site Work	1	10 to 11
Concrete, Plain and Reinforced	2	12 to 15
Masonry	4	15 to 16
Carpentry	5	16 to 20
Millwork and Glazing	6	20 to 25
Roofing	7	25 to 27
Finishing Hardware	8	28 to 34
Sheet Metal Work	9	35 to 36
Painting	10	37 to 38
Mechanical and Heating	11	39 to 40
Electrical	12	41 to 45

they have to be followed closely so that no misinterpretation can occur. Any change from the original must be agreed upon by all parties concerned and must be put in writing to become binding.

Government and institution funding of homes supply the largest percentage of monies in the Canadian housing market. Builders using this money are required to adhere to the National Building Code—"Residential Standards". Since the residential standards are in themselves *specifications* for houses built under the loaning agent, the builder is not required to write a complete set of specs. He will instead use what is commonly known as a fill-in form. This form lists information that would be pertinent to the house in question (Table 3-3).

BUILDING CODE

A building code is a set of **minimum** regulations regarding the safety of buildings, specifically with reference to public health, fire protection and structural considerations.

In Canada, the most complete and up to date building code available is the National Building Code of Canada. This code was compiled by the Associate Committee on the National Building Code, assisted by a number of standing committees. Each of these is responsible for one major part of the code. The Division of Building Research of the National Research Council of Canada has its head office in Ottawa and has numerous regional offices across Canada. It supplies the technical and secretarial support staff required to do research into new ideas, materials and methods of construction. This is vital to keeping the building code current. The Division reports back to the Associate Committee on the National Building Code, and amendments are then made to the code to supplement the existing code as a *Revision Series*. The amend-

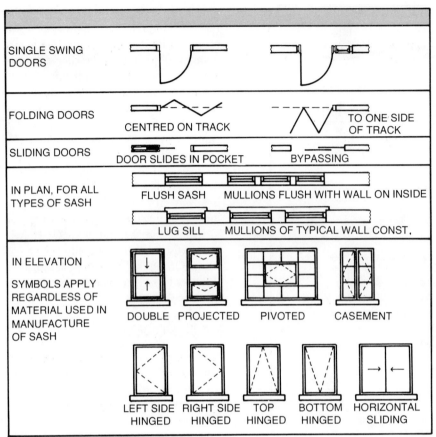

Figure 3-13   Door and window symbols.

Courtesy National Research Council

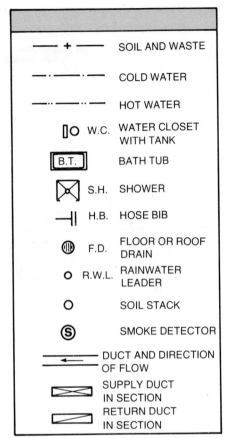

Courtesy National Research Council

Figure 3-14   Symbols used for plumbing and heating.

TABLE 3-3   PORTION OF A STANDARD FILL-IN FORM

Specifications

Job No. _____

Customer _____   Address _____

City/Town _____   Telephone _____

Site Location-Lot _____   Block _____   Plan _____

Section #3—Structural

Floor Construction

Beam(s) Douglas Fir Constr. Grade 3 ply 38 mm × 235 mm

Telepost(s) Conform to "Residential Standards" Table 10-B

Joists Douglas Fir Constr. Grade 38 mm × 235 mm @400 mm O.C.

Bridging Cross bridging 38 mm × 38 mm @2100 mm max.

Sub Floor Douglas Fir plywood—15 mm

Finish Floor Firlock—6 mm

Basement Stairs Housed Stringer—30 mm tread, 11 mm ply. riser

ments are then included in the succeeding editions of the National Building Code.

It is accepted practice today for the provincial government to adopt the National Building Code or use it as a base for their building requirements, upgrading the minimum requirements where applicable to their area. The cities and municipal governments within the province are then required to conform to the code as laid out by the provincial authorities.

A supplement to the National Building Code is published because of the many and varied buildings requiring coverage. This supplement-Residential Standards

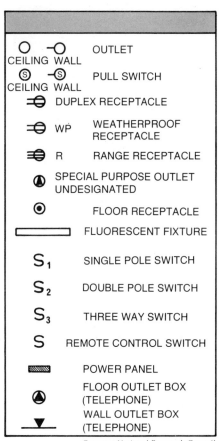

O —O	OUTLET
CEILING WALL	
Ⓢ —Ⓢ	PULL SWITCH
CEILING WALL	
⊐●	DUPLEX RECEPTACLE
⊐● WP	WEATHERPROOF RECEPTACLE
⊐● R	RANGE RECEPTACLE
⊕	SPECIAL PURPOSE OUTLET UNDESIGNATED
⊙	FLOOR RECEPTACLE
▭	FLUORESCENT FIXTURE
S_1	SINGLE POLE SWITCH
S_2	DOUBLE POLE SWITCH
S_3	THREE WAY SWITCH
S	REMOTE CONTROL SWITCH
▨	POWER PANEL
⬤	FLOOR OUTLET BOX (TELEPHONE)
▼	WALL OUTLET BOX (TELEPHONE)

Courtesy National Research Council

Figure 3-15 Electrical symbols.

Canada–is Part 9 of Housing and Small Buildings of the National Building Code. In general, *Residential Standards Canada* contains requirements for residential buildings only. All other types of buildings must be referred to the National Building Code.

Canada Mortgage and Housing Corporation

The government agency that controls the funding of homes under the National Housing Act is called the Canada Mortgage and Housing Corporation (CMHC). If the structure is being financed or insured through CMHC, the owner will have to build according to *Residential Standards Canada*. It is the responsibility of CMHC to see that work conforms to the construction standards as laid out in *Residential Standards* and also with the plans and specifications the owner submitted to them when applying for the loan. Other lending agencies such as banks, credit unions and trust companies finance home construction. It is common practice for them to adopt the Residential Standards and require the home to be constructed according to the code.

Building inspectors for CMHC make a minimum of three inspections of the house during construction. The inspector's job is not one of policing or supervising the construction of the home. It is designed to ensure security for the National Housing Act loan. This security is two-fold—first that every house will be built within the prescribed minimum standards (or better) of good construction practices and that the home constructed meets the approved plans and specifications as provided in support of the owner's loan application.

Local Building Requirements

The local building code or building bylaw is administered by the city, township or municipality to provide for the erection, classification, alteration and repair of buildings within its jurisdiction. It is common practice for the governing body to adopt the National Building Code as set forth and to add any necessary regulations that would be conducive to good building practices in that area. It also sets out the zoning regulations for the area under its jurisdiction. Zoning regulations specify the types of structure allowed in a proposed building area. Examples of these would be residential, commercial, industrial, agricultural and special building areas.

Like CMHC, the local governing bodies appoint building inspectors whose job it is to process and administer building inspections during the construction in their areas. Building permits are required before the owner can begin construction or renovation (over a certain dollar value) to an existing building. When applying for a permit the owner usually has to comply with some or all of the following requirements:

(a) Use the official building permit form for that jurisdiction. See Table 3-4 for a typical building permit.

(b) State the proposed use of the building (this is required for zoning regulations)

(c) File for the local Governing Board:
   1. A complete set of drawings for the building
   2. Specifications for the building
   3. Dimensions of the property along with the position of the building on the property
   4. Necessary grade levels of street, sewer and water lines and of the proposed building
   5. Any other information as required by the governing body

Once the Building Permit has been accepted by the agent (Building Department) of the local government and the owner begins construction, it is then a matter of administering and enforcing the local building code. As with CMHC inspectors, the local building inspector has the right at any time during or after construction to inspect the building site. This ensures that the owner/builder is conforming to the local building code. If contraventions to the code are found inspectors have the authority to issue a *stop work order* until such time as the violations are corrected to the department's satisfaction.

While the use of building codes may seem restrictive, they have and will continue to serve a worthwhile purpose. Urban and rural developments do not just ''happen'' today. To ensure orderly growth it is necessary to restrict or promote development of residential and commercial areas in our cities and rural areas.

If planning authorities are conscientious in laying out new developments within cities and municipalities, and if the builders, owners and population in general are aware of these ideas, then well-planned, publicly accepted developments should be the result.

TABLE 3-4   TYPICAL BUILDING PERMIT
City of Your Choice—Building Department

Application No. _____

Legal   Assessment
Description   No. _____
                     Block _____
                     Plan _____

                     General _____
                     Plumber _____
Contractor   Electrician _____
                     Architect _____

                     Width _____
Size of   Length _____
Building   Floor Area _____ m³

Building Specifications

Foundation Material
   a) Footing _____
   b) Wall _____ etc.
Floor joist _____ @ _____ o.c.

Building _____
      Address _____ Plan No. _____

               Name _____
               Phone No. _____

Owner   Building Permit Sent
                     to _____
                     _____

Type   New _____ Demo.
                     and/or _____
of   Alterations _____
                     Removed _____
Work   Addition _____
                     Other _____

                     Frontage _____
                     Depth _____
Zoning   Site Area _____ m³
Information Front yard _____ m
                     Side yard _____ m

Request for Building Permit

I the undersigned verify the above
information as correct in accordance with the
      building and zoning bylaws of the city

                                              signed

City of Your Choice—Building
   Department

Permission been granted

                              Building Inspector

Date _____

REVIEW QUESTIONS

3—BLUEPRINT READING

Answer all questions on a separate sheet

A. Write full answers

1. Describe the two types of drawings used by architects to portray their ideas to owners and builders. What purpose does each type of drawing serve?

2. List the type of drawings found in a "set" of working drawings.

B. Select the correct answer

3. Views taken in the horizontal plane are called
   (a) Elevation views
   (b) Plan views
   (c) Sectional views
   (d) Detailed views

4. The location of the house on the lot is usually found on the
   (a) Plot plan
   (b) Elevation plan
   (c) Floor Plan
   (d) Foundation plan

5. Views showing the exterior of the home are called
   (a) Elevation views
   (b) Plan views
   (c) Sectional views
   (d) Detailed views

6. The shape and arrangement of rooms are found on the
   (a) Plot plan
   (b) Sectional plan
   (c) Floor plan
   (d) Detailed plan

7. Border lines and lines used to outline the house are shown as
   (a) Break lines
   (b) Dotted lines
   (c) Solid lines
   (d) Section lines

8. The size of the first floor joist is shown on the
   (a) Foundation plan
   (b) Main floor plan
   (c) Plot plan
   (d) Elevation plan

9. Where information given in the specifications of the home differs from that shown on the blueprint, the builder should
   (a) Make a compromise between the two
   (b) Use the information given on the blueprint
   (c) Use the information given in the specifications
   (d) Make a personal judgement

10. A vertical cross-section of a house shown on a blueprint is known as a(n)
   (a) Detail view
   (b) Sectional view
   (c) Elevation view
   (d) Schedule

C. Define the following:

11. (a) Object line
    (b) Dimension line
    (c) Equipment line
    (d) Extension line
    (e) Break line
    (f) Invisible line
    (g) Section line
    (h) Centre line

D.

12. List the common metric scale(s) used on a set of blueprints for:
    (a) Site plans
    (b) Plan views
    (c) Elevation views
    (d) Sectional views
    (e) Detailed views

E. Mark as either TRUE or FALSE

13. The practice of scaling directly off the blueprint to obtain dimensions is quite acceptable.

14. Schedules list recurring items in a house.

15. Symbols are used to indicate the types of materials used in the construction of the house.

F. Replace the Xs with the correct word

16. Written instructions which accompany the set of blueprints are called XXXXXX.

17. XXXXXXX is the government agency designated to handle funding of home construction in Canada.

18. To obtain a building permit, the XXXXX must submit a copy of the house plans and specifications to the local government agency.

19. The most complete and up-to-date building code in Canada is the XXXXXX, under the responsibility of the XXXXX.

G. Write a full answer

20. Visit your local building inspector's office to obtain copies of the local building code and other information regarding the construction of homes in your area. From this information prepare a written report on the erection, classification, alteration and repair of buildings within your area.

4 SITE PREPARATION AND EXCAVATION

SITE PLANNING

When beginning to plan a home, the *proposed site* should be evaluated before finalizing the blueprints to see what effect it will have on the design and function of the home. Not normally taken into consideration by most home planners or builders are such things as orienting the home to capitalize on the daily quantity of sun, the prevailing winds in the area, the topography of the site, both natural and artificial, the presence of trees and shrubbery and surrounding buildings. These should all have a direct influence on the design. All of these factors should be kept in mind because of the potential to adapt them to the overall scheme and to minimize the energy requirements of buildings.

Employed wisely, the features of the building site can be used to the homeowner's advantage to reduce energy requirements, particularly those associated with heating, cooling and lighting. In any energy-efficient building the sun's energy can be used to advantage with a little forethought. Knowing that during the winter months the sun rises in the southeast and sets in the southwestern sky, and in summer rises in the northeast and sets in the northwest; the most ideal building site is one with a southern exposure. This does

not mean to say that all buildings should be placed on east-west streets. However, it does mean that buildings running along north-south streets, or crescents, etc. should be oriented to obtain as much solar energy as possible through the high energy consumption periods of late fall-winter and early spring (Fig. 4-1).

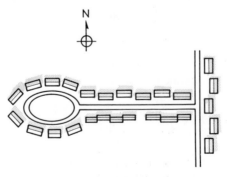

Figure 4-1   Street orientation using solar exposure

Prevailing winds can be either advantageous or detrimental in establishing a building on a site. Ideally, the site and building should work in harmony to cool the building in the summer and offset cold winds in the winter. Wind direction and velocity can be affected by surrounding hills, valleys, wooded areas and large buildings.

Trees offer a natural windbreak. Coniferous trees placed on the north and north-western exposures will not block out the solar radiation during the winter months, but will protect the site from prevailing north-west winds. Deciduous trees and shrubbery can be used to advantage during the summer. If they are positioned to the south and west of the site they can provide the required summer shade for cooling. In fall, the loss of leaves permits

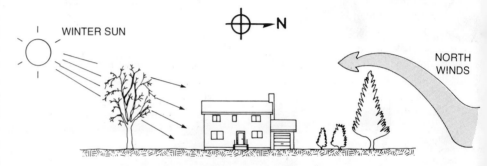

Figure 4-2   Trees used to shelter the home but not to restrict the winter sun

the penetration of sunlight, thereby allowing solar radiation and natural lighting to be used in abundance through the winter months (Fig. 4-2).

Naturally, not all conditions can be achieved on every building site, although decisions that are made should be of a long term rather than short term and should be made on the basis of energy conservation.

LOT OR PROPERTY LAYOUT

Once the building site has been chosen, blueprints finalized and drawn, and the necessary permits obtained from local governing bodies, the preliminary work can begin.

Site preparation can result in hours of wasted time unless the owner or builder is familiar with local requirements. Many municipalities require that both the property and position of the building on that property be certified as stated on the building permit. The only accepted certification is the one obtained from a registered engineer or company involved in land surveying. Depending on the knowledge and equipment available, the owner/builder can go ahead and stake out the lot and have it certified at a later date. The alternative is to contract this work out to a land surveyor before beginning the layout of the building.

Assuming that the building site has a reasonably flat profile (no steep slopes involving a hill or valley), and that the lot is rectangular in shape and the house is also a rectangle or series of rectangles, the following method of layout can be used for both the lot and the building. This method requires the use of the following hand tools: wood stakes (38 × 38 × 300 mm) or steel pins, hammers (claw and small hand sledge), nails, several lengths of nylon building line and measuring tapes (15 or 30 m).

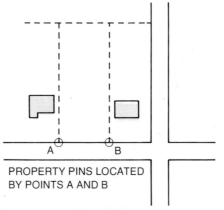

PROPERTY PINS LOCATED BY POINTS A AND B

Figure 4-3 Typical building site with the front corner pins known

Step #1: Locate the corner pins on the property. This could include all four corners or only the front corners of the lot. Figure 4-3 indicates a lot with only two points known.

If the two front corners of the lot are the only available points to work from, it is up to the builder to establish the two back corners of the property. To locate these points a third point at 90° to the existing known points must be established. To establish the third point at a right angle to the two existing property pins A and B, the *Pythagorean theorem* or *Right Angle* law is used. Rather than using the formula of $a^2 + b^2 = c^2$ and calculating the hypotenuse of the right angle triangle, it is common practice to use the base numbers of 3, 4, and 5. The carpenter then uses the largest multiple possible of the base numbers to lay out the lot, while staying within the boundaries of the lot.

Example: A lot measures 19 × 38 m. Using a multiple of 6, the right angle formed would measure 6(3, 4, 5) or 18, 24, and 30 m (Fig. 4-4).

For a more detailed explanation of the Pythagorean theorem and its application see Chapter 9—Roof Construction.

Step #2: Between the property pins A and B erect the lot line using stakes and a nylon line (Fig. 4-5). Measuring along

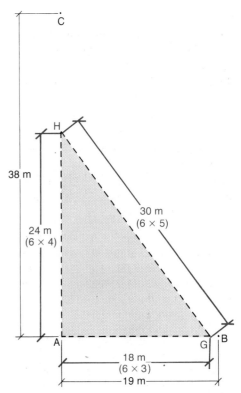

Figure 4-4 Layout of the right triangle using the property pins

Figure 4-5 Layout of the lot using two property pins

the line 18 m from point A, establish a point G. **Note:** when using wood stakes, if a nail is driven into the stake after it has been placed at approximately the correct distance, it will speed up the layout and increase the accuracy.

Step #3: Using two measuring tapes, measure out 24 m from point A, and 30 m from point G, establishing a point H at the intersection of the two measurements. Points A, G, and H will then form the right angle triangle required to establish the back property pin (Fig. 4-5).

Step #4: From point A extend a line the depth of the lot (38 m) and establish point C. The other corner of the lot (point D) is obtained by measuring 19 m from point C, and 38 m from point B.

Step #5: Check the layout of the lot and ensure that all corners are square. To do this, measure the diagonals. If diagonals AD and BC are equal then the corners are right angles and the layout is correct (Fig. 4-6).

Layout of the Building

To establish the position of the house on the property the following information is needed; the size and shape of the house (obtained from the blueprint), and the distance the house is from the front and side property lines. This distance is termed setback and must conform to the local building bylaws in your area.

Using the house dimensions (13 000 mm × 9000 mm) from chapter three, Figures 3-4 and 3-5, and establishing the position of the house at 7 m from the front property line and 2 m from the side property line, the layout of the building on the property can then proceed.

Step #1: Establish the front and side property lines by running a line from point A to B and from A to C (Fig. 4-7).

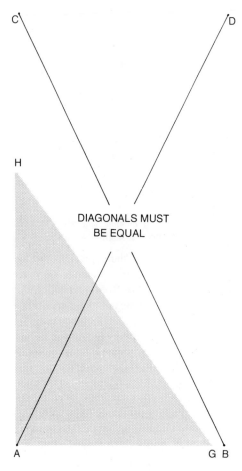

Figure 4-6   Completed layout of the lot with all corners established

Step #2: Measure the setback along AC (7 m) plus 1200 mm the cantilevered portion, and establish a point E. From point B measure up the same distance and find point F.

Most local governing authorities establish the minimum setback and side yard requirements and measure to that part of the building nearest the property line. Depending on the style of home this measurement could be taken from the basement wall, the frame wall of the house or even the roof line. In the example we are measuring to those portions of the building that are cantilevered over the base-

ment wall (Fig. 4-7). This will move in the basement wall more than the established 7 and 2 m.

Step #3: Measure along AB the side yard (2 m plus 750 mm) for the cantilevered portion and establish point G. From point C measure the same distance and establish point H.

Step #4: Run lines from points E to F and from points G to H. The intersection of the two lines becomes the corner of the basement wall–point J (Fig. 4-8).

Step #5: Measure along line EF the length of the basement (13 000 mm) and establish a point K. Using the line GH and the width of the basement (9 000 mm), establish point L.

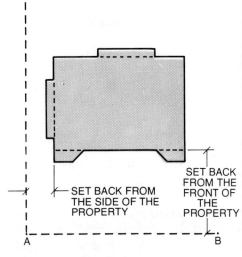

Figure 4-7   Proposed home situated on the property

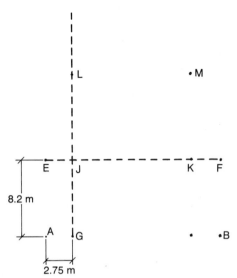

Figure 4-8   Front and side property lines

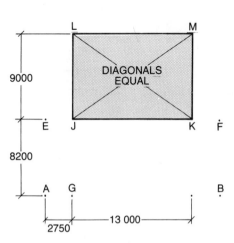

Figure 4-9   Completed layout of the building on the property

Step #6: The last corner of the basement, point M can be found by measuring from point L (13 000 mm) and K (9 000 mm) simultaneously.

Step #7: To check your accuracy in laying out the basement, measure the diagonals LK and JM. They must be equal (Fig. 4-9).

If the layout of the house requires an *offset* (either an **L** or **T** shaped home), or has a number of smaller offsets, it is generally accepted practice to lay out the larger rectangle first and then add or omit the offset afterwards (Fig. 4-10).

INSTRUMENT LAYOUT

The alternate method for site and building layout is to use surveying instruments. Today the carpenter uses surveying instruments both to speed up the work and

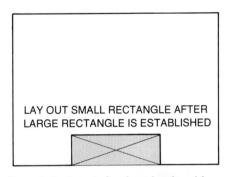

Figure 4-10   Layout of an irregular shaped home using the larger rectangular layout first

to ensure its accuracy. Surveying simply means making precise measurements, which are a must no matter what stage of construction you are working on.

Types of Instruments

The simplest of the surveying instruments is the *builder's level* (referred to sometimes as the dumpy level). In this instrument, a telescope in a fixed position turns a complete 360° (Fig. 4-11). Although layout work is not possible with this instrument, it can be used for testing level-

Courtesy Faber-Castell Canada Ltd.

Figure 4-11   Builder's level

ness, transferring points and measuring angles on a horizontal plane.

The *transit-level* (Fig. 4-12) is a combination of the engineer's transit and the builder's level. It is becoming increasingly popular in the construction field, because it can perform all the operations of the builder's level and can also measure vertical angles up to 45° from the horizontal plane. The vertical mobility of the telescope allows it to be used for both lot and building layout. Figure 4-12 shows the transit-level and its component parts.

In addition to the features contained on the other two instruments described, an *engineer's transit* has a telescope which will rotate 360° both horizontally and vertically. It is also equipped for measuring directions (it has a built-in compass) as well as distance measurement. Although nice to have, these features are not necessary for most layout work. Therefore, all other information on surveying equipment will pertain to the transit-level.

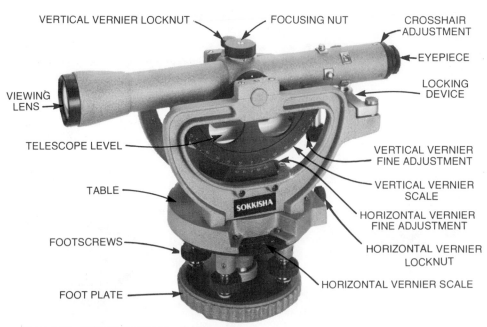

Figure 4-12   Builders transit-level and component parts

Courtesy Faber-Castell Canada Ltd.

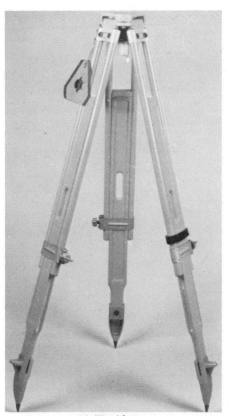

(a)   Wood frame

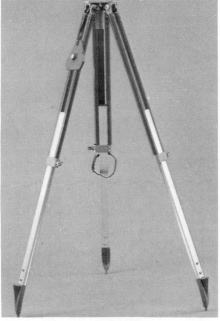

Courtesy Faber-Castell Canada Ltd.
(b)   Metal frame
Figure 4-13   Instrument tripods

Operation of the Transit Level

In operating the instrument the first and most important step is knowing how to assemble and set it up. The following procedure is used for four point leveling:

Step 1: Set up the tripod, a three legged wood or metal stand (Fig. 4-13) securely to prevent collapse of the tripod and the instrument.

Step 2: Place the instrument on the tripod, ensuring that the threads are properly meshed and that the foot plate is tightened onto the tripod.

Step 3: After snugging up the footscrews to the footplate (Fig. 4-12), rotate the telescope until it is directly over one pair of opposite footscrews (Fig. 4-14).

Step 4: Adjust the footscrews to centre the level bubble using the "Thumbs In—Thumbs Out" method (Fig. 4-14) **Note:** The level bubble always moves in the direction of the **left** thumb.

Step 5: Once the telescope level is centred, the telescope is turned 90°, which places it directly over the opposite pair of foot screws. The leveling process in step 4 is then repeated (Fig. 4-15).

Step 6: The telescope is returned to its original position and any necessary adjustments made. Bring it back to the position in step 5. It should now be level. Final adjustment is complete when the telescope can be turned in a complete circle without any change occuring in the position of the bubble.

Step 7: Periodic checks may be necessary to ensure the instrument remains in adjustment.

Note: On instruments which have three point leveling footscrews with a circular level bubble, leveling is a simple operation. Footscrews are adjusted in pairs until the bubble is centred perfectly.

Layout of a Lot and House using a Transit-Level

Previously, a method was described for laying out the lot and house by measure-

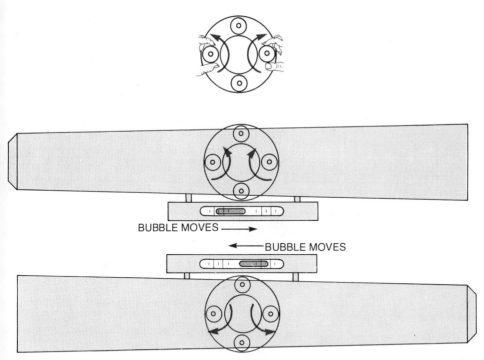

Figure 4-14 "Thumbs In—Thumbs Out" method of leveling the instrument

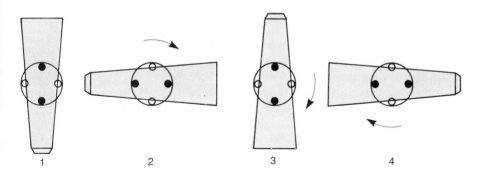

Figure 4-15 Positioning the telescope over the leveling footscrews.

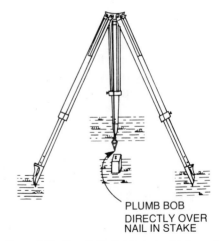

Figure 4-16 Transit Level set up over the property pin

Step 1: Locate and set up the instrument over one of the property pins. Use a plumb bob fastened to the bottom of the instrument to locate the centre of the pin in Figure 4-16.

Step 2: Adjust the instrument until it is level. Release the locking mechanism which secures the telescope in the horizontal position.

Step 3: Sight in, or locate the other property pin (Point B) and lock the transit level head in position.

Step 4: Using the same lot, house and setback dimensions as before, establish points C (2.75 m) and D (15.75 m). The position of the two points is obtained by tilting the telescope and running a straight line from point A to point B. Measure along the line of sight from A. The two points are established at their respective distances from point A (Fig. 4-17).

ment given certain conditions. If these conditions cannot be met or if other factors restrict their use, the transit level becomes the principle instrument used in the layout.

When using a transit level it is quite common to lay out both the lot and the building simultaneously to save repetition of some operations. The following is an accepted procedure:

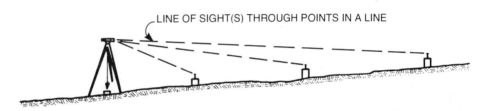

Figure 4-17 Establishing points along the line sight.

Step 5: Move the instrument to point C, setting it up directly over that point using the plumb bob.

Step 6: Adjust the horizontal scales to 0° and sight in point B locking the instrument in this position.

Step 7: Release the scale lock and rotate the telescope a full 180°, picking up point A in the line of sight.

Note: This is a check to make sure you have set up over point C correctly. If point A is in the line of sight you can proceed with the layout.

Step 8: Release the horizontal scale lock and rotate the telescope 90°, locking it again in this position (Fig. 4-18).

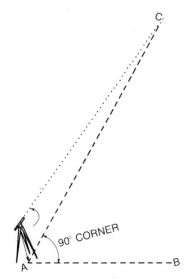

Figure 4-18   Turning a 90° corner

Step 9: Measuring out 8.2 m from point C and 17.2 m in the line of sight, establish points E and F as the two corners of the house.

Step 10: Move to point D and repeat steps 5 to 9, establishing points G and H as the remaining two corners of the house (Fig. 4-19).

A measurement of diagonals will indicate the accuracy of the layout. Diagonals must be equal. If it is necessary to obtain

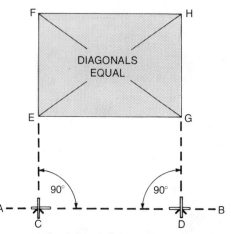

Figure 4-19   Completed layout of the house on the building site

the other corners of the lot, steps 5-9 can be repeated to establish the required points.

THEORY OF LEVELING

The most common use of the transit-level is to obtain a perfectly level line of sight. Measurements are then taken to determine the difference in elevation (height) between any number of points. The line of sight is perfectly straight and therefore every point on the line of sight will be exactly level with every other point. Using a reference point of known elevation, the difference in elevation of points above

and below the known point can be found with this instrument (Fig. 4-20).

Leveling Rod

To check the difference in elevations on the site, it is necessary to know how to use a level rod. Manufactured rods come in two styles, the one on the right in Figure 4-21 is referred to as the **G style-E pattern**, while the one on the left is an **F style** rod. Most collapsible rods are available in lengths up to 5 m.

The G style-E pattern rod is graduated to the nearest centimetre and marked in consecutive decimetres. The distance between 07 and 08 in Figure 4-21 is one decimetre. That distance is broken into ten equal parts (centimetres–cm) shown as white or black divisions. Consequently, reading to either the bottom or top of a black division gives a reading to the nearest cm. For finer measurements the target plate (Fig. 4-21) is adjusted to coincide with the crosshairs on the telescope of the instrument. Readings to the nearest millimetre are possible.

The graduations on the F style rod are identical to the G style E pattern rod, with the exception of the dot, which appears above the number to indicate the height in metres. One dot therefore represents one

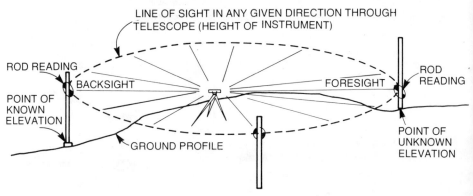

Figure 4-20   Theory of leveling

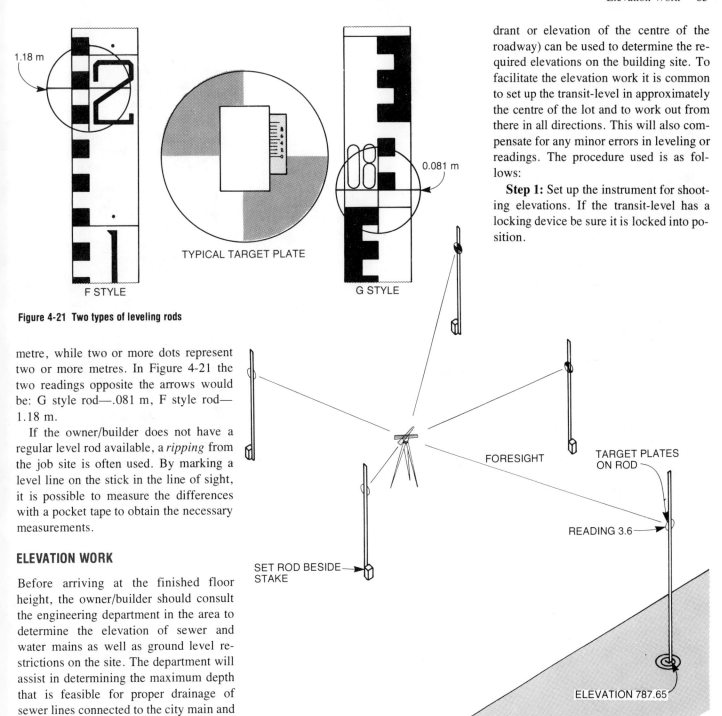

Figure 4-21 Two types of leveling rods

Figure 4-22   Determining the elevations on the building site

drant or elevation of the centre of the roadway) can be used to determine the required elevations on the building site. To facilitate the elevation work it is common to set up the transit-level in approximately the centre of the lot and to work out from there in all directions. This will also compensate for any minor errors in leveling or readings. The procedure used is as follows:

Step 1: Set up the instrument for shooting elevations. If the transit-level has a locking device be sure it is locked into position.

metre, while two or more dots represent two or more metres. In Figure 4-21 the two readings opposite the arrows would be: G style rod—.081 m, F style rod—1.18 m.

If the owner/builder does not have a regular level rod available, a *ripping* from the job site is often used. By marking a level line on the stick in the line of sight, it is possible to measure the differences with a pocket tape to obtain the necessary measurements.

ELEVATION WORK

Before arriving at the finished floor height, the owner/builder should consult the engineering department in the area to determine the elevation of sewer and water mains as well as ground level restrictions on the site. The department will assist in determining the maximum depth that is feasible for proper drainage of sewer lines connected to the city main and if this would facilitate proper surface drainage (Figs. 4-22 and 4-24).

Once established, the known elevation or *bench mark* (manhole cover, fire hy-

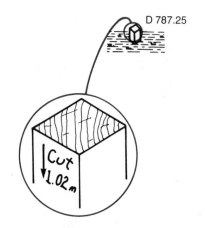

D 787.25

C 790.20

B 789.35

(b)   Grade stake showing depth of excavation

A 788.10

Figure 4-23 (a)   Site layout indicating grade stakes for excavating

mine the depth of the basement excavation.

From Figure 4-24:

Bench mark (B.M.) = 787.65 m
Backsight (B.S.) + B.M. =
Height of Instrument (H.I.)
3.60 + 787.65 = 791.25
Height of Instrument (H.I.) −
Readings (Fig. 4-24(a)) =
Elevation of points

Point A: 791.25 − 3.15 = 788.10
       B: 791.25 − 1.9 = 789.35
       C: 791.25 − 1.05 = 790.20
       D: 791.25 − 4 = 787.25

Elevation of Points − Elevation of the bottom of the excavation = **Depth of excavation**

Point A: 788.10 − 786.23 = 1.87 m cut
       B: 789.35 − 786.23 = 3.12 m cut
       C: 790.20 − 786.23 = 3.91 m cut
       D: 787.25 − 786.23 = 1.02 m cut

After calculations are complete, individual stakes are then marked for the amount of cut required (Fig. 4-23 (b)). This allows the excavator to proceed with excavating. Before removing the transit level it is common practice to set the height of the batter boards with it as well.

Step 2: Hold the rod on the established bench mark (B.M.)—787.65 m (Fig. 4-22), obtaining a reading on the rod of 3.60 m. This reading is termed the *backsight*. Record this reading which will be added to the bench mark to determine the *height of the instrument* (H.I.). B.M. 787.65 + backsight 3.60 = H.I. 791.25 m.

Step 3: Moving the rod to one corner stake of the house, take a reading at a marked point on the stake (Fig. 4-23(a) and (b)). This is termed the *foresight*. The reading 3.15 m is then subtracted from the height of the instrument to obtain the elevation of the point. H.I. 791.25 − foresight 3.15 = elevation point A 788.10 m.

Step 4: Obtain readings at the other corners of the house—points B, C, and D. Repeat the operations in step 3 to obtain readings of 1.9, 1.05, and 4 m respectively.

Step 5: If any other excavating is to be done at this time, stakes are driven in at appropriate intervals and readings obtained as in step 3.

Using Figures 4-23 (a) and 4-24 and the data obtained from the elevation work, it is now possible to catalogue the points and obtain measurements to deter-

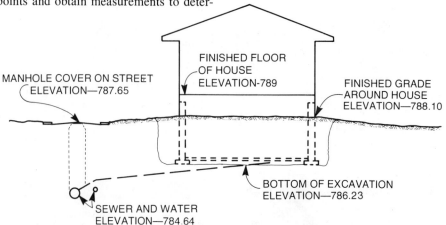

Figure 4-24   Elevations of Street, sewer and water mains and proposed building

ERECTION OF THE BATTER BOARDS

Once the exact position of the building has been established on the lot, the points or corners must be retained for future reference in the construction of the foundation wall. To accomplish this *batter boards* are erected at each corner of the building as well as at offsets or jogs in the foundation. The batter board is usually set back a minimum of 1200 mm from the building line (Fig. 4-25). It is common however, to increase this distance if more room is required for excavating equipment. Unstable soils tend to cave-in during digging and if more room is required for the forming, pouring and exterior work on the foundation wall before backfilling, this distance is vital.

Batter boards consist of 38 × 89 mm stakes driven firmly into the ground. Nailed horizontally to the stakes are 19 × 140 mm boards or 38 × 89 mm dimensional stock. The stakes should be driven in parallel to the building line. To set the height of the batter board, an elevation mark is established on the corner stake (Fig. 4-25). This level is either the top of the foundation wall or the top of the finished floor. In Figure 4-24 the elevation of 789 m would be used for this point. The top of the batter board is then leveled and nailed to the remaining corner stakes.

Transfer of the Building Points

When the batter boards are complete, a nylon line is stretched over the points on the corner stakes, running from one corner to the other and extending out to the batter boards. Either a plumb bob attached to the line, or a hand level held plumb is used to position the line directly above the corner points (Fig. 4-25). Where the line crosses the batter board, either a nail is driven in or a *saw kerf* is made to position and fasten the line. This

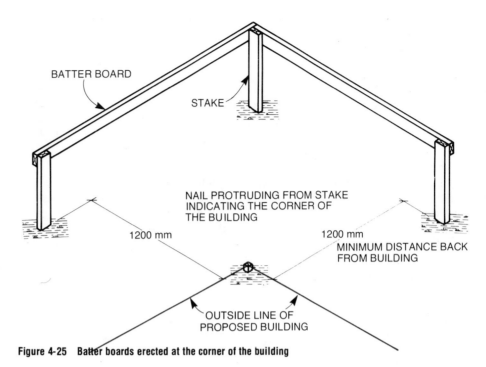

Figure 4-25   Batter boards erected at the corner of the building

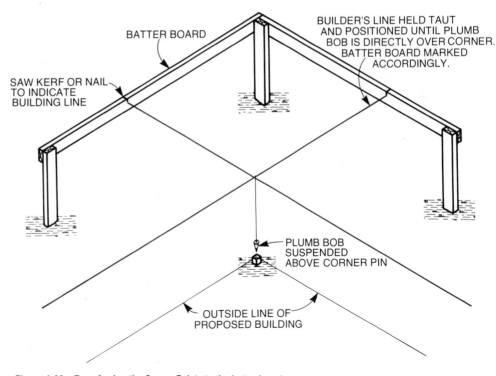

Figure 4-26   Transferring the Corner Points to the batter boards

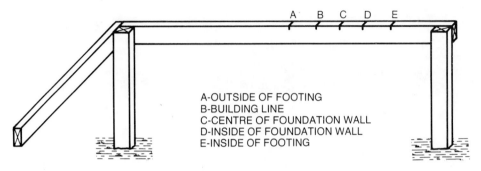

A-OUTSIDE OF FOOTING
B-BUILDING LINE
C-CENTRE OF FOUNDATION WALL
D-INSIDE OF FOUNDATION WALL
E-INSIDE OF FOOTING

Figure 4-27   Layout of the batter boards

procedure is repeated on all sides of the building as each exterior wall line is transferred to a set of batter boards. After transferring these points, the building lines are set up to check the accuracy of the points. The length and width of the building is established between the given lines. Diagonals are again checked to make sure they are equal (Fig. 4-26).

Once all the measurements are checked and confirmed with those on the blueprint, the remainder of the batter boards can be marked out. Figure 4-27 shows a typical layout indicating the position of a number of points in relationship to the building line. Always make sure that the points are clearly indicated on the batter board so that no mistake will be made during the construction of the house.

SOIL

Before beginning to excavate, it is important to have an understanding of the soil, both surface and subsoil. Soils can be divided into three major categories; coarse grained, fine grained and organic.

Coarse grained soils are those grouped together and defined as particles made up of individual pieces which are visible to the naked eye. Broken down according to size there are:

1. **Boulders**—those particles larger than 250 mm in diameter
2. **Cobbles**—particles which range from 75 to 250 mm in diameter
3. **Gravel**—particles smaller than 75 mm but larger than 6 mm in diameter

4. **Sand**—particles ranging from 6 mm in diameter to those particles just visible to the naked eye.

Fine-grained soils are those which are invisible to the naked eye. The consistency of the fine grained soils varies with the water content and density. Its ability to change or retain shape when under pressure is termed as its *plasticity*. The degree of plasticity depends on the amount of water that can be held and still allow the soil to be moulded. The size of a particle and its ability to change shape can only be determined by specific laboratory tests which determine the shrinkage or swelling that will take place under different moisture conditions. In the two categories of silt and clay, some or all of the following characteristics are visible:

1. **Silt**—when moisture is present, if a sample of soil is broken, the fractured edge will appear dull. In a dried state the soil powders and breaks down easily. It tends to wash away easily and brushes off your hands when dry.
2. **Clay**—if moisture is present, blading a sample of soil will leave moisture on its surface. This shiny surface indicates clay content. When dry, some clays can be broken down easily, indicating silty clay of medium plasticity. Samples which cannot be broken down by hand indicate a highly plastic soil. Unlike silt, clays have a tendency to stick to the hands when wet and do not wash off easily.

Organic soils are generally defined as those containing organic material and

supporting forms of plant life. Broken into two groups they are:

1. **Partly organic**—are soils made up of organic material containing varying amounts of silt and sand particles. Partly organic soils have some of the same characteristics as fine grained soils.
2. **Organic material**—contains fiberous material, usually black or dark brown in color. Soils in this category include muskeg, peat and peatmoss.

Topsoil

Topsoil is that layer of soil which will support plant life. It contains both fine-grained soils and organic soil. On the building site the type and amount of topsoil will determine its value in the final landscaping of the site. It is common practice to strip off the topsoil on the site, stock-piling it so that it may be used later.

Fill Soil

Fill soils are generally found in areas where depressions have been brought up even with surrounding levels. Fill areas are usually identified by waste materials and natural soils which are placed over an original layer of topsoil. If fill soil is encountered while excavating for a basement, it should be removed entirely until proper subsoil is reached. If this is not possible, pilings should be considered to stabilize the foundation of the house.

Methods of Testing Soils

Due to the fact that the majority of foundations are selected to meet only minimum requirements, and that the foundation relies on the stability of the subsoil, it is imperative that soils be examined for type and characteristics.

Two methods of identifying soils are normally used. The *open pit method* is used where a shallow excavation is used. It is used mainly on house construction and permits the owner/builder to visually inspect the subsoil upon which the foundation will be constructed. In cases where there is any doubt about the conditions of the subsoil under the footing area, further testing should take place.

The other method used for testing the subsoil is called an *auger* or *core test*. A truck mounted drill rig is usually brought onto the building site and a series of holes drilled at predetermined intervals. Samples of the subsoil are taken at specific depths. When a sample is taken a hollow pipe is fastened to the drill rig, lowered and forced into the subsoil. When brought to the surface it is removed as a ''core,'' packaged and the position on the site and the depth recorded. Samples are then moved to a laboratory where a series of tests determine the type of soil, moisture content, plasticity, consistency and density. From the information obtained, it is possible to determine the bearing quality of the entire site. If a problem exists in the stability of the soil, corrective measures must be taken in the design of the foundation.

WATER TABLE

Defined as a level at which the ground is saturated with water, the water table in any area is usually not a problem when first constructing a building. However, its presence must be taken into consideration because as urban development takes place, changes occur below the ground as well as above.

Ground water levels vary according to the amount of rainfall or snow accumulation. A change in the water table also occurs when land that was formerly underdeveloped is taken up with buildings and streets. Often 30% to 60% of the original land area is taken up with construction causing the water table in the remaining area to change drastically.

If too pronounced, this change may lead to problems in the bearing quality of the subsoil, such as the increased depth to which the ground may freeze in the winter and the liquefying of mineral salts (alkali) in the surrounding soils.

Excess water may have two different effects on the soil, which in turn affects the bearing strength of the subsoil. Depending on their consistency, clay soils can shrink or expand. The foundation therefore may undergo movement which can reach damaging proportions, and which can cause maintenance problems throughout the life of the building.

For frost to affect the home, water must first be present in the soil and weather conditions such that they cause the soil and water to freeze. Generally, the coarse grained soils such as gravel or sand do not heave or swell when frozen. Clay, silt and fine grained sands however, do tend to expand or contract to varying degrees. A high water table will draw frost.

If mineral salts exist in the soil, counteraction is generally taken when constructing the foundation itself. If concrete is used for the foundation, *alkali-resistant cement* is a must. All foundations must be carefully coated on the outside with a waterproof compound or other acceptable material.

To control or minimize the problems with ground water levels, a number of steps may be taken. Many of these steps are dealt with fully in later chapters in this book. To minimize ground water problems:

1. Place drainage tile and/or crushed stone below floor level at the bottom of the footing. Water collected in the drain tile is then disposed of through the sewer line or drywell.
2. Damp-proof the outside foundation and footing.
3. Provide drainage from the ground surface to the drainage tile to eliminate surface water buildup.
4. Slope the grade away from the building.
5. Control the runoff from buildings, driveways, etc.

For large buildings, determining soil types, characteristics, and load restrictions, etc. is the responsibility of the engineer. On smaller jobs such as individual housing projects, if any doubt exists as to problems which might arise it is advisable to enlist the help of soil analysis professionals. A few dollars spent on preventative measures at the foundation stage of the construction will be beneficial in later years in terms of maintenance costs.

EXCAVATION

Once the position of the house and the grade levels have been established, and once the batter boards have been erected, the excavating for the foundation can begin. The style of building and the type of foundation to be used, along with the other points mentioned in the chapter, dictate the depth to which the excavating will take place. The equipment commonly used in excavating for buildings today are either four-wheeled or track-type front end loaders or machine equipped backhoes (Fig. 4-28).

If a good layer of topsoil exists, the excavator's first job will be to strip this away. It is either stockpiled on the site or removed for use when the final grading of the site takes place. When excavating for

Courtesy Caterpillar Americas Co.

(a) Track-type Front-end Loader

Courtesy JCB Excavators Ltd.

(b) Wheel-type Backhoe and Loader

Figure 4-28   Excavating equipment

the foundation, the operator must be careful not to disturb the batter boards, as they are the only record the builder has of the position of the house once excavating begins. The *working room* or extra excavation which may be required outside the building line depends on the following variables:

1. Type of foundation
2. Forming method to be used
3. Type of soil encountered
4. Depth of the foundation

   The amount of working room allowed for varies from 600 to 1000 mm. Special circumstances require more or less working room depending on the particular

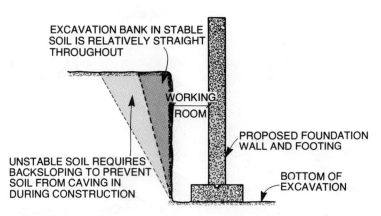

Figure 4-29   Allowance for working room

problem. For instance, if the stability of the soil is a concern the operator of the machine will slope the earth bank back to prevent a possible cave-in, which could result in a great deal of lost time during the construction of the foundation. (Fig. 4-29).

As the operator nears the desired depth, the owner/builder will set up the transit-level and, working from the reference point and the blueprint, check the elevation at the bottom of the excavation. The finished depth is determined by the type of foundation and what the builder wants. The builder will have the operator excavate to either the bottom of the footing or to the bottom of the gravel fill (pages 94-116). Working to a *close tolerance* (plus or minus 25 mm), at the bottom of the excavation will pay off when constructing the footings and foundation wall. An excavation which is too shallow will result in a considerable amount of hand digging, while too-deep an excavation requires extra gravel fill and concrete which means extra dollars.

DETERMINING THE NUMBER OF CUBIC METRES OF EARTH IN AN EXCAVATION

To determine the number of cubic metres of earth that would have to be removed from the excavation to allow for a foundation, three factors must be known:

(a) **Building size** (13 000 × 9000 mm)
(b) **Excavation depth** (2 m)
(c) **Working room** (600 mm)

Note: that the space for working room is added to all four sides of the building, allowing for setting up and stripping of forms, foundation erection and drainage systems to be installed (Fig. 4-30).

Using the volume formula in which:
Length × width × depth = cubic metres, the following calculations take place:

Step 1.   L × W × D = m³ All measurements have to be changed from mm to m.

Step 2.   Length = 13 + 1.2 (working room)
Width = 9 + 1.2 (working room)
Depth = 2

Step 3.   L × W × D = m³ or
14.2 × 10.2 × 2 = 289.68 m

Step 4.   If the resulting answer—289.68 m³—comes out to a fraction of a cubic metre, round off your answer to the next largest cubic metre, or 290 m³.

With the excavation now complete the building lines can be replaced on the batter boards and the house is ready to progress to the next phase of construction.

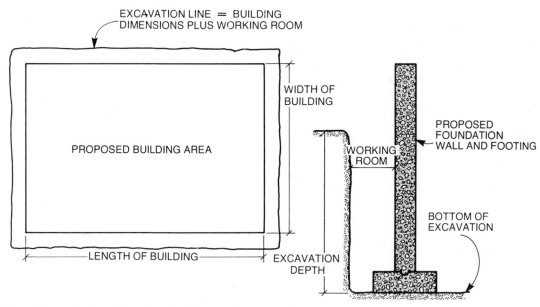

EXCAVATION LINE = BUILDING DIMENSIONS PLUS WORKING ROOM

WIDTH OF BUILDING

PROPOSED BUILDING AREA

PROPOSED FOUNDATION WALL AND FOOTING

WORKING ROOM

BOTTOM OF EXCAVATION

LENGTH OF BUILDING

EXCAVATION DEPTH

Figure 4-30   Calculating the number of cubic metres in an excavation

REVIEW QUESTIONS

4—SITE PREPARATION AND EXCAVATION

Answer all questions on a separate sheet.

A. Write full answers

1. List the factors to consider when choosing a lot upon which to build a house. What effect can this have on the design on the house?

2. What method can the carpenter/builder use to lay out the house on the building lot? Explain its application.

B. Select the correct answer

3. If the length of two sides of a right triangle are 28.5 and 38 m respectively, the length of the third side is
   (a) 40.5 m
   (b) 44.5 m
   (c) 47.5 m
   (d) 50.7 m

4. The distance that the house is from the front property line is referred to as the
   (a) Setback
   (b) Offset
   (c) Contour
   (d) Backsight

5. One of the following operations **cannot** be accomplished with the builder's level.
   (a) Transferring points
   (b) Measuring angles
   (c) Testing levelness
   (d) Layout work

6. Of the following operations, one is unique to the engineer's level.
   (a) Measuring angles
   (b) Measuring distance
   (c) Layout work
   (d) Vertical layout

7. With the transit-level set and the leveling rod held on the B.M., it reads 781 mm. Moving the rod to a point on the foundation we obtain a reading of 1400 mm. If the B.M. is 10 000 mm, what elevation is the point?
   (a) 12 181
   (b) 2181
   (c) 9381
   (d) 619

8. If the transit-level is set at 1456 mm above a B.M. (elevation 11 200 mm), what reading would the target plate be set at to establish a finished elevation of 9100 mm at the bottom of the foundation excavation?
   (a) 3556
   (b) 7644
   (c) 21 756
   (d) 10 556

9. Positioning the transit 1434 mm above the B.M., where will the target plate be set to establish the elevation (750 mm) above the B.M.?
   (a) 2184
   (b) 750
   (c) 2384
   (d) 684

C.

10. Rearrange the following six steps so that they are in the correct order for setting up the builder's level:

(a) Snug up the footscrews against the footplate.
(b) Turn the telescope 90° to the right or left and adjust the other set of footscrews.
(c) Set up the tripod keeping the head approximately level.
(d) Using the ''thumbs in-thumbs out'' method, adjust the footscrews.
(e) Set the telescope parallel with a ''set'' of diagonally opposite footscrews.
(f) Fasten the builder's level to the tripod.

D.

11. Define the following:
(a) Bench Mark
(b) Backsight
(c) Height of Instrument
(d) Foresight
(e) Line of Sight
(f) Elevation

E. Replace the Xs with the correct word

12. When adjusting the footscrews of a leveling instrument, the bubble moves in the direction of the XXXXXXX thumb.
13. Grade stakes are used to indicate the XXXXX on the building site.
14. XXXXXX are set up at the corners of the building before the excavating for the foundation begins.
15. Diagonals of a square or rectangular layout are always XXXXXX.

F. Write full answers

16. Plasticity of soil refers to a certain group of soil. What does the term mean and to what group of soils does it refer?
17. List and define the other groups of soils encountered when excavating.

G. Replace the Xs with the correct words(s)

18. Two methods of testing for the type and characteristics of soils are XXXXXX and XXXXXX.
19. A XXXXXXX is defined as the level at which the ground is saturated with water.
20. List the factors which may cause the level of ground water to vary.

21. What effects does a high water table produce on the bearing strength of soil?
22. XXXXXXX present in soil requires the use of special ''alkali-resistant'' cement in a concrete foundation.

H. Write a full answer

23. List the variables and explain why the working room will vary on each construction site.

I. Replace the Xs with the correct word

24. When excavating for the foundation, the XXXXXX is usually stripped and stockpiled for later use.

J.

25. Find the number of cubic metres in the following excavations:
(a) Building size = 8.5 × 12 m
Excavation depth = 2.15 m
Working room = 900 mm
(b) Building size = 8.8 × 10.2 m
Excavation depth = 1350 mm
Working room = 600 mm
(c) Building Size = 10.8 × 14.2 m
Excavation depth = 1200 mm
Working Room = 450 mm

5 CEMENT AND CONCRETE

PORTLAND CEMENT

Portland cement is not a brand, but a type of cement, consisting of certain chemical compounds and characterized by its greyness and extreme fineness. When mixed with water, portland cement forms a paste which binds materials such as sand and gravel tightly together. Therefore, there is really no such thing as a "cement" footing, "cement" floor, or a "cement" driveway. Once mixed with water and an aggregate it becomes **concrete.**

Joseph Aspdin, the inventor of portland cement, patented the process in 1824 and called it "portland cement" because it produced a concrete very similar in colour to the limestone quarries on the Isle of Portland in England. The first plant to produce cement in Canada was built in Hull, Quebec in 1889. Today, cement plants are found from coast to coast, manufacturing portland cement under carefully controlled conditions and producing a product of high quality.

MANUFACTURE OF PORTLAND CEMENT

The manufacture of portland cement involves four separate operations—quarrying, grinding, burning and another grinding of the final product.

Quarrying is simply the gathering of the raw materials that go into the manufacture of portland cement. The materials used must contain the right proportions of lime, silica, alumina and iron components. The selected raw materials are then transported to the plant. Table 5-1 gives examples of the raw materials used to extract the four essential ingredients in this process.

Once quarried, the raw materials are fed into primary crushers which reduce the raw materials down to less than 125 mm in size. From there they go to the secondary crushers where further reductions to 20 mm size take place; then on to

storage areas to wait for the second step in the process.

Grinding-Blending: In this step the raw materials are proportioned and ground to a powdery consistency. After the grinding (in either a tube or ball mill), the raw materials are tested and proportioned so that the resulting mixture has the desired chemical composition. This process can be as simple as combining two types of rock, but often requires the combination of three or more different types.

This final step before burning is accomplished either wet or dry. In the wet process water is added to the mix while

TABLE 5-1 TYPICAL SOURCES OF RAW MATERIALS USED IN THE MANUFACTURE OF PORTLAND CEMENT

Components			
Lime	**Silica**	**Alumina**	**Iron**
Cement rock	Sand	Clay	Iron ore
Limestone	Traprock	Shale	Iron calcine
Marl	Calcium silicate	Slag	Iron dust
Alkali waste	Quartzite	Fly ash	Iron pyrite
Oyster shell	Fuller's earth	Copper slag	Iron sinters
Coquina shell		Aluminum ore refuse	Iron oxide
Chalk		Staurolite	Blast-furnace
Marble		Diaspore clay	flue dust
		Granodiorite	
		Kaolin	

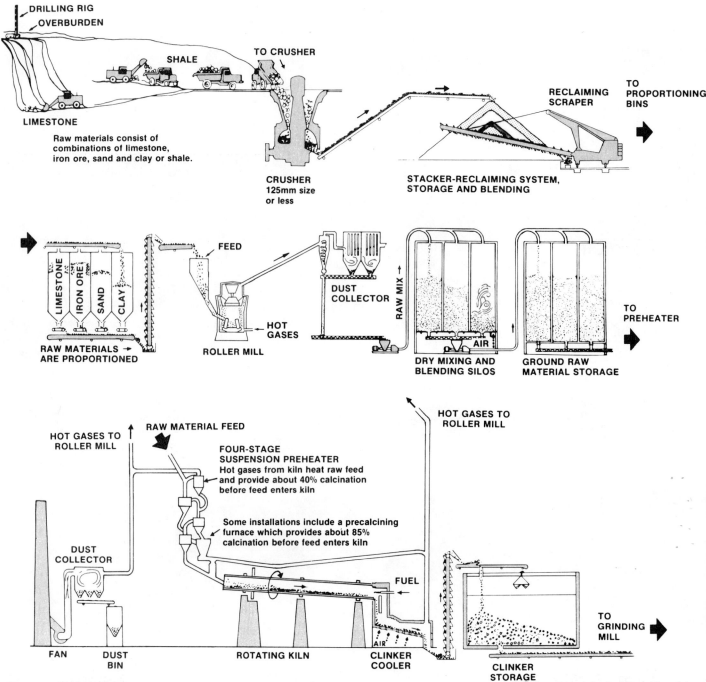

Figure 5-1   Flow chart showing the steps in manufacturing portland cement

Courtesy Portland Cement Association

grinding. The resulting mixture is called *slurry*. This slurry is then piped to open storage tanks where additional proportioning can take place before burning.

In the dry process the material is fed into the same type of grinding mill but no water is added during the process. The resulting powdery material is then stored in silos where additional blending and mix-

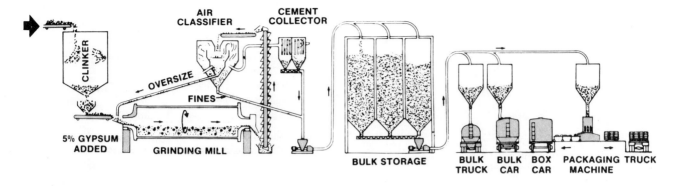

Figure 5-1 (cont'd)

ing can take place in preparation for the third step.

Burning: Burning is the most spectacular step and is the key process in the manufacture of portland cement. It is here that the raw materials are chemically combined to form a clinker called portland cement. The kiln is usually 3 m or more across the inside diameter and more than 150 m long. It averages one revolution per minute and reaches temperatures of 1400 to 1650°C. It is during this exposure to heat that the raw mix changes to a burnt clinker. The clinker is then cooled and stored for the forth and final stage in the manufacturing process.

Final Grinding: Before the final grinding begins a small amount of gypsum is added to the clinker. This regulates the setting up time of cement when mixed with other materials to make concrete. Once the grinding is complete, the resulting finished product—"portland cement"—is finer than flour or face powder (Fig. 5-1).

The finished product is then either shipped by bulk carriers or bags to its destination. In bulk form it is sold by the metric tonne; while bagged portland cement is sold in 40 kg units. Masonry cement used for brick, block and stone work comes in 30 kg bags.

TYPES OF PORTLAND CEMENT

Different types of portland cement (12 or more) are manufactured to meet the specific needs of the user. The following are the most commonly used types.

Type 10 Normal—this is a general purpose cement used for most standard concreting jobs. It is used where the concrete is not subjected to sulphate attack or objectionable heat build up due to hydration. It is used for sidewalks, driveways and foundation work of all kinds.

Type 30 High Early Strength—This is used where high strengths are required of the concrete at an early stage (one week or less). Its uses include concreting where form removal must be early, or in cold weather to reduce the curing time.

Type 50 Sulphate Resistant—This type is necessary where cement is subjected to sulphate (alkaline) conditions caused by soils and/or ground water having a high sulphate content. It is a slower setting cement than Type 10 Normal, and therefore requires a longer curing time than normal portland.

Masonry—is a special type of cement used for making mortar for brick, block or stonework construction. It contains additives that allow it to work smoothly with a trowel. A mortar mix is made by adding water and sand.

CONCRETE

Concrete is a mixture of portland cement and water (the binding agent), fine aggregate (sand) and coarse aggregate (gravel or crushed stone), which harden through a chemical reaction called "hydration". Hydration turns the cement and water into a rock-like mass. Fine aggregate consists of particles up to 10 mm in size. Coarse aggregate consists of particles which range from 5 to 40 mm or larger. As indicated in Figure 5-2, the volume of cement used is usually between 7% and 15% with the water being 14% to 18% in any given volume of concrete. With air-entrainment ranging up to 8% maximum, the remaining 59% to 75% consists of the fine and coarse aggregate which must supply the adequate strength and durability of the concrete.

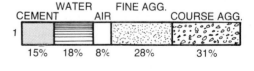
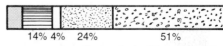

Courtesy Portland Cement Association

Figure 5-2   Range in proportions of materials used in air-entrained concrete by volume

Concrete has many properties which make it one of the most widely used building products today. Among those properties are its ability to:

1. Set up under water
2. Be formed into almost any shape (flexibility)
3. Resist fire
4. Be unaffected by hot and cold weather
5. Be impervious to decay, corrosion and insects

Steps in Obtaining Quality Concrete

Although the person using the concrete may not design the concrete mix, it is helpful to understand some of the fundamentals of quality concrete. To ensure that the mix produces a high quality finished product, it is important that the steps for making quality concrete be followed carefully.

1. Select the proper ingredients
2. Proportion correctly
3. Mix the ingredients thoroughly
4. Place the concrete correctly
5. Finish the fresh concrete correctly
6. Cure the concrete correctly

Selecting the Proper Ingredients:

(a) **Portland Cement**–the types of portland cement and its uses were explained earlier in this chapter. It should be inspected before use to see that it is free of all lumps which indicate moisture has been absorbed and could weaken the resulting concrete mix. If it contains lumps that may be broken down by crushing in the hand, the cement is acceptable for use. Cement stored in a dry place will last for a long period of time without deteriorating.

(b) **Water**–the mixing water performs two functions. It reacts with the cement to produce a bonding agent to hold the aggregates together called "hydration," and it permits the mixing and ease of placement of the concrete into the mould or form. Almost any natural water that is drinkable is suitable for mixing with cement to make concrete. Water that is questionable should be tested to make certain that setting time is not affected or impurities do not have an effect on the quality of the finished product.

(c) **Aggregates**–occupy from 60% to 75% of the total concrete mass. The most commonly used aggregates such as sand, gravel, crushed stone and slag produce concrete from about 2150 to 2550 kg/m³. They therefore strongly influence the mix proportions and the economy of the concrete.

Aggregates should be clean, hard, durable particles free of clay and other fine materials which can affect the bond of the cement paste. The shape and texture of the aggregate will also influence the freshly mixed concrete. Rough, irregularly shaped particles tend to require more water and cement to make a more workable mix; but the bond between the cement paste and rough textured aggregate is better than with round, smooth materials. As well as having quality and the proper shape, aggregates need to be properly proportioned (Fig. 5-3).

In Figure 5-3, a beaker with 28 mm particles has a given void area; a beaker with 10 mm particles has the same void space. However, in the third beaker where both sizes have been mixed, the void volume is considerably less. Therefore, it requires less cement paste to bond it together. Additional sized aggregates would decrease this void area even more.

It is more economical to grade and combine the aggregates, thus keeping the void space to a minimum. This requires a minimum amount of cement paste to fill the remaining area and bond the aggregates together.

(d) **Admixtures**–are classified as any additives used with the three basic ingredients, cement, water and aggregate to make concrete. In almost all cases concrete can be made workable, finishable, strong, durable, watertight and wear-resistant through the use of a well designed mix made of only the basic ingredients. Where setting time needs to be accelerated or retarded, where increased strength through the use of a water-reducing or strength-

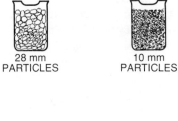

28 mm PARTICLES   10 mm PARTICLES   28 and 10 mm PARTICLES

THE AMOUNT OF LIQUID REQUIRED TO FILL THE VOID AREAS IN EACH OF THE ABOVE BEAKERS.

Courtesy Portland Cement Association

Figure 5-3   The level of liquid in beakers where aggregate size is constant and where sizes are combined

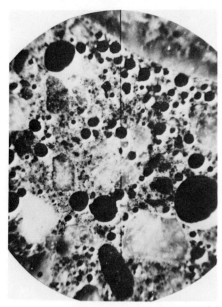

Figure 5-4   Air-entrained concrete as seen through a microscope

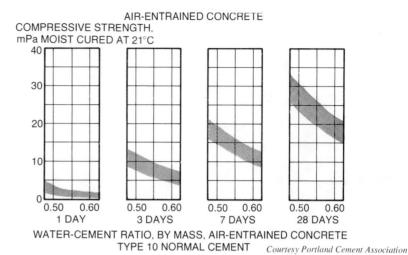

AIR-ENTRAINED CONCRETE
COMPRESSIVE STRENGTH,
mPa MOIST CURED AT 21°C

WATER-CEMENT RATIO, BY MASS, AIR-ENTRAINED CONCRETE
TYPE 10 NORMAL CEMENT   *Courtesy Portland Cement Association*

Figure 5-5   The approximate relationship between water-cement ratio and compressive strength by mass

increasing admixture is necessary, or to simply make the mix more workable. A superplasticizing admixture or an entrained air admixture can be used.

The effectiveness of admixtures depends upon the type of cement used, water, aggregate shape and graduation along with proper handling and placing of the concrete. However, an admixture should not be considered as a substitute for good concreting practices.

Air-entrainment is the one admixture used quite extensively in concrete. It produces billions of microscopic air bubbles. By volume it constitutes 4% to 6% of a cubic metre of concrete (Fig. 5-2). The use of this admixture improves the:
1. Durability of exposed concrete to freezing and thawing
2. Resistance to surface sealing and dusting
3. Workability of the fresh concrete (it acts like a lubricant), while reducing segregation and bleeding through cracks in the forms

The admixture can be either in the form of air-entrained cement or a liquid which is added to the concrete at the time of mixing (Fig. 5-4).

Proportioning the Concrete Mix

The purpose of proportioning a concrete mix is to determine the most economical and practical combination of materials available, with the design mix meeting the following requirements:
1. That it is workable
2. That it is strong, durable and has a good appearance when properly placed and cured.
3. That it is economical

Workability is used to describe how easy or difficult the concrete is to place, consolidate and finish. Strength, durability and good appearance depend on a suitable cement paste known as cement-water *ratio*. A suitable cement-water ratio is one with a low ratio of water to cement with the right proportion of entrained air. The quality of the mix is dependent on the correct cement-water ratio, therefore water should be minimized to reduce the amount of cement required, but not below a point that would affect its performance.

Put another way, as long as the mix is workable and shows *consistency* and the aggregates are made up of strong, sound material; the less water used, the stronger the mix.

As can be seen in Figure 5-5, by reducing the cement-water ratio, the compressive strength of the concrete with less water is stronger than that with more water. This increase in strength becomes more established as time increases. With all other factors being equal, such as placing and curing the concrete, the least amount of water which will satisfy the three requirements should be selected. Economy of the mix is obtained when proper proportioning minimizes the cement to water ratio without sacrificing the quality of the concrete. In Figure 5-6, examples are shown of mixes that are poorly proportioned for various reasons. In all cases except number five the concrete would fail because of one or more of the points outlined.

Trial batch mixes are normally pre-established in laboratories where proper control and exacting measurements can ensure accurate results. Once an economical, workable and strong mix has been designed using the field materials, a re-

TOO WET: This mix is too wet because it contains too little sand and coarse aggregate for the amount of cement paste. Such a mix would not be economical or durable, and it would have a tendency to crack.

TOO DRY: This mix is too dry and stiff because it contains too much sand and coarse aggregate. It would be difficult to place and finish properly.

TOO SANDY: This mix is too sandy because it contains too much sand and not enough coarse aggregate. It would place and finish easily, but it would not be economical and would be very likely to crack.

TOO STONY: This mix is too stony because it contains too much coarse aggregate and not enough sand. It would be difficult to place and finish properly and would result in honeycomb and porous concrete.

JUST RIGHT! This workable mix contains the correct amount of cement paste, sand and gravel. With light trowelling, all spaces between coarse aggregate particles are filled with sand and cement paste.

Courtesy Inland Cement Industries Ltd.,
A Division of Gemstar

Figure 5-6 Proportioning concrete mixes

port goes out to those concerned giving them the necessary data. If any doubt exists as to the quality of the concrete to be obtained, it is suggested that quality overrule economy.

Two tests are used to control the proportioning of materials used in concrete and to control the resulting finished product at the job site. One of these is called the *slump test*. This test is a measure of consistency of the mix, not a measure of workability. Different jobs call for different slumps. Normally the slump required will be indicated in the job specifications. If it is not specified it should be checked before proceeding. Table 5-2 may be used as a guide for slump ranges where the concrete is to be vibrated.

In conducting the test, the *slump cone* as it is called (Fig. 5-7) is placed on a flat surface and held down by the foot pieces.

The mould is then filled to approximately one-third its height with concrete. Then it is rodded 25 times with the bullet-pointed rod. The following two layers are then placed in the cone, each layer again

TABLE 5-2   SLUMP RANGES FOR VIBRATED CONCRETE

Type of Construction	Slump, mm	
	Maximum	Minimum
Reinforced foundation walls and footings	80	20
Unreinforced footings, caissons, and walls	80	20
Reinforced slabs, beams, and walls	100	50
Building columns	100	50
Bridge decks	50	20
Pavements	50	20
Sidewalks, driveways, slabs on ground	80	20
Heavy mass construction	50	0

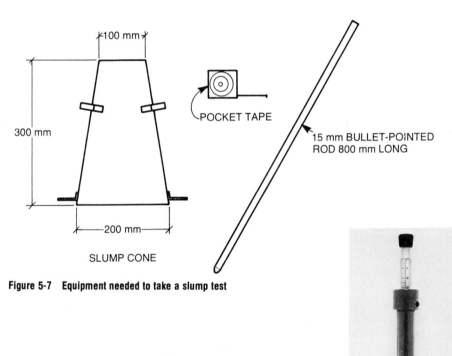

Figure 5-7   Equipment needed to take a slump test

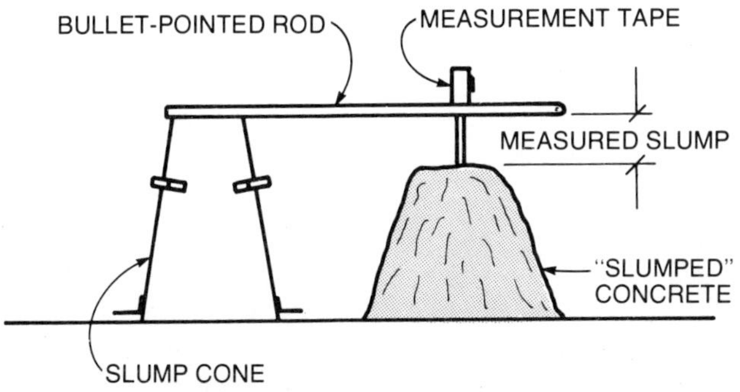

Figure 5-8   Measuring the slump of a concrete mix

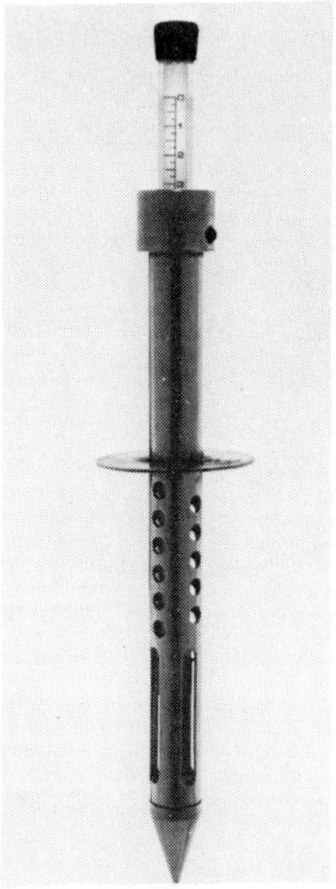

Courtesy K. Nasser

Figure 5-9   K slump tester

rodded 25 times uniformly within the cone. The top is then struck off level with the top edge of the mould and the spilled concrete removed from around the base. The mould is removed immediately by raising it vertically and placing it beside the slumped concrete. The measurement is taken from the top of the cone to the top of the concrete as shown in Figure 5-8.

Another testing device which is gaining acceptance is known as the **"K" slump tester** (Fig. 5-9).

Basically a probe, the K slump tester is a 300 mm hollow tube 20 mm in diameter. To take a slump test it is inserted into the fresh concrete to the depth of the penetration ring. After leaving it in the concrete for sixty seconds, during which time the wet concrete has entered through the openings in the lower portion of the probe, the graduated rod is lowered until it rests on the concrete which has flowed into the hollow portion. The slump of the concrete can be then read directly off the graduated scale.

Slump tests should be taken whenever concrete appears to change in consistency and whenever compression test cylinders are to be taken.

Compression testing is taken to determine (1) if the mix has the proper proportion for the designed strength, thus promoting quality control and (2) when form removal can begin and when the structure can be put into service. Measured in megapascals (MPa), normal concrete strengths range from 15 to 35 MPa although individual designs may dictate lower or higher strengths than those mentioned. The commonly used strengths for house construction are 20 and 25 MPa.

The standard cylinders used in testing are either made from a waxed cardboard form or a split steel form which is bolted together during the actual pouring (Fig. 5-10).

The procedure for taking samples in compressive testing is to sample the con-

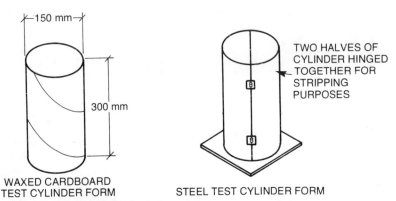

WAXED CARDBOARD
TEST CYLINDER FORM

STEEL TEST CYLINDER FORM

Figure 5-10   Test cylinders for compressive testing

Courtesy Portland Cement Association

AIR-ENTRAINED CONCRETE

COMPRESSIVE STRENGTH
MPa MOIST-CURED AT 21°C

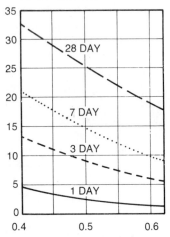

WATER-CEMENT RATIO, BY MASS
CEMENT–TYPE 10

Figure 5-11   Testing and test results of compression tests taken on 150 × 300 mm cylinders

Courtesy K. Nasser

Figure 5-12   K 5—strength tester

crete at three or more intervals during the unloading of a given batch of concrete. Samples are not normally taken at the beginning or end of the batch to allow for a more representative sample. Usually only three cylinders are required at any given time. To fill the test cylinders, the procedure is identical to the one used for obtaining a slump test, in which three layers of concrete are rodded 25 times. Once filled, they are covered to prevent evaporation and left for twenty-four hours.

After the initial curing period the samples may be field or laboratory cured, or both, depending on the requirements of the job. After curing for seven days, one or two cylinders are tested for their compressive strength. At this time the concrete should have attained at least two-thirds its designed strength (Fig. 5-11). Assuming a concrete mix was designed for 30 MPa, it should therefore test out at a minimum 20 MPa after the initial seven days.

The remaining cylinders are then cured for twenty-eight days at which time the concrete should have reached its designed strength of 30 MPa or better.

Like the K slump tester, the relatively new **K-5 strength tester** is shortening the curing period from days to hours. Developed by Professor K. W. Nasser of the University of Saskatchewan, Saskatoon, (inventor of the K slump tester), the K-5 strength tester is able to give compression test readouts the same day the concrete is poured.

Concrete samples are placed in specially designed K-5 cylinders which are equipped with a heating element. Once the samples have been obtained the cylinders are placed in the compression tester (Figs. 5-12), pressurized and heat cured for five hours, which accelerates the curing time. At the end of five hours the concrete cylinders are removed from the moulds and tested to determine the potential strength of the concrete.

Mixing the ingredients

All concrete should be thoroughly mixed. Each and every particle of sand, gravel and/or crushed rock must be completely coated with the cement paste to ensure a strong, durable product. Mixers come in all sizes from small gasoline or electric-powered machines, which can be rented in most areas, to the large truck-mixers available from ready-mix concrete plants across the country (Fig. 5-13).

Courtesy Goldblatt Tool Co.
(a) Batch Mixer

Courtesy Portland Cement Association
(b) Truck Mounted Mixer
Figure 5-13 Mixing of concrete in a small batch mixer and large truck mounted mixers.

No matter what the size of the mixer, it should never be loaded above its capacity or operated above its designed speed. If increased output is required, larger or additional mixers should be used. Generally, a minimum of one minute of mixing time is required for mixers of up to one cubic metre capacity, with an increase of twenty seconds for each additional cubic metre. When transit mixers are hauling concrete from a central batching plant, the maximum time allowed between the batching and complete discharge of the

concrete is about one and a-half to two hours.

In charging the mixer the ingredients are normally mixed in the following proportions:
1. About 10% of the specified amount of water is first added to the mixer.
2. Water and solids are added uniformly along with any admixtures such as air-entrainment.
3. About 10% of the water is left to be added after the other ingredients are charged into the mixer.
4. Mixing is continued until all materials are thoroughly coated with the cement paste.

Placing Concrete

The transportation and handling of concrete should be carefully controlled to make sure that the quality of the concrete

(a) Wheelbarrow

Courtesy Portland Cement Association
(b) Concrete Pump
Figure 5-14 Transporting concrete by wheelbarrow (a) and concrete pump (b)

is maintained. Wheelbarrows operated over runways are generally used for transporting concrete on small jobs (Fig. 5-14).

On larger jobs movement of concrete can take place in a variety of ways including hoppers, chutes, buckets, belt conveyors and pumps (Fig. 5-14). The method of transporting concrete does not, however, necessarily guarantee a quality concrete. Excessive handling and movement may cause separation of the fine and coarse aggregate, resulting in a poor concrete.

Prior to placing the concrete a number of preparatory steps must take place.

Courtesy Portland Cement Association
Figure 5-15 A sidewalk slab being prepared for pouring

These include the erection of footing or wall forms, and making sure that they are clean, oiled, tight and well braced. For slab construction, compaction and leveling of the subgrade must be finished, along with outside forms and intermediate *screeds* if required (Fig 5-15).

If the outside forms are too far apart for the straight edge to reach, intermediate screeds are set. These are generally at the same level as the outside form unless a slope is required. The subgrade is moistened prior to pouring so that the moisture is not drawn out of the concrete. If required, reinforcing steel is also placed in

Courtesy Portland Cement Association

Figure 5-16   Pouring a large concrete slab

the forms before pouring. Make sure it is kept clean so that the concrete makes a good bond when forming around it.

Concrete should be deposited as near as possible to its final position. In slab construction, placing should begin at one end with the edges poured first and the centre filled as the work progresses (Fig. 5-16).

When spreading and spading the concrete ensure that the material is worked into the corners and edges as well as around the reinforcement.

Jobs requiring a considerable depth of concrete, such as the foundation wall, should be poured in horizontal lifts of 300 to 500 mm around the entire wall area. Each layer should be consolidated before the next layer is placed. When pouring walls, the first batch should be placed at each end. Placing then progresses toward the middle of the form. Avoid moving the concrete horizontally in either slab or form work as the cement paste will tend to separate from the coarse and fine aggregates, resulting in poor quality concrete.

Consolidating concrete is necessary in forms, around embedded parts and reinforcements to eliminate *voids* other than the entrained air. Plastic flowing mixes (slump of 75 to 100 mm) can be tamped by hand using a steel rod or wood tamper long enough to reach the bottom of the

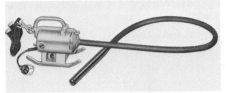

Courtesy Goldblatt Tool Co.

Figure 5-17   Internal electric vibrator

form and small enough to pass by the reinforcement.

Mechanical *vibrators* are used for stiffer mixes of concrete. The internal vibrators most commonly used on foundations are driven by air, electricity, or gas, and consist of a flexible shaft with a vibrating head. Inside the head is an unbalanced mass which when rotated at a high speed causes the vibration (Fig. 5-17).

Proper use of the mechanical vibrator is important. In wall forms the vibrator head should be lowered vertically at regular intervals, allowing it to penetrate the previous layer of concrete. In slab work the head should be held at an angle or horizontally so that it will be completely submerged. The vibrator is left in the concrete until a film of cement paste appears. It is then drawn out at the same rate of penetration. A word of caution—do not use the vibrator for moving concrete horizontally in forms as separation will occur. Also, do not over-vibrate as this may cause undue pressure on the forms which could result in them giving away.

Finishing Concrete

Finishing concrete is normally associated with the finishing of concrete slabs. Any finishing that is required on foundation walls can be achieved through the use of special forms, forming techniques, or finishing the exposed portions after the forms have been removed.

Concrete slabs can be finished in a number of ways depending on the effect desired. In most cases the finish is floated, troweled or broomed. As soon as

Courtesy Portland Cement Association

Figure 5-18   Placing and consolidating the concrete

the concrete is placed and spread in the forms, finishing can take place. In Figure 5-18, once the fresh concrete has been deposited in the forms, it is spread, consolidated and the excess struck off at the edge of the form.

Screeding is the first step in finishing the concrete slab. A *straight edge* (wood or metal) is used in a sawing action to level the concrete between the forms or, on a large pour, between the edge form and intermediate screeds (Fig. 5-19). As the straightedge is moved back and forth it is advanced slowly with a small surplus of concrete ahead of it to fill in any depressions.

Contractors specializing in concrete work quite often use the vibrating strike off or power screed (Fig. 5-20). This machine combines the action of the vibrator with the straight edge; thus speeding up the operation of consolidating and leveling the concrete slab.

Figure 5-19   Striking off the concrete to the proper level

Figure 5-20   Power screed

(a) Darby

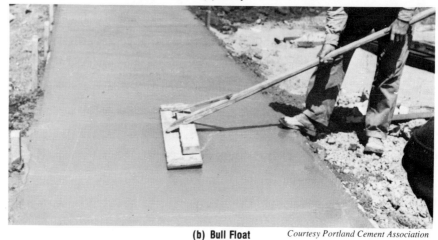

(b) Bull Float

Figure 5-21   Darby and bullfloat

Once struck off to the level of the forms, the darby or bullfloat is used to remove any ridges left by the screeding operation and to fill any low spots (Fig. 5-21). This also helps to embed any coarse aggregate that may be left on the surface of the slab.

The darby is substituted in this operation for the larger and longer reach of the bullfloat used on large slab work. Both tools can be made of wood, magnesium or aluminum.

All the finishing operations on the slab to this point must be completed before any excess *bleed water* appears on the surface. Bleeding is caused by the heavier materials consolidating and settling and

Figure 5-22   Edging and jointing a sidewalk slab

forcing excess water to the top of the slab. It is critical that all finishing operations **cease** until the bleed water disappears. The continuation of the finishing operations at this point will lead to serious dusting and scaling of the concrete slab at a later date.

When the bleed water and sheen has disappeared from the surface of the concrete, finishing of the slab can resume. On exterior slabs such as sidewalks, driveways and patios, edging of the slab is quite common. Edging removes the sharp exposed edges, preventing them from breaking away during form removal or chipping at a later date (Fig. 5-22).

Care should be taken when using the edger. Make sure to work in the coarse aggregate, leaving the edge smooth, while watching that the impression left from the edger is not too deep. Once edging is completed or even while edging, control joints may be put in the slab (Fig. 5-31). *Control joints* (explained on page 87), are quite often spaced at intervals equal to the width of the sidewalk or slab which gives a squared effect. They should never be more than 6 m apart.

When the edging and jointing are completed the main area of the slab is ready for floating. This can be the last step in the finishing of the concrete if a good slip resistant surface is required, or it may be the preliminary step to steel troweling if a smooth or brushed finish is necessary. Floating accomplishes the following: it helps to embed particles of aggregate and cover them with a layer of cement paste; it helps to eliminate any remaining high spots or dips in the surface of the slab; and as previously mentioned, it compacts the surface to prepare it for further finishing.

If the resulting floated surface is to be the finished texture of the slab, a metal float is often used in place of the wood one as it tends to give a slightly finer finish (Fig. 5-23).

Courtesy Portland Cement Association

Figure 5-23   Floating the surface of the concrete slab

A troweled finish is quite often required for the basement floor slab in a house. It produces a smooth, hard finish which is quite often left exposed as is or it may be painted, or covered with carpet or linoleum as the home is developed at a later date. To achieve this finish the floating is followed by steel troweling. This is often done at the same time as the floating. (Fig. 5-24).

As the concrete sets up it may be necessary to repeat the steel troweling a number of times to produce the desired smoothness. Power troweling is often

Courtesy Portland Cement Association

Figure 5-24   Floating and steel troweling simultaneously

used to complete the last two steps in finishing. The power trowel (Fig. 5-25) has three or four rotating steel blades, which can be tilted as required and is gas or electrically powered.

Because of its size and ability to cover a larger area more quickly, machine troweling is quite common on slab work other than sidewalks.

If a non-slip surface is desired once the floating or steel troweling is completed, it is common to broom the surface (Fig. 5-26) on exterior slabs. Generally broomed at right angles to the flow of traffic, a coarse or fine texture can be achieved by using the appropriate broom. If it is necessary, edges and control joints can be gone over again to maintain clean lines.

Courtesy Goldblatt Tool Co.

Figure 5-25   Power trowel

Courtesy Portland Cement Association

Figure 5-26   Broom finish on a sidewalk

Curing

One of the most important and yet most neglected steps in pouring concrete is the curing operation. No matter how carefully the previous steps of mixing, transporting and finishing were followed, if poorly cured, concrete will not attain the desired properties inherent in it. Concrete's watertightness, wear resistance and volumetric stability all increase as hydration continues between the cement and water.

The strength of concrete grows quite rapidly for the first twenty-eight days (the period of time compression tests are based on), then levels off and continues increasing in strength for an indefinite period of time. The main purpose of curing is to have the exposed surfaces of the concrete kept moist and warm so that hydration can take place. An excessive water loss will cause the concrete to shrink rapidly and cracks will appear as tensile stresses are relieved due to shrinkage. With a temperature at or near 0°C the hydration is slowed to a point where little or no strength is acquired. It is therefore important to maintain a balance between moisture loss and the proper temperature to ensure proper curing during the initial setting.

Methods of Curing Concrete

Methods of curing concrete once it is poured are normally broken down into three areas. It is not uncommon to see more than one of these methods being used depending on the circumstances involved. In the first method, water is added to the surface of the concrete once the finishing has been completed and damage to the surface cannot occur. This can be done by flooding the area using earth or sand as a retainer around the edge of the concrete, sprinkling using a system of hoses to flood the area, or using sand or sawdust spread to about 50 mm deep and kept moist by periodic wetting.

The second method involves surface sealing. This is accomplished by covering the exposed area with waterproof paper or polyethylene sheets, leaving the forms in place (usually in wall construction) or by spraying on a sealing compound. The use of waterproof paper or polyethylene sheets is common on large areas. The major problem lies in securing it so that it will stay in place (Fig. 5-27).

In vertical pours, forms left in place do an excellent job of protecting concrete, as long as the top is kept damp. A soaker hose used with the forms left on works well. The one disadvantage with this system is that the forms are unavailable for a period of time.

Sealing compounds are used immediately after the finish on the slab is complete. They are applied by sprayer, and care should be taken to ensure complete coverage of the exposed area. In Figure 5-28 a concrete slab is sprayed with a curing compound using a hand sprayer.

The third method is generally used under the controlled conditions of a precasting plant. Steam and heat curing ac-

Courtesy Portland Cement Association

Figure 5-28   Use of a curing compound

Figure 5-27   Curing by covering with polyethylene

Courtesy Portland Cement Association

celerates the strength of the concrete by supplying heat and moisture, thereby shortening the curing time.

The total curing time required for concrete is dependent on a number of variables—the type of concrete, mix proportions, required strength, size and shape of the concrete mass as well as the weather conditions. A curing period of five to seven days should be sufficient unless adverse conditions warrant other measures.

Hot and Cold Weather Concreting

Concrete is sometimes poured in what can be described as less than ideal concreting weather. If this occurs, precautionary measures must be taken to avoid problems with the finished product.

During hot weather, the major problem is placing and finishing the concrete before excessive evaporation of water takes place. Some steps that can be taken include wetting down the forms, reinforce-

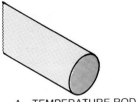

A   TEMPERATURE ROD

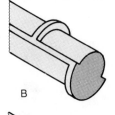

B

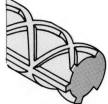

C   B & C–DEFORMED ROD (RE-BAR)

Figure 5-29   Reinforcement used in concrete

ment and subgrade periodically during the pour; using more manpower and speeding up the operation; or covering the exposed areas between the steps in slab finishing thereby retaining any surface moisture. Curing of the slab should begin immediately after the finishing is complete.

With cold weather concreting the setting time is slowed down, therefore the strength of the concrete is also set back. In planning concrete pours during cold weather, precautionary measures include heating of the water, aggregates and area to be concreted; using a high early cement (Type 30) or additives like calcium chloride which are used to increase the setting time; and when the pour is complete to keep the area above freezing for an extended period of time to ensure sufficient strength.

Plain and Reinforced Concrete

Concrete is strong in compression but weak in tension; that is, it can withstand large amounts of pressure placed directly on it but is very weak when pressure is applied that it tends to bend or pull it apart. The reverse is true of steel rods and when concrete and steel are used together, one makes up for the deficiency in the other. Concrete and steel work well

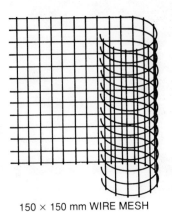

150 × 150 mm WIRE MESH

together because (1) concrete bonds well to steel rods (2) concrete protects steel from corrosion and fire and (3) their coefficients of expansion are almost identical. If the coefficients of expansion were different, the concrete would be ripped apart by the internal stresses put upon it during extreme temperature changes.

Concrete that has no reinforcement or has either wire mesh or smooth rods up to 10 mm in diameter is quite often referred to as *plain concrete* (Fig. 5-29). This type of concrete is used for parts of a structure which carry relatively light loads, such as sidewalks, patios, driveways and basement or garage slabs. In other words, most concrete slabs which are supported by a sub-base of gravel (or equivalent) beneath them. The temperature rods are placed in the concrete not to increase the tensile strength, but rather to help control the expansion or contraction of the slab due to temperature changes.

When tensile stresses are a concern, a *deformed bar* (re-bar) is used to replace the wire mesh or smooth bar (Fig. 5-29). By deforming or increasing the surface area of the rod, a greater bond is achieved between the concrete and steel. Increasing the bond strength between the two materials allows the steel to be placed where the greatest tensile stress occurs, thereby transferring the stress to the steel rod (Fig. 5-30). Tension normally occurs opposite the load being placed on the concrete. In the case of footings or concrete walls in the basement this usually occurs at the bottom of the footing and on the inside face of the wall. The re-bar is generally placed both vertically and horizontally in the case of walls while footing steel is horizontal. When placing it in the forms, corners are simply bent to suit the requirements of the building while end laps (where two pieces of re-bar join) are normally lapped 24 to 36 times the rod diameter.

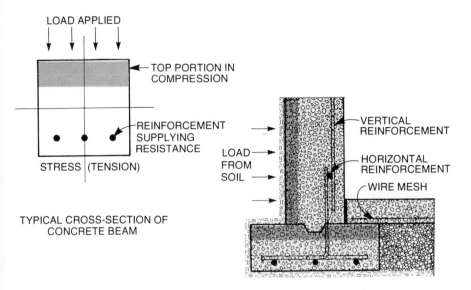

Figure 5-30   Placement of reinforcing steel

The number of rods required, spacings necessary, and the size and special shapes should all be referred to an engineer who has experience in reinforced concrete design. He or she understands the forces on the structure and knows the limitations of the materials.

Controlled Cracking in Concrete

Cracking of concrete is caused by one of two things; stress due to loading or changes due to shrinkage. With the use of properly placed reinforcement and control joints, unsightly cracking can be minimized. In foundation walls, *control joints* are required if the wall is over 25 m in length with the joints spaced not more than 15 m apart. Control joints are scored into the sidewalk, driveway, etc. so that cracking will tend to follow the grooves cut into the surface (Fig. 5-22).

Isolation joints separate two parts of a structure. In Figure 5-31 the slab is separated from the wall with the use of a compressible filler such as a strip of insulating board.

Construction joints are used when two separate pours are required at different times during the construction period. Construction joints appear between the footing and foundation and are quite often keyed as shown (Fig. 5-31).

Quality control of materials used in the construction field is taken for granted, but in the manufacture of cement and the subsequent proportioning, mixing, pouring, finishing and curing of concrete, it is one more key ingredient. Continuous sampling and testing ensures the quality of the product arriving at the job site. The owner/builder must then assume this responsibility and see that the materials are properly handled to guarantee a finished product of the highest quality.

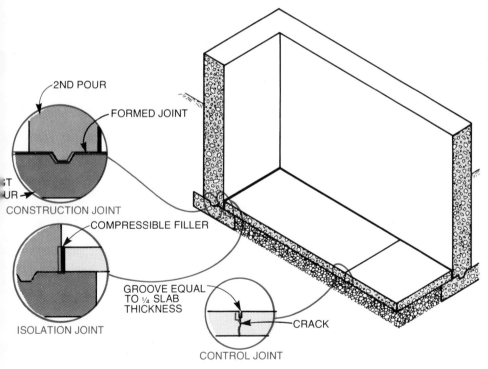

Figure 5-31   Types of joints used in concrete

REVIEW QUESTIONS

5—CEMENT AND CONCRETE

Answer all questions on a separate sheet

A. Replace the Xs with the correct word

1. Portland cement is known for its ability to form a XXXXXXX which binds aggregates together.
2. "Quarrying," one step in the process of manufacturing cement is the gathering of XXXXXX.
3. "Grinding-blending" of materials for manufacturing cement is accomplished by either XXXXXXX and/or XXXXXX grinding.
4. When burnt, the raw materials combine to form a XXXXXXX called portland cement.
5. Bagged portland cement comes in XXX kg bags.
6. Concrete, a mixture of four essential ingredients, should contain the following: XXXXXX, XXXXXX, XXXXXXX and XXX XXX.

B.

7. Arrange the following steps for obtaining quality concrete in their proper sequence.
   3(a) Thoroughly mix the ingredients
   5(b) Finish the exposed surface
   2(c) Prepare the correct proportions
   6(d) Cure the concrete
   4(e) Place the concrete in the form
   1(f) Use the correct materials for quality concrete

C. Select the correct answer

8. Cement is designed for various purposes. One type, designed to resist mineral salt found in some soil is called
   (a) Type 10
   (b) Type 30
   (c) Type 50
   (d) Masonry

9. Water added to a concrete mix provides
   (a) Mixing qualities and strength
   (b) Mixing qualities and ease of placement
   (c) Strength and ease of placement
   (d) Mixing qualities and hydration

10. A sieve analysis of the aggregate being used will determine if
    (a) The aggregate will make strong concrete
    (b) The aggregate is workable
    (c) The aggregate is properly proportioned
    (d) The aggregate has an adequate amount of moisture

11. Strength, one of concrete's many properties, is primarily determined by the
    (a) Use of admixtures
    (b) Type of aggregate used
    (c) Cement-water ratio
    (d) Proportion of aggregate and cement

12. The test for the consistency of concrete is called a
    (a) Slump test
    (b) Core test
    (c) Cylinder test
    (d) Moisture test

13. For design purposes, concrete's compressive strength is based on a
    (a) 5 day test
    (b) 7 day test
    (c) 21 day test
    (d) 28 day test

14. When performing tests for slump and compressive strength on concrete, moulds are filled up in specific layers. Each layer is rodded
    (a) 20 times
    (b) 25 times
    (c) 30 times
    (d) 35 times

15. Standard cylinders used for compressive testing measure
    (a) 127 mm in diameter by 254 mm high
    (b) 150 mm in diameter by 300 mm high
    (c) 150 mm in diameter by 254 mm high
    (d) 127 mm in diameter by 300 mm high

16. The "slump" of a concrete mix is established by removing the cone form and
    (a) Measuring the distance from the top of the cone to the concrete mass
    (b) Measuring the total height of the concrete mass
    (c) Measuring the diameter of the concrete mass at the base
    (d) Measuring the diameter of the concrete mass at the top

17. Once mixing water has been added, concrete should be placed within
    (a) One hour
    (b) One-half to one hour
    (c) One to one and a-half hours
    (d) One and a-half to two hours

18. Vibrating of concrete
    (a) Is used to distribute concrete in the form
    (b) Separates the coarse and fine aggregate
    (c) Consolidates the mass
    (d) Causes voids to appear
19. A hand tool commonly used when leveling and compacting the surface of concrete is called a
    (a) Float
    (b) Trowel
    (c) Vibrator
    (d) Tamper
20. Curing of concrete is essential to
    (a) Prevent hydration and shrinkage cracks
    (b) Prevent water loss and hydration
    (c) Prevent early use of the concrete
    (d) Prevent rapid water loss and shrinkage cracks

D. Replace the Xs with the correct word(s)

21. An additive, XXXXXX or the use of a XXXXXX type of cement speeds up the setting time of concrete during cold weather concreting.
22. The purpose of positioning reinforcing steel in concrete is to increase its XXXXXX strength.
23. Two types of reinforcing rods used are XXXXXX and XXXXXX.
24. Wire mesh, which is often used in concrete is known as "XXXXXX concrete."
25. To control cracking due to shrinkage and pressure exerted on concrete, three commonly used joints are XXXXXXX, XXXXXX and XXXXXXX.

6 FOOTINGS AND FOUNDATIONS

With the excavation of the house now complete work can begin on the footings and foundation walls. Footings are the enlarged portions of the foundation wall which act as a base, transmitting the load of the structure to the undisturbed soil beneath. The footing must be constructed in such a manner that it will minimize any settlement or movement when the structure is completed.

The foundation walls form the enclosure upon which the house frame sits and is fastened. Depending on the type of foundation, the enclosed area can be used as (a) a full basement with a total usable area (b) a partial basement and crawl space, or (c) non-usable area under the building. They must be able to carry the load from the superstructure and any additional loads when occupied, and to transmit it to the footing and soil below. Normally, the foundation is the same size as the house structure unless the building design calls for some areas to be cantilevered out over the foundation wall. It is imperative that the foundation wall and footings be built level, square and to the dimensions given in the blueprints for the structure. Poor quality work at this point will result in constant problems throughout the rest of the construction period and will plague the homeowner in the future.

TYPES OF FOOTINGS

Acting as the support for the foundation wall and superstructure above, the footing transmits loads down into the subsoil below. The footing must be constructed so that it is (a) suitable for the soil conditions upon which the structure is to be built and (b) far enough into the ground to ensure that it is below the frost level for that area. Concerns about energy usage today along with energy conserving measures being used in the foundation area makes this last point extremely important. Whether the foundation wall is insulated inside or outside, the ground adjacent to the foundation is colder and if certain conditions exist (moisture and cold temperatures), the danger of the soil freezing down to and under the footing is increased. Heaving of the footing, wall and floor slabs can cause unmeasurable damage if freezing does occur. (Fig. 6-1).

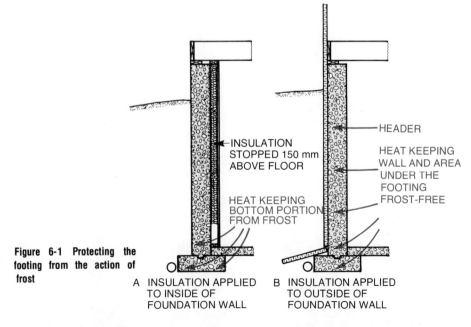

Figure 6-1 Protecting the footing from the action of frost

A INSULATION APPLIED TO INSIDE OF FOUNDATION WALL

B INSULATION APPLIED TO OUTSIDE OF FOUNDATION WALL

INSULATION STOPPED 150 mm ABOVE FLOOR

HEAT KEEPING BOTTOM PORTION FROM FROST

HEADER

HEAT KEEPING WALL AND AREA UNDER THE FOOTING FROST-FREE

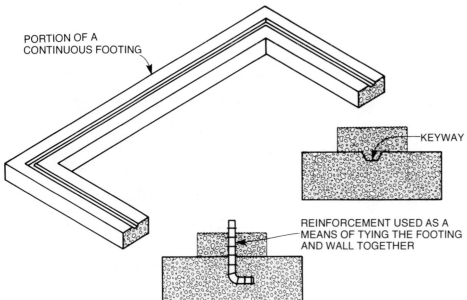

PORTION OF A
CONTINUOUS FOOTING

KEYWAY

REINFORCEMENT USED AS A
MEANS OF TYING THE FOOTING
AND WALL TOGETHER

Figure 6-2   Continuous or strip footing

Continuous or Strip Footing

A continuous or strip footing is one which supports the perimeter of a building. If the design of the building is such that an interior partition is used in place of a beam to support the inner ends of the floor joists, then the footing also runs under the partition. The footing is unbroken throughout its length, and the width and depth must conform to the job specifications which in turn have to at least meet the NHA code minimum requirements. Minimum footing sizes are set out for stable soils, but when unstable conditions exist, adjustments must be made to increase the width and depth. Keyways or reinforcing dowels are used to tie the footing and foundation wall together as they are usually poured at separate times. In Figure 6-2, a portion of continuous footing is shown.

Stepped Footings

Stepped footings are used when differences in elevation are encountered.

This difference could be a product of the house design or style (split-level house), a house built into the side of a slope, or where unstable soil conditions in one part of the excavation warrant a deeper footing placed on more stable soils. In forming and pouring a stepped footing the width of the vertical portion between the different elevations must be maintained. The vertical portion must be 150 mm thick (minimum) and the total distance not more than 600 mm in height. If more than one step is required between the two levels, the horizontal distance should be 600 mm or more (Fig. 6-3).

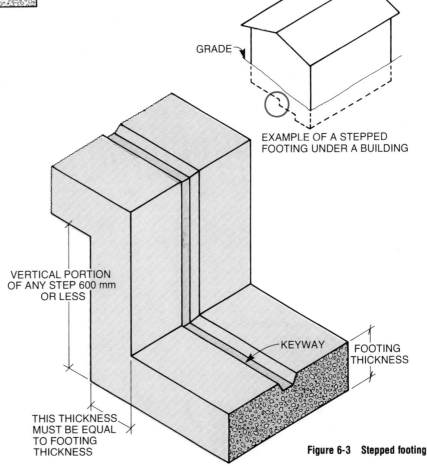

GRADE

EXAMPLE OF A STEPPED
FOOTING UNDER A BUILDING

VERTICAL PORTION
OF ANY STEP 600 mm
OR LESS

THIS THICKNESS
MUST BE EQUAL
TO FOOTING
THICKNESS

KEYWAY

FOOTING
THICKNESS

Figure 6-3   Stepped footing

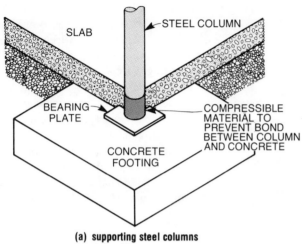

(a) supporting steel columns

(b) supporting wood columns

Figure 6-4   Independent footings

Independent Footings

Pad-type footings or independent footings, support posts or columns which in turn carry the inner ends of the floor joists and loads upon them. The pad footing should be centred under the load being transmitted; its size conforming to the code requirements or job specs.

As shown in Figure 6-4(a), steel columns are normally positioned on the footing so that the floor slab is poured around them. With wood columns it is good building practice to set the post on a pier which is part of the footing and which rises slightly higher than the floor slab. The pier has either straight or sloped sides and is poured with the footing. A dowel pin may be inserted into the fresh concrete when pouring it to anchor the column. Before placing the column on the pier a damp-proof membrane must be placed between the concrete and the column to prevent moisture from causing future decay (Fig. 6-4(b)).

Footings for *pilasters*, (enlarged–portions of the foundation wall that carry concentrated loads) chimneys, and fire-

places are usually incorporated into the continuous footing. The length, width and thickness of the portion is specified in the blueprint. In certain cases free-standing chimneys and fireplaces may be located within the enclosure requiring an independent footing (Fig. 6-5).

In construction, if conditions other than the norm are encountered, the footings should be specially designed by a professional engineer using practices proven in the area. Basic rules governing the design and construction of footings under normal conditions include: (a) Minimum thick-

ness of footings should be 150 mm but never less than the projection out past the foundation wall in an unreinforced footing (b) Never backfill under footings, they must rest on undisturbed soils (c) Well built forms mean square, level footings (d) Footings must extend below the frost levels of the area (e) Any diggings under the footing (sewer and water lines) should be backfilled with concrete and the footing bridged with reinforcement (f) Freshly poured concrete must be protected from hot or cold temperature extremes.

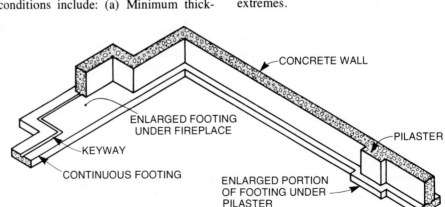

Figure 6-5   Continuous footing with an enlarged portion for a fireplace and pilaster

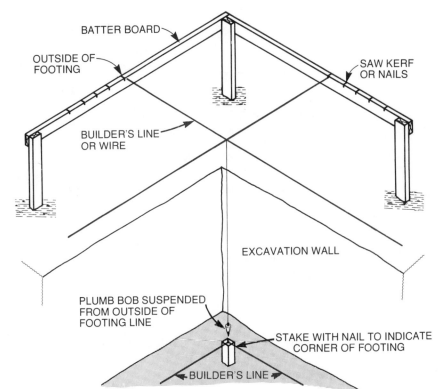

BATTER BOARD

OUTSIDE OF FOOTING

SAW KERF OR NAILS

BUILDER'S LINE OR WIRE

EXCAVATION WALL

PLUMB BOB SUSPENDED FROM OUTSIDE OF FOOTING LINE

STAKE WITH NAIL TO INDICATE CORNER OF FOOTING

BUILDER'S LINE

Figure 6-6(a) Establishing the corners of the building in the excavation using a plumb bob

plumb bob or a level and straight edge. When using a plumb bob always attach it to the bottom line so that lines hang free; simply lifting the line with the plumb bob attached will indicate its exact position in relationship to the other line. A slip knot will make easy the adjustment of the plumb bob to the correct height above the bottom of the excavation (Fig. 6-6(a)).

Once the approximate corner has been established a stake (38 × 38 mm) is driven directly under the plumb bob to the elevation of the top of the footing. This elevation is determined from the blueprint and can be measured from the builder's lines if the batter boards are set at a predetermined elevation or it can be obtained by using the transit-level working from the bench mark or reference point. When the stake is at the correct height, re-position the plumb bob, marking the corner

Forming the Concrete Footing

Once the type and size of footing has been determined, the first step in setting up the footing form is to locate the position of the building at the bottom of the excavation. Using the batter boards erected during the initial layout, builder's lines are stretched from one batter board to another. Where the lines cross (Fig. 6-6), that point then becomes the corner of the building. If the batter board was laid out as discussed in Chapter 4, Figure 4-27, the builder's line could be positioned on either the inside or outside footing line, to indicate the exact position of the footing.

With the lines positioned on the batter board the corner points can be transferred down to the excavation floor by using a

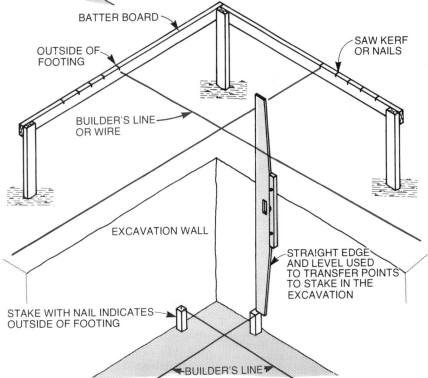

BATTER BOARD

OUTSIDE OF FOOTING

SAW KERF OR NAILS

BUILDER'S LINE OR WIRE

EXCAVATION WALL

STRAIGHT EDGE AND LEVEL USED TO TRANSFER POINTS TO STAKE IN THE EXCAVATION

STAKE WITH NAIL INDICATES OUTSIDE OF FOOTING

BUILDER'S LINE

Figure 6-6(b)   Establishing the corners of the building in the excavation using a straight edge and level

on the top of the stake, and drive a nail in at this point. The procedure is then repeated at the other corners of the foundation.

Transfer of points using the level and straightedge is similar to the plumb bob method except that two stakes are often used at each corner. The line position is then plumbed down and marked on the stake (Fig 6-6(b)). Diagonal measurements of the corners in the excavation will verify the accuracy of the transfer of the corner points.

With the corners now established, attach builder's line to the stakes. Assum-

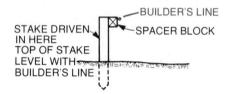

A SETTING FOOTING FORM STAKES

(a) Using a spacer block

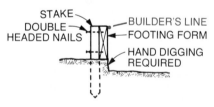

B BOTTOM OF EXCAVATION UNDERCUT

(b) Shallow excavation, hand digging required

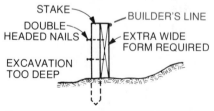

C BOTTOM OF EXCAVATION CUT TOO DEEP

(c) Excavation is deeper than required, therefore forming must be the full depth

Figure 6-7 Footing forms in place

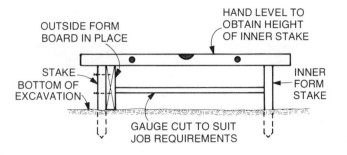

Figure 6-8 Use of a gauge and hand level to set the inner form stake(s)

ing that the outside footing line was used we are now ready to set the outside footing form in place. Stakes of 38 × 38 mm or 19 × 64 mm are driven in perpendicular to the excavation floor every 600 to 800 mm on centre outside the line to the thickness of the forming material. Using a block of wood the same thickness as the form will aid in gauging this measurement (Fig. 6-7(a)), and driving the stakes to the height of the line eliminates the need to mark and cut the stakes off later.

Once the stakes are in, the outside form boards can be set in place. Choose forming material as close to the correct width as possible to eliminate concrete spillage. If the bottom of the excavation has been undercut, narrower material will be required. The full depth of the footing is achieved by hand digging the last bit of soil from between the forms once they are fastened in place (Fig. 6-7). If the excavation has had too much soil removed you will have to increase the footing depth. Remember fill soils **cannot** be placed under the footings to bring them to the correct elevation.

When doing any concrete work requiring the use of forms, always be sure to fasten the forming materials adequately but do not over-fasten as the forms will have to be stripped when pouring of the concrete is complete. With footing forms, the use of double headed nails and the practice of nailing from the outside will speed up the stripping time required (Fig. 6-8).

Having formed up the outside of the footing we are now ready to set the inside form to the correct width and depth. Cut a piece of material to the width of the footing plus the thickness of the forming material; this gauge will be used in establishing the position of the stakes for the inner form (Fig. 6-8). Using the gauge to determine the proper spacing, drive the stakes in at predetermined centres and to the level of the outside form. The use of a hand level will ensure that the stake is at the correct height. With the inner stakes in place, installing the inner form is a simple matter (Fig. 6-8).

If stepped footings are required in the continuous footing, they have to be formed in such a way as to ensure a continuous concrete pour. Figure 6-9 shows two ways of forming steps of differing depths. Awareness of the pressures ex-

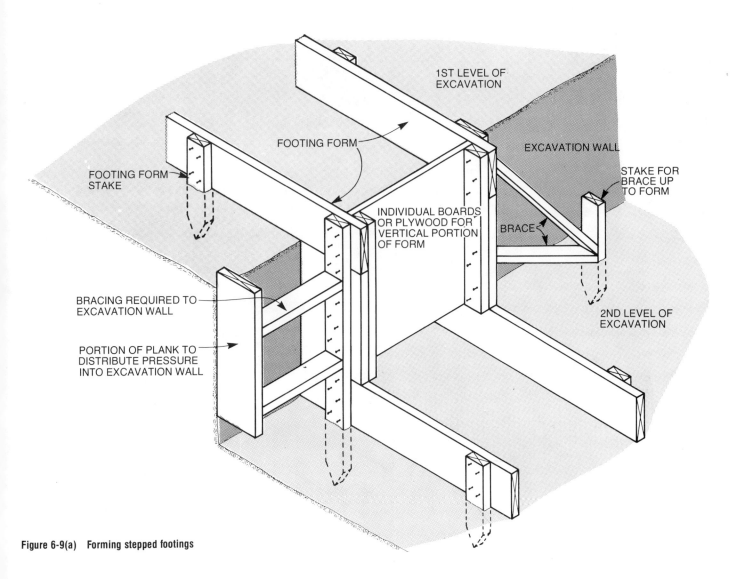

1ST LEVEL OF
EXCAVATION

FOOTING FORM

EXCAVATION WALL

FOOTING FORM
STAKE

STAKE FOR
BRACE UP
TO FORM

INDIVIDUAL BOARDS
OR PLYWOOD FOR
VERTICAL PORTION
OF FORM

BRACE

BRACING REQUIRED TO
EXCAVATION WALL

2ND LEVEL OF
EXCAVATION

PORTION OF PLANK TO
DISTRIBUTE PRESSURE
INTO EXCAVATION WALL

Figure 6-9(a)   Forming stepped footings

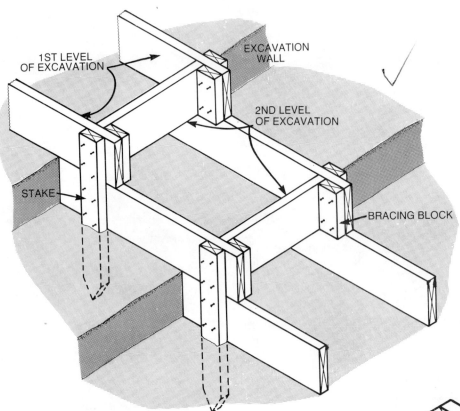

Figure 6-9(b)   Forming stepped footings

column just slightly above the concrete floor when it is poured.

If the centre beam which supports the inner ends of the floor joist is to be replaced with a bearing wall, a continuous footing will have to be constructed in place of the independent footing normally used. The footing will have to have a projection to support the bearing wall slightly above the basement floor slab. The lower portion of the footing is formed into the ends of the perimeter footing and constructed in the same manner. To form up the raised projection on which the bearing wall will rest, use longer stakes of about 1 m on centre (o.c.), and suspend braces from them to hold the form in place during pouring (Fig. 6-11).

With the forming complete, unreinforced footings are ready to be poured. If required, reinforcing bars should be positioned in the forms before pouring takes place. Here strict adherance to the job

erted by freshly poured concrete will result in adequate forms, yet eliminate the need for excessive forming materials and bracing.

Enlarged sections of the continuous footing required to support pilasters, chimneys or fireplaces are built into the footing as the form work progresses. Size and placement are dependent on the job specs.

To obtain the height and position of the independent footings, centre lines can be run off the outside footing form. Forming of the independent footing is quite simple if steel columns are used. When a pier footing is required to support a wood column, the top portion can be suspended as shown in Figure 6-10(b). The pier should be constructed high enough to place the

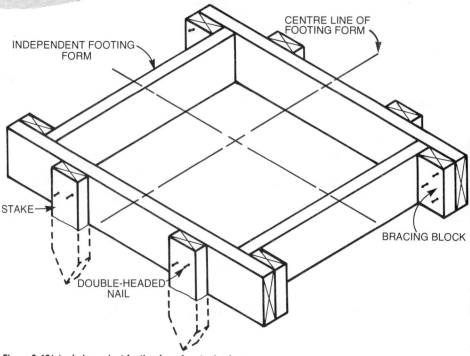

Figure 6-10(a)   Independent footing form for steel column

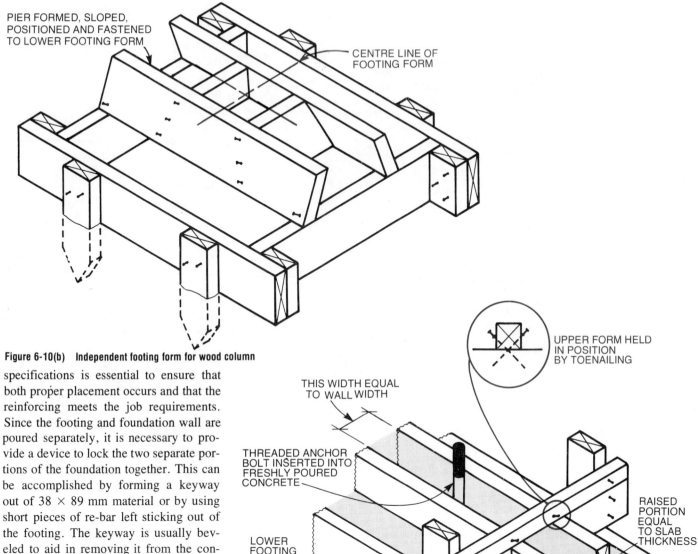

PIER FORMED, SLOPED,
POSITIONED AND FASTENED
TO LOWER FOOTING FORM

CENTRE LINE OF
FOOTING FORM

Figure 6-10(b)   Independent footing form for wood column

UPPER FORM HELD
IN POSITION
BY TOENAILING

THIS WIDTH EQUAL
TO WALL WIDTH

THREADED ANCHOR
BOLT INSERTED INTO
FRESHLY POURED
CONCRETE

LOWER
FOOTING
FORM

RAISED
PORTION
EQUAL
TO SLAB
THICKNESS

STAKE

Figure 6-11   Footing for a bearing wall

specifications is essential to ensure that both proper placement occurs and that the reinforcing meets the job requirements. Since the footing and foundation wall are poured separately, it is necessary to provide a device to lock the two separate portions of the foundation together. This can be accomplished by forming a keyway out of 38 × 89 mm material or by using short pieces of re-bar left sticking out of the footing. The keyway is usually beveled to aid in removing it from the concrete after it has hardened. (Fig. 6-2, 6-12)

FOUNDATION WALLS

Referred to earlier in the chapter as enclosures for the basement, and used to support the loads of the structure, walls are usually made of concrete, concrete block or pressure-treated wood. The wall thickness, height and type of construction used for them depends upon the local code restrictions, the function of the area under the structure, the depth below grade, and

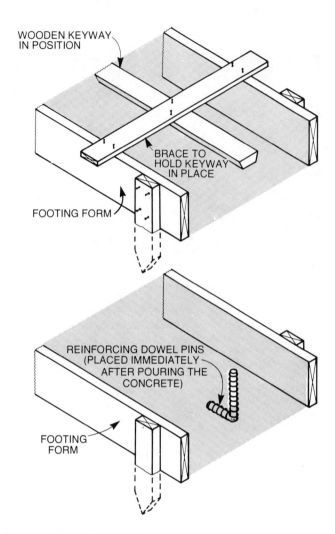

WOODEN KEYWAY
IN POSITION

BRACE TO
HOLD KEYWAY
IN PLACE

FOOTING FORM

REINFORCING DOWEL PINS
(PLACED IMMEDIATELY
AFTER POURING THE
CONCRETE)

FOOTING
FORM

Figure 6-12   Keyway and dowels in the footing

the lateral support supplied by the floor framing system.

Where full basement areas are utilized, the basement height from the finished basement floor to the underside of the supporting beams or floor joists should be at least 1950 mm to allow for adequate headroom (Fig. 6-13). Where foundations provide only limited access under the floor joist, the clearance can vary from 300 to 600 mm minimum depending on the required services (Fig. 6-14).

Full Foundation Wall

As indicated in Figure 6-13, a full foundation is one where the basement area under the house frame is a *liveable* one. This area is used for the distribution of services coming into the home, such as sewer and water lines, and electrical and heating fuels (natural gas, oil, etc.). The remainder of the basement is then used by the occupants, be it for storage, work or living.

Grade Foundation Wall

With a grade foundation, the area under the structure is not a liveable one. However, it is used to distribute services throughout the home. It is generally referred to as a *crawl space*. As shown in Figure 6-14, the foundation wall may be built in the same way as a full foundation, and like a full foundation must extend below the frost line. Trench-type excavations are generally used for this type of construction; however, if the only reason for the deep wall is to penetrate below frost, a combination of the grade foundation and pilings can be used. This type of foundation is quite common with split-level homes (see Chapter 2) where a portion of the home has a full basement under it and the other portion a crawl space. With a crawl space there is no need for a concrete floor. Ground cover is usually provided in the form of polyethylene covered with gravel or sand to protect and weigh it down.

Slab Foundation

This consists of a concrete slab poured over granular fill. Generally referred to as a *slab-on-grade*, it requires a minimum of 120 mm of compacted granular material beneath a 100 mm minimum thick slab. The slab may or may not be supported at the perimeter by a grade foundation resting on a spread footing or piling. Dampproofing of the slab is achieved by placing a polyethylene sheet over the fill just before pouring. The slab design is dependent on the loads being applied, stability of the soil and frost penetration (Fig. 6-15).

Foundation Piles

Bearing pilings or piers are vertical members designed to transmit applied loads from the structure into the surround-

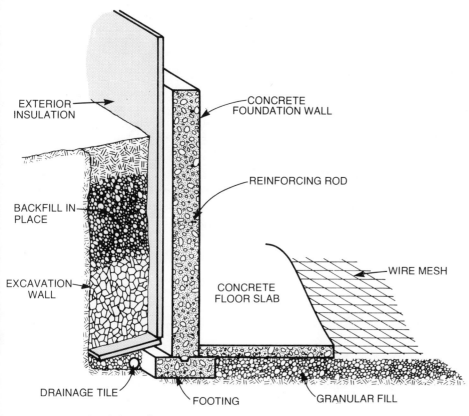

EXTERIOR INSULATION

CONCRETE FOUNDATION WALL

REINFORCING ROD

BACKFILL IN PLACE

EXCAVATION WALL

CONCRETE FLOOR SLAB

WIRE MESH

DRAINAGE TILE

FOOTING

GRANULAR FILL

Figure 6-13   Full foundation wall

ing soil. Pilings are vertical members, usually found below ground supporting the foundation, while piers are vertical members found above grade level, which support the principal framing members or beams of the structure. Although uncommon, it is possible to have the piling and pier combined, thus eliminating the foundation.

In house construction, pilings and piers are usually made of concrete. However, concrete block, pressure-treated wood or steel may be used instead. Their principal use is to support the structure or sections of the structure that do not require a full basement, or structures where a full basement is not feasible due to design, location or cost.

The size, spacing and depth of the piles will vary according to the load imposed on them and the soil conditions of the site. Like all footings, bearing piles must be below the depth of frost, so that heaving or shifting of the foundation is eliminated. Engineer-designed for each individ-

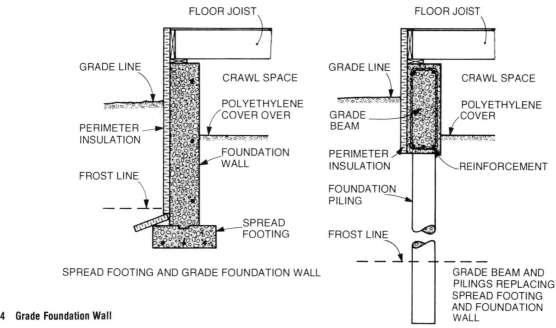

FLOOR JOIST

GRADE LINE

CRAWL SPACE

PERIMETER INSULATION

POLYETHYLENE COVER OVER

FOUNDATION WALL

FROST LINE

SPREAD FOOTING

SPREAD FOOTING AND GRADE FOUNDATION WALL

FLOOR JOIST

GRADE LINE

CRAWL SPACE

GRADE BEAM

POLYETHYLENE COVER

PERIMETER INSULATION

REINFORCEMENT

FOUNDATION PILING

FROST LINE

GRADE BEAM AND PILINGS REPLACING SPREAD FOOTING AND FOUNDATION WALL

Figure 6-14   Grade Foundation Wall

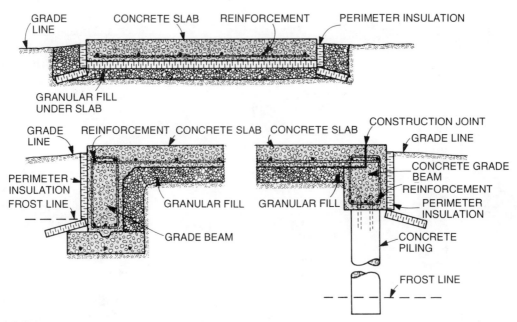

Figure 6-15   Slab foundation

ual structure, the simplest pilings are known as *friction piles*. Usually constructed on concrete which is poured-in-place, they adapt themselves readily to house construction (Figs. 6-14, 6-15).

Friction-type pilings can be dug with an auger if only a few pilings are required. If a large number are needed or hard digging is encountered, machines for boring the holes can be rented. Unless resting on extremely hard soil called hardpan, or on rock, the friction pile obtains most of its bearing capacity from the soil surrounding its circumference and a small percentage from the bottom of the pile. If sufficient bearing cannot be obtained in this manner, the adaptations to the friction pile are used. One such adaptation is the *belled pile*, shown in Figure 6-16. Special equipment must be used in forming the bell.

Pilings driven into the ground are made of precast, reinforced concrete, or pressure-treated wood or steel. They support the foundation above, and function the same way as does the concrete friction

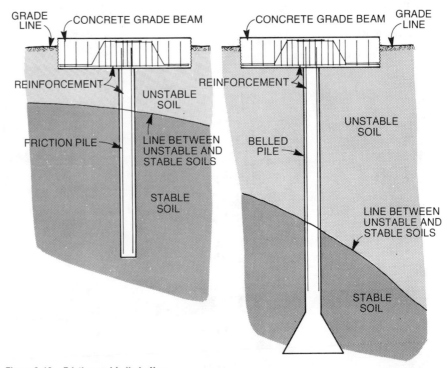

Figure 6-16   Friction and belled piles

pile. In the case of pressure-treated wood pilings, the area extending below the ground level will not be vulnerable to decay. The portion above the ground level must be taken into consideration though, as it may deteriorate and decay if subjected to wet-dry conditions. It is common practice when using pressure-treated wood to cap the top of the piling with a concrete pad, beam or pier.

Sometimes it is necesssary to elevate buildings above ground to ensure that the soil remains in its original state, thus retaining its stability and bearing qualities (Fig. 6-17). The portion of the piling above ground is referred to as the pier. In areas of permafrost, pilings-piers can be used in this manner. Beams made of wood, concrete or steel are then positioned over the piers and the structure is fastened to them.

FOUNDATION WALL FORMS

Forms for concrete foundation walls are available in many different types which require different methods of assembly. Although all wall forms contain the same basic parts, they may vary as to design and to the type of hardware used in their assembly. A basic part of the wall form is the *sheathing*, which gives the concrete the desired shape. Sheathing may or may not be backed up with an external framework for support. The forms are held apart at the correct width by a *metal tie* which holds the wall *together* while the concrete is in its plastic form. Fastened to the metal ties are horizontal or vertical members called *walers*, which resist the pressure of the concrete. The forms are secured by bracing where required to align and support them during the pouring of the wall (Fig. 6-18).

Forming Materials

The sheathing material used for concrete forms is usually solid stock lumber or

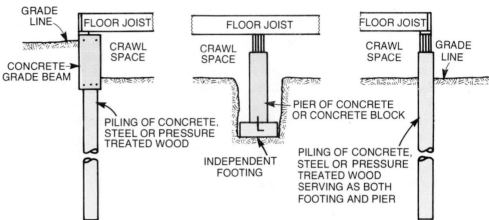

Figure 6-17   Piling and pier foundations

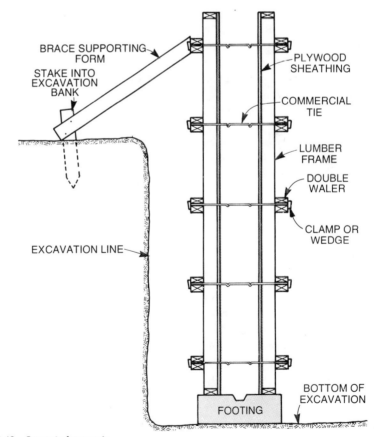

Figure 6-18   Concrete form parts

plywood. Lumber sheathing 19 mm thick can be any width depending on the job situation and is normally made of shiplap or common board.

Plywood panels used for concrete forms are available in standard 1200 × 2400 mm sheets. For on-site use the panels are cut to sizes ranging from 300 to 1200 mm depending on the type of tie used, and whether a framework is used on the back of the panel for support. Factory prepared panels are available with coatings either impregnated into or applied to the surface of the panels. This coating protects the wood and extends the life of the forms. A *release agent* (form oil) applied to the forms before pouring will facilitate the stripping operations. Other materials used in forming foundation walls include steel, aluminum and fiberglass, but these are in limited use.

Framed form panels usually consist of 11 to 19 mm plywood panels placed on a framework of 38 × 89 mm dimensional stock. The panel may also be applied to a metal frame if desired. Sizes are usually 1200 × 2400 mm or 600 × 2400 mm with numerous odd sizes used as fillers. By their very nature they are a much stronger form and therefore tend to take more abuse and last longer than the unframed panel form.

Form Hardware

Form hardware refers to all the devices used in setting up and securing the forms in position, before and during the concrete pour. Many different types of form hardware have been developed, but we will only refer to those which apply to residential construction. Ties are an inexpensive method of form support. The number of rows and the spacing of ties should be obtained from the manufacturer's specifications, since the strength

of the ties must be calculated according to numerous variables, including: the width of the wall, the height of the pour, the rate of pour, etc. Ties also have a *break-back* feature which allows the ends of the tie to be broken off a set distance inside the wall surface, either as the form is being removed or later on. Other hardware required includes, clamps, waler bars and break-back tools. Figure 6-19 shows three common types of ties used.

Forming Systems

The type of forming system used will depend largely on the following factors:

1. Type of forms available (in many areas rental forms are at your disposal).
2. Height and width of the concrete wall to be poured.
3. Type of tie used (this may be governed by the type of form available).
4. Desired quality of the finished foundation wall.
5. Expected use of the forms after the concrete pour is complete.

Since formwork consists of a good per-

centage of the overall cost of the foundation, it is quite obvious that pre-planning and the correct choice of forming materials will result in a significant saving. The intent in this section is to show commonly used systems in residential construction. Which system best suits the job requirements should be decided by the owner/builder.

Built-up Forms

Built-up forms make extensive use of lumber sheathing. Although not used as extensively today as it has been in the past, it can still be used where the convenience and economy of using lumber forms exists and the material is designated for later use elsewhere in the structure. Built-up forms can be used where it is uneconomical to cut plywood forms, where stepdowns (differences in elevation) occur in the footing or where forms cannot be removed after pouring.

Assembly time is slower as the individual boards have to be fastened to the frame backing. Proper construction techniques will, however, result in forms as strong and tight as these produced by

STRAP TIE

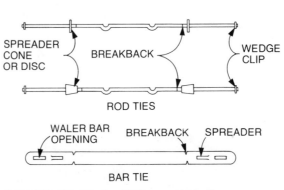

ROD TIES

BAR TIE

Figure 6-19   Concrete form ties

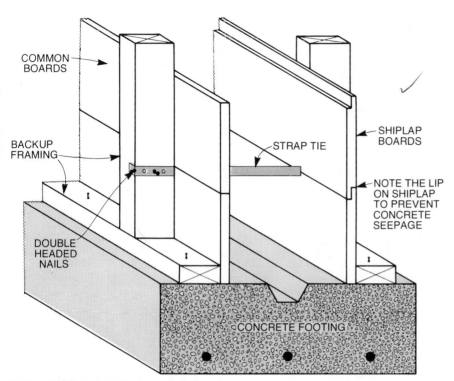

COMMON BOARDS

BACKUP FRAMING

DOUBLE HEADED NAILS

STRAP TIE

SHIPLAP BOARDS

NOTE THE LIP ON SHIPLAP TO PREVENT CONCRETE SEEPAGE

CONCRETE FOOTING

Figure 6-20 Built-up wall forms

other methods. Factors which can be used to reduce time and overall costs of the forming system include: fabrication of the form on the ground, making sections large enough to be easily handled by the available manpower; the proper use of shiplap and/or common boards (Fig. 6-20); and keeping nails to a minimum.

The type of tie used will determine how the walers are to be placed. With the strap tie shown in Figure 6-19 a backup frame is needed, however the use of walers is eliminated. The strap tie is notched into the sheathing next to opposing studs in the framework. It is fastened at the desired width by nailing through the pre-drilled holes with double headed nails into the studding on the form. It becomes quite rigid, requiring a minimum number of braces to set and align it. After the concrete is poured and the forms removed, the strap tie ends are broken off on the outside and inside of the wall. The inside ties can be left so that future framing can be fastened to the ties (Fig. 6-20).

PREFABRICATED PANEL SYSTEMS

Panel systems are the most widely used system today, and can be adapted to most structures. Pre-drilling of tie holes in the panels makes it essential that the footings are level and that the panels are placed correctly. Panel systems can be broken into two groups; unframed systems (having no external framework) and framed systems (supported on an external frame).

Unframed Panel Forms

In the section on forming materials it was mentioned that plywood 19 mm thick is the most widely used panel form. Factory designed forms for foundation work are coated with a preservative to extend the use of the material and aid stripping.

They will also improve the quality of the finished product. If the form panels are new, it will be necessary to drill or slot them depending on the type of ties to be used. One of the most widely used ties for this system is the flat bar tie shown in Figure 6-19. However, the rod tie may be used as an alternate.

Framing is kept to a minimum as a bottom plate and top waler of 38 × 89 mm material is all that is necessary for positioning and aligning the form panels. The flat bar tie is inserted into the preformed slots and held in place by steel bars of various lengths. These steel bars (walers) are all that are required to resist the force of the concrete until it sets. Inside panels are sometimes cut 50 mm shorter for joist filled construction (Fig. 6-27) and must be placed directly opposite the exterior panels, inserting the ties and locking them in place with the waler bars as assembly continues.

With pouring of the concrete completed, stripping is simply a matter of breaking off the bar tie ends and lifting the panels up out of the excavation. Figures 6-23 to 6-28 show this type of form system being used on a basement foundation. Because the components are light and easy to handle, and because assembly time is reduced to a minimum, it is probably the most widely used system in residential foundation building today.

Framed Panel Forms

Framed panel forms consist of a panel face (usually plywood) with a wood or metal reinforcing frame, and are normally 600 × 2400 mm in size. The rod tie (Fig. 6-19), which is only one type of tie available, is adaptable to all types of residential and commercial construction. The framed panel system lends itself to gang forming quite well. In *gang forming,* individual forms are fastened together in large sections and hoisted into position

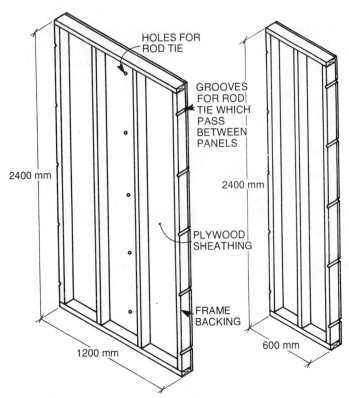

2400 mm

HOLES FOR
ROD TIE

GROOVES
FOR ROD
TIE WHICH
PASS
BETWEEN
PANELS

2400 mm

PLYWOOD
SHEATHING

FRAME
BACKING

1200 mm

600 mm

Figure 6-21(a)   Framed panel forms

with the use of cranes or similar equipment.

The forms are designed so that the rod ties either fit between the panels (Fig. 6-21), or the panels are drilled and the rods inserted. Like the unframed panel, both inside and outside forms have to be matched for size so that the ties fit properly. Odd sized panel sections are made up as required to meet the job specs. Horizontal walers (usually 38×89 mm material) are placed above and below the rod tie. Walers and forms are then tightened against the spreader collar with a metal wedge. A system similar to this is available for unformed forms which uses a combination metal waler holder and clamp assembly. With this system a single 38×89 mm waler is used.

As this system secures the inner and outer form tightly together, it requires little bracing except for alignment. However, a considerable amount of material is necessary for walers. When stripping, the clamps and walers (along with any necessary bracing) are removed first. Panels are then stripped away, leaving the tie rod

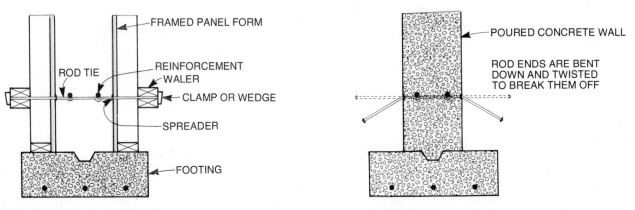

FRAMED PANEL FORM

ROD TIE

REINFORCEMENT
WALER

CLAMP OR WEDGE

SPREADER

FOOTING

POURED CONCRETE WALL

ROD ENDS ARE BENT
DOWN AND TWISTED
TO BREAK THEM OFF

Figure 6-21(b)   Before pouring

Figure 6-21(c)   After pouring

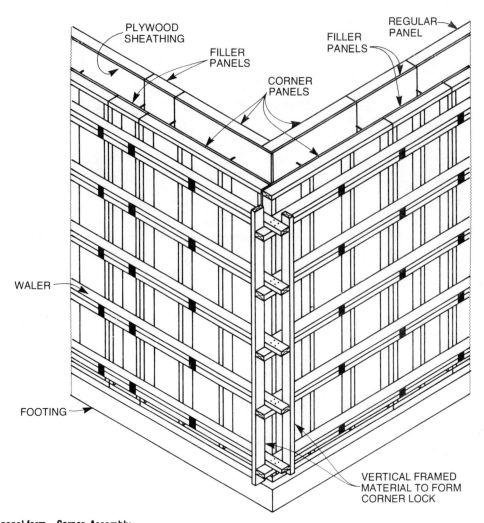

Figure 6-21(d)   Framed panel form—Corner Assembly

ends (minimum 100 mm) sticking out of the concrete wall. These ends are then bent back against the concrete wall and twisted in a circular motion snapping off at the weakened section of the rod inside the wall (Fig. 6-21).

Forming the Concrete Wall

With the footing pour completed, construction of the wall forms can begin. Back at the batter boards, builder's lines are set up again; this time the outside building line is used. Using the same procedure as was described to obtain the footing line, the corners of the foundation are now established on the footing (Fig. 6-22).

Once the corners of the foundation have been obtained, use a chalk line to snap a line along the footing to indicate the outside of the concrete wall. We will assume we are going to form a 200 mm thick concrete wall using the unframed panel forms with the flat bar tie.

With the outside building line snapped on the footing, measure back 19 mm (thickness of the panel) and secure a 38 × 89 mm plate to the footing. If the concrete footing is still *green* (one or two days old) it is possible to simply nail the plate to the footing. The alternative method is to leave the outside footing form in place and use a plate wide enough to allow nailing through it into the footing form (Fig. 6-23).

Once the plate is in position, the outside corner panels can be positioned, plumbed and nailed secure. A top plate of 38 × 89 mm is placed flat slightly below the form top, joining the outside corner panels. Be sure to keep the distance from inside of corner to inside of corner the

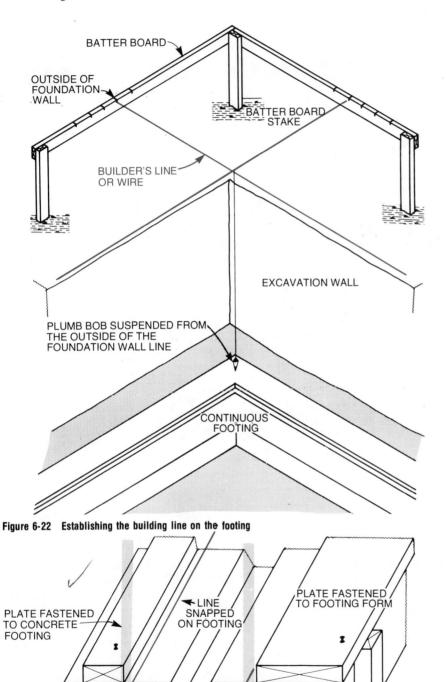

BATTER BOARD

OUTSIDE OF
FOUNDATION
WALL

BATTER BOARD
STAKE

BUILDER'S LINE
OR WIRE

EXCAVATION WALL

PLUMB BOB SUSPENDED FROM
THE OUTSIDE OF THE
FOUNDATION WALL LINE

CONTINUOUS
FOOTING

Figure 6-22   Establishing the building line on the footing

PLATE FASTENED
TO CONCRETE
FOOTING

LINE
SNAPPED
ON FOOTING

PLATE FASTENED
TO FOOTING FORM

PLATE SET BACK
THE THICKNESS
OF THE FORM
PANEL

FOOTING FORM
AND STAKES

REINFORCEMENT

CONCRETE
FOOTING

same at both the top and bottom of the form. The outside form is then completed by nailing on the remaining panels. When odd dimensions are encountered, filler panels of plywood or lumber can be cut to fit (Fig. 6-24).

Bar ties designed for a 200 mm wall are then inserted into the preslotted form and the 6×19 mm steel locking bars (walers) are positioned. Due to the fluid pressure exerted by the fresh concrete, it is important to watch the positioning of the bars. No joins should be made in succeeding rows of bars, and bars that span joins in individual panels should be held by at least one and preferably two form ties (Fig. 6-24).

Before the inside form is positioned, any door or window bucks, offsets for pilasters, beam pockets or reinforcing steel should be set (Figs. 6-29 to 6-32). Inside forms are then positioned using the same procedure as the outside form. Corner panels, and then panels to match the outside forms are set. No bottom plate or top plate is necessary on the inside form, but a top plate does aid in holding and positioning the panels. When putting in any fillers, measure the overall distance of the form and subtract the concrete wall thickness on both ends; then make the cutting accordingly. Remember the outside form is fastened solid so any movement will occur inwards when the concrete pour commences.

With the forms completed, aligning and bracing of the outside form is necessary. A builder's line is strung on the inside of the outside wall form and blocked out as shown in Figure 6-25. Braces are then nailed to the top plate and the wall moved in or out according to the line and

Figure 6-23(a)   Securing the plate to the footing

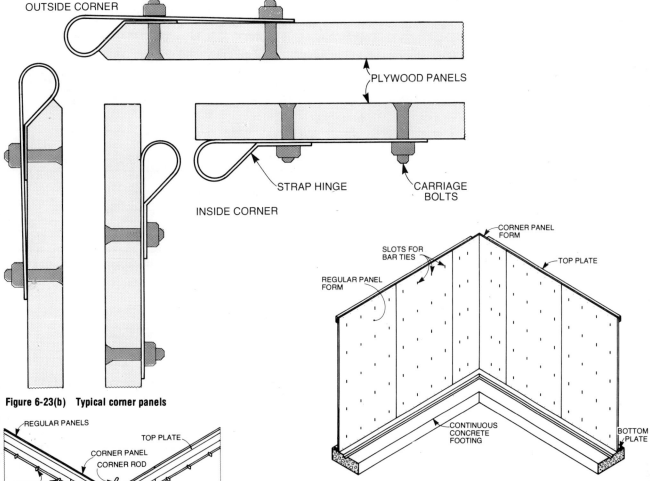

OUTSIDE CORNER

PLYWOOD PANELS

STRAP HINGE

CARRIAGE BOLTS

INSIDE CORNER

Figure 6-23(b)   Typical corner panels

CORNER PANEL FORM

SLOTS FOR BAR TIES

REGULAR PANEL FORM

TOP PLATE

CONTINUOUS CONCRETE FOOTING

BOTTOM PLATE

Figure 6-23(d)   Wall panels set in place

REGULAR PANELS

TOP PLATE

CORNER PANEL CORNER ROD

BAR TIES

WALER BARS

CORNER HINGE

BOTTOM PLATE

CONTINUOUS CONCRETE FOOTING

Figure 6-23(c)   Unframed wall panel complete with ties and waler bars

test block being used. The outer ends of the brace are nailed to stakes secured to the excavation bank.

Depending on the connection used between the foundation and the floor frame system, the concrete wall must be struck off at a certain level. With **box sill construction**, a level grade must be established and anchor bolts set. With **joist fill construction**, the floor joist will have to be set on the form wall before pouring. Figures 6-26 and 6-27 give more detail on these construction procedures, as well as Chapter 7 (Floor Framing).

Assuming that the box sill method is to be used and that the sill is bolted to the top of the wall, a 19 × 38 mm strip of wood or grade nails can be fixed to the inner face of the outside form indicating the correct height of the concrete (Fig. 6-26). This height is obtained from the blueprint and positioned with the use of the transit-level from the bench mark. If a wood grade strip is used it can be either left in and removed with the concrete forms or can be fastened in permanently by driving nails in the back at an angle, thus securing it into the concrete wall. When the concrete is leveled, anchor bolts (12.7 mm minimum diameter) are placed not more than 2400 mm o.c. and embedded at least 100 mm into the concrete. The wood sill is attached to these bolts when the concrete has set.

Joist filled construction requires the beam and joist to be set into position on the forms before concreting takes place. This technique has superseded the box sill type of construction in many areas. To achieve this, the inner forms are usually cut 50 mm shorter than the outer form allowing the header joist to be nailed to the outer form at the correct height (Fig. 6-27). The height is established using grade

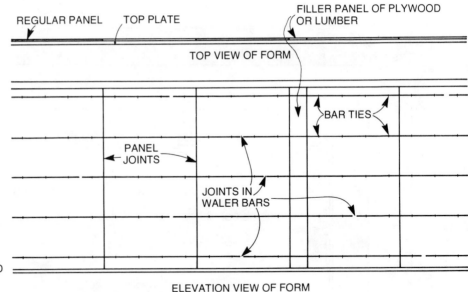

Figure 6-24   Location of panels, ties and waler bars

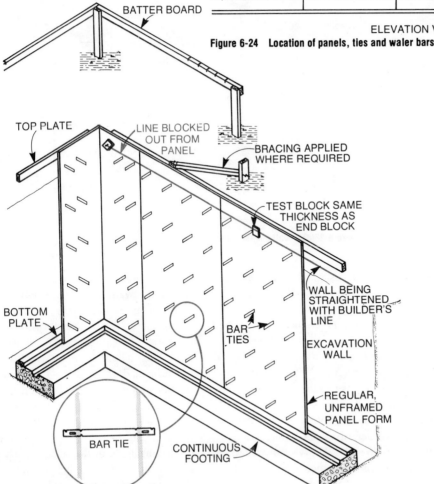

Figure 6-25   Aligning and bracing the outside form

nails. With the headers set at the correct height, the beam is positioned and temporarily braced. Joist layout takes place as indicated in Chapter 7 (Floor Framing), with the joist end nailed to the header and either butted or lapped at the beam. Blocking is installed between the joists over the inner wall form to prevent concrete spillage (Fig. 6-27).

When pouring, the concrete is leveled off about 25 mm below the top of the joist, embedding them into the concrete wall. Advantages of this type of assembly are that the foundation and floor systems are identical size with the floor joist. In place, a perfect platform is available for working on while pouring the concrete.

After allowing the concrete to cure properly, stripping of the form begins by reversing the setting up procedure. The bracing, plates and wooden walers are the first to be removed; make sure to pull out all the nails as you proceed. To remove the waler bars, strike the tie from above or below with a hammer on the bar tie end (Fig. 6-28). This fractures the bar tie at the designed break back point inside the

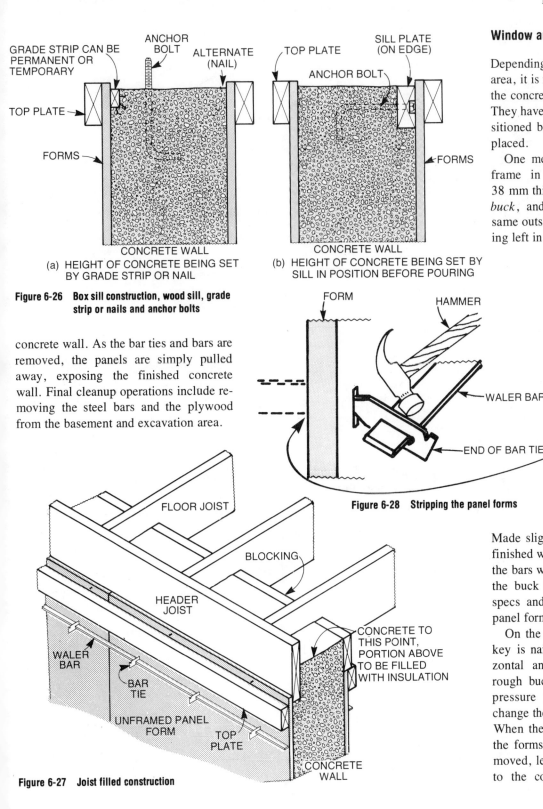

Figure 6-26   Box sill construction, wood sill, grade strip or nails and anchor bolts

(a) HEIGHT OF CONCRETE BEING SET BY GRADE STRIP OR NAIL

(b) HEIGHT OF CONCRETE BEING SET BY SILL IN POSITION BEFORE POURING

Figure 6-27   Joist filled construction

Figure 6-28   Stripping the panel forms

concrete wall. As the bar ties and bars are removed, the panels are simply pulled away, exposing the finished concrete wall. Final cleanup operations include removing the steel bars and the plywood from the basement and excavation area.

Window and Door Openings

Depending on the design of the basement area, it is not unusual to have openings in the concrete wall for windows and doors. They have to be built into the wall and positioned before the inner form panels are placed.

One method which can be used is to frame in the complete opening using 38 mm thick material. It is called a *rough buck*, and when constructed, it has the same outside dimensions as a rough opening left in a framed wall.

Made slightly narrower (3 mm) than the finished width of the concrete wall so that the bars will slide easily into the tie ends, the buck is positioned according to the specs and nailed into place through the panel form.

On the outside of the buck, a tapered key is nailed on (Fig. 6-29). Both horizontal and vertical bracing inside the rough buck are essential to ensure that pressure from the concrete does not change the size and shape of the opening. When the concrete has been poured and the forms stripped, the rough buck is removed, leaving the tapered key anchored to the concrete. The window or door

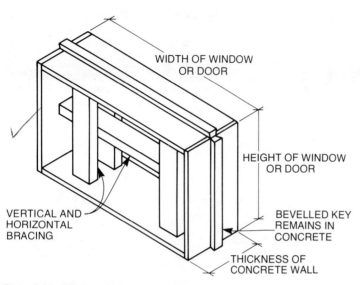

Figure 6-29   Window and door rough bucks

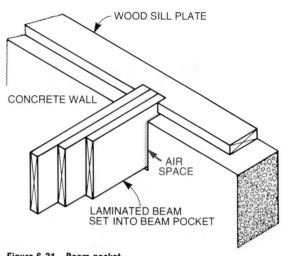

Figure 6-31   Beam pocket

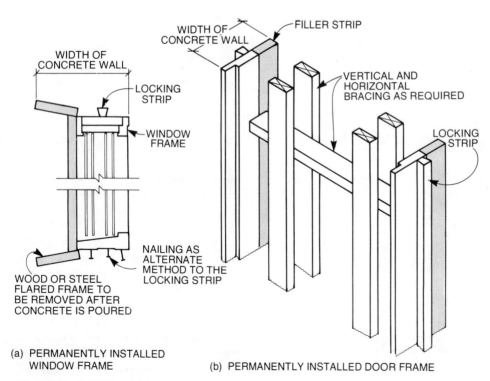

(a) PERMANENTLY INSTALLED WINDOW FRAME

(b) PERMANENTLY INSTALLED DOOR FRAME

Figure 6-30   Installing a window or door frame in a concrete wall

frame is then positioned in the opening and fastened to this key.

The other method which can be used is to position and nail the actual window or door frame onto the form panel before the concrete pour. To compensate for the difference in thickness between the frame and wall, a filler buck is added. If desired, the buck may be tapered to achieve a flared look around the opening, removing the sharp, square edge and making it easier to strip. Bracing is required both within the unit and the rough buck to prevent movement, but the locking key is not required. As an alternative, nails are driven into the back of the frame and left protruding into the concrete wall, forming the locking device (Fig. 6-30).

Beam Pockets and Pilasters

A beam pocket is necessary when the box sill construction method is used. A beam pocket is a recessed opening left in the end walls of the foundation in which the main support beam rests. The pocket is formed by making a box 24 mm wider than the beam width by the beam thickness and 100 mm deep. The box is then

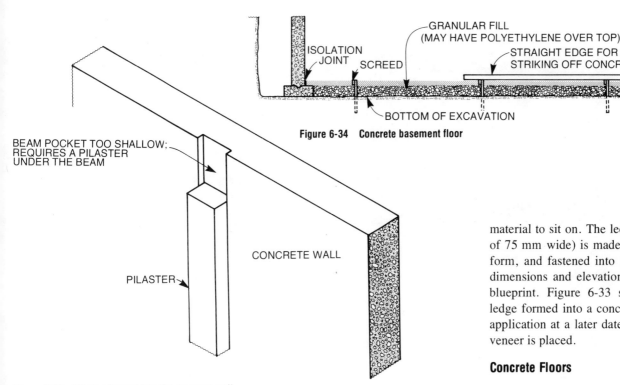

Figure 6-34   Concrete basement floor

Figure 6-32   Pilaster formed into the concrete wall

placed in position on the inner form with the top of the box 38 mm above the top of the concrete. This sets the top of the beam level with the sill. Removed after the concrete has hardened and the wall forms are removed, it forms a pocket for the end of the beam to sit in (Fig. 6-31).

Pilasters are required where the beam pocket which has been formed into the wall is considered inadequate. The pilaster is simply an enlargement on the inner face of the concrete wall where the beam sits and the concentrated load is then transferred down into the soil below. Formed in as the inner wall forms are set up, it provides the necessary support (Fig. 6-32).

Brickledge

If the brick or stone veneer is used as an exterior finish on the structure, it may be necessary to form a ledge for the masonry

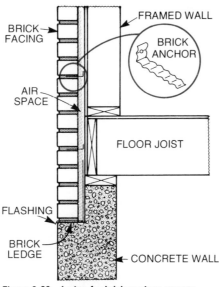

Figure 6-33   Ledge for brick or stone veneer

material to sit on. The ledge (a minimum of 75 mm wide) is made up, set into the form, and fastened into place, using the dimensions and elevations shown on the blueprint. Figure 6-33 shows a typical ledge formed into a concrete wall and its application at a later date when the brick veneer is placed.

Concrete Floors

After the necessary ground work has been completed in the basement area, such as the sewer and water lines connected, and drain tile and other plumbing roughed-in, the basement floor can be poured. With concrete foundation walls this can be before or after the backfilling of the foundation takes place. With concrete block and wood foundations however it is advisable to place the floor before backfilling.

To prepare the floor, a layer of coarse granular fill not less than 120 mm thick is placed over the excavation floor, then leveled and compacted. After compacting, the screeds are set to the level of the finished floor. Screeds are spaced so that a 3 to 4 m straight edge can be used to strike off the concrete surface. Pouring, finishing and curing operations are then carried out as discussed in Chapter 5. Figure 6-34 shows an accepted method of placing the screeds and the position of the concrete floor in relation to the wall and footing.

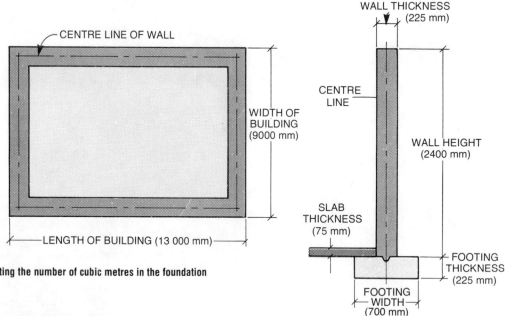

Figure 6-35   Calculating the number of cubic metres in the foundation

Calculating the Number of Cubic Metres of Concrete Used in a Foundation, Footing, Wall and Floor Slab.

To determine the number of cubic metres of concrete that would have to be ordered to pour the footing, wall and floor slab, the following factors would have to be known:

(a) **Building size**—13 000 × 9000 mm
(b) **Footing size**—700 × 225 mm
(c) **Wall height**—2400 mm
(d) **Wall thickness**—225 mm
(e) **Floor slab thickness**—75 mm

Note:

(1) Use the volume formula: Length × Width × Height = cubic metres
(2) To find the total wall length use the centre line length. The *centre line length* (CLL) of the building is equal to the perimeter of the building (L + W)2 minus 4 times the wall thickness. The centre line length for the wall and footing are the same unless the footing is offset.
(3) When ordering concrete always take your answer to the next largest 0.25 m³.

Step 1. Determine the CLL:
CLL = Perimeter (L + W)2 minus 4 times wall thickness
  = (13 + 9)2 − 4 × 0.225
  = (22)2 − 0.9
  = 44 − 0.9
  = 43.1 m

Step 2. Determine the number of cubic metres of concrete in the footing:
L × W × H = cubic metres
43.1 × 0.7 × 0.225 = 6.79 or rounded off 7 m³

Step 3. Determine the number of cubic metres of concrete in the wall:
L × W × H = Cubic metres
43.1 × 0.225 × 2.4 = 23.27 or rounded off 23.5 m³

Step 4. Determine the number of cubic metres of concrete in the floor slab:

For this calculation we are unable to use the CLL, however the length and width are obtained by subtracting 2 times the wall thickness from the length and width of the building.
L = length of building − 2 × wall thickness
  = 13 000 − 2 × 225
  = 13 000 − 450
  = 12 550
W = width of the building − 2 times wall thickness
  = 9000 − 2 times wall thickness
  = 9000 − 450
  = 8550
using the volume formula L × W × H = cubic metres
12.55 × 8.55 × 0.075 = 8.05 or round off 8.25 m³

MASONRY FOUNDATIONS

Concrete masonry blocks can be used as an alternative to concrete for foundation walls. Masonry walls should be planned out to make the fullest use of full and half blocks. This means that wall lengths should work out to modular panels (Fig. 6-36) and that doors and windows should also conform to the modular layout. By pre-planning and working to the modules, the time required to cut and fit odd sized blocks will be kept to a minimum. In Figure 6-36, the standard block sizes are

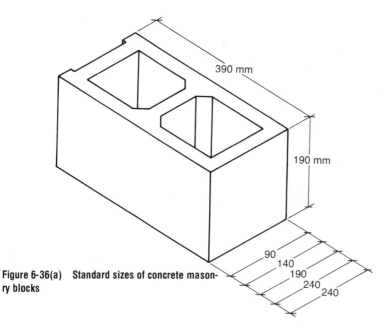

Figure 6-36(a)   Standard sizes of concrete masonry blocks

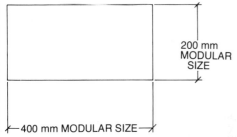

Figure 6-36(b)   Typical metric block

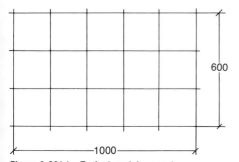

Figure 6-36(c)   Typical modular panel

shown along with a modular panel and layout of a door and window using the module. As well as the standard concrete block shown, other blocks available include; half, corner, jamb, lintel and sill blocks.

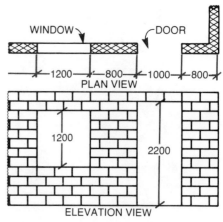

Figure 6-36(d)   Fully modular layout

Courtesy Portland Cement Association

Figure 6-37(a)   Layout of a concrete block foundation

Courtesy Portland Cement Association

Figure 6-37(b)   Anchor bolts set in poured concrete

Masonry work, performed by the bricklayer (mason) begins after the footing has been poured. Locating the corners of the building (Fig. 6-22) from the batter boards, the mason begins the layout by setting the first row of blocks on the footing to adjust the spacing. After the first row (course) is established, the corners of the foundation wall are built up (Fig. 6-37).

Once the corners are established, the mason uses a builder's line strung from corner to corner to maintain a straight wall line, and lays up the section of wall between the corners. Openings for doors, and windows, pilasters, etc., are all built into the wall as the courses of block are laid. "Tooling" is the term used for compacting and smoothing the mortar joint. It produces clean neat lines on all exposed surfaces.

When building with a masonry foundation, it is common practice to use the box sill method of construction. This means that anchor bolts must be set once the basement wall height is reached. Using concrete block the accepted method is to

Courtesy Portland cement Assoc.

Figure 6-37(c) Parging a block foundation

fill the top two courses of block with concrete and set the anchor bolts into this (Fig. 6-37).

To ensure a watertight basement, the exterior side of the foundation wall is parged using a cement plaster coat at least 6 mm thick. This parged coat can be discontinued at the ground level or can continue to the top of the foundation wall. If the parging is taken above the ground it is often swirled or pebbled to give a decorative effect, enhancing the exterior decor of the house.

PRESERVED WOOD FOUNDATIONS

One of the most recent systems to be developed is the **pressure-treated preserved wood foundation (PWF)**. Since the first experimental homes were built in the early 1960s, the interest in this system has grown until today many areas in Canada have accepted the PWF system. Provided it is designed and built in accordance with the National Building Code of Canada, the PWF system can be expected to function as well as any other foundation system.

Engineering procedures which have been developed have demonstrated that in designing a PWF system, there are some special considerations. *Lateral loads* imposed on the foundation from the surrounding soils and the connection of the basement floor and the first floor joist with the foundation wall are of prime importance. Most municipal building codes require engineer-designed and inspected PWF systems. It is therefore imperative that before proceeding with a preserved wood foundation, local building codes are consulted and adhered to.

Materials Used in Preserved Wood Foundations

To ensure top quality and satisfactory results, it is essential that the selection of materials and their application be carefully monitored. PWF lumber, like all framing lumber must follow the guidelines used for grading lumber under the National Building Code of Canada. In addition, all lumber used for wood foundations must be treatable and only those species approved and stamped with the Canadian Standards Association certification mark are acceptable. The **size** and **grade of species** accepted will be dependent on the *spacing* of the framing members, the *type* of construction used, the *backfill depth* and whether the structure is *one* or *two stories*.

The plywood used for PWF is either select sheathing or standard sheathing grade. The **thickness** required is dependent on the *spacing* of the studding, the *direction* of application and *height* of the backfill. There are tables available which enable the engineer, builder, or owner to select the correct dimensional lumber and thickness of plywood for each job.

Treatment of lumber for a PWF must conform to the Canadian Standards Association guidelines and is either Chromated Copper Arsenate or Ammoniacal Copper Arsenate—the two currently accepted treatments. Due to the fact that the lumber used for pressure-treating must be dry before undergoing the treatment, plus the high cost of treating, results in pressure-treated wood being more expensive than regular construction lumber. Make sure that pressure- treated wood is used only where required. This includes all areas where the wood is in contact with the soil or where a decay hazard exists. In areas where it is not needed, switch to regular construction grade lumber to reduce the overall cost. Any on-site cutting of pressure-treated materials must be protected by repeated brushing or dipping until the exposed area has thoroughly soaked up all the preservative it will hold. A good building practice here is to keep all cut ends or edges away from contact with the soil and moisture if possible.

Nails used in PWF must conform in *size* and *number* to the job specs. The type of nail required will vary according to moisture conditions of the area. The minimum is a hot dipped galvanized nail with the stainless steel nail being used as an alternative. In some cases, stainless steel nails are required by local building codes. If used, staples must be stainless steel. All other types of **metal fastening devices** must be made of *hot dipped galvanized* material.

Sealing the plywood to prevent moisture from penetrating the wall is accomplished by *caulking* all joints between the panels with a high performance caulking compound. A watertight seal of butyl or silicone-based caulking compound is suitable, providing the caulking remains flexible, has a long service life and is compatible with the pressure-treated wood.

A continuous cover over the P.T. plywood is required, to deflect the surface water away from the foundation wall. This **moisture barrier** usually consists of 150 μm (micrometre) *polyethylene* applied to the exterior of the wall from the bottom of the foundation up to the finished grade level. Care must be taken with the polyethylene, to avoid tearing it. Protection can be achieved with either a rigid panel board or cardboard placed over the polyethylene to protect it while backfilling of the excavation.

Types of Preserved Wood Foundations

Three basic types of PWF have been designed for residential houses. Presently, most PWF foundations are limited to residential buildings up to two stories plus a basement area. If other adaptations of the PWF are required, they must be designed to suit the building which it supports.

The three accepted types of PWF shown in Figure 6-38 include (a) the **traditional** one-or two-storey home placed on a PWF normally 2400 mm in height and set into a *full* excavation. The basement floor consists of either a *wood sleeper* floor or a *concrete* slab. The *suspended wood* floor shown in (b) allows for 3000 mm foundation walls with the suspended floor approximately 600 mm above the gravel bed. This system is adaptable to split-entry homes, where footings must be extended to reach below the level of frost penetration, or where unstable soils must be excavated to obtain a firm bearing. Where a full basement (c) is not required, an excavated trench and crawl space allows for shallow PWF walls to be erected. A suspended floor over the crawl space allows services to be run under the floor system.

Building a Preserved Wood Foundation

Like all building systems, there are some practices common to all and others which are unique to the system being used. Because the materials in a PWF are made of wood they are totally different from other foundation materials commonly used in the past. Therefore the installation procedures are somewhat different, but in the end, all foundations must function in the same manner.

In the excavation, a layer of clean granular material or crushed stone a minimum

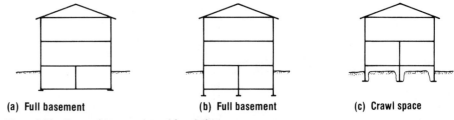

(a) **Full basement**   (b) **Full basement**   (c) **Crawl space**

Figure 6-38   Types of preserved wood foundations

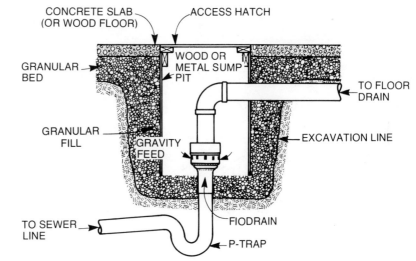

Figure 6-39   Application of the sump in the gravel bed   *Courtesy Fiodrain Limited*

of 125 mm thick is placed over the undisturbed soil on the excavation floor. This gravel layer must extend under the footing and 300 mm out past the footing plate. The entire bottom of the excavation drained by this system must be graded into a sump pit, where it drains either by gravity or is pumped into a sewage system using a sump pump (Fig. 6-39).

Pressure-Treated Footings

Footings for a PWF can be made of *either* pressure-treated wood *or* poured concrete described earlier in the chapter. It is recommended that the wood footing plate be used in conjunction with the wood foundation wall providing the minimum bearing capacity of the soil is 75 kN/m^2 or

better. This is because the total weight of the wall is much less in comparison with concrete, and the wood plate sitting on top of the gravel pad does not impede the flow of water through it.

The size of the footing plate must conform to the job specs. Installation may take place before the foundation wall is set into place or may be fastened to the bottom of the wall prior to setting the wall up. In either case, joints in the footing plate must be **offset** a minimum of 600 mm away from those in the bottom wall plate to ensure a strong joint. By using different lengths of footing material, cutting can be avoided to a large degree. Rather than cutting to the exact length, ends can extend beyond the corner

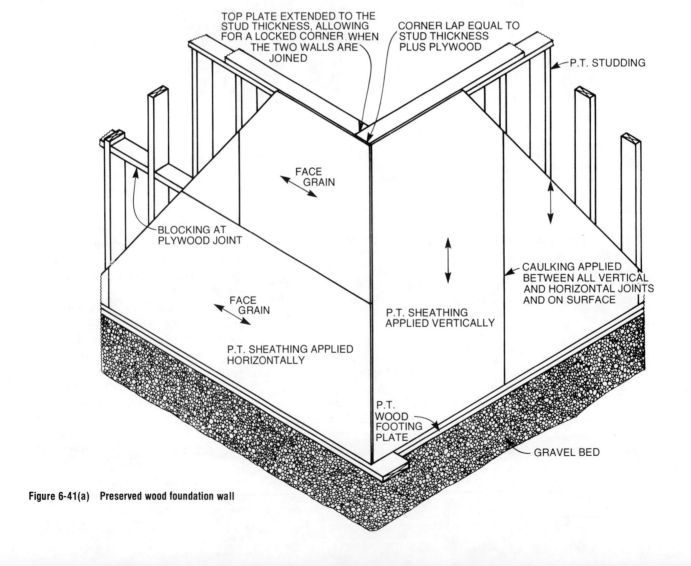

EXCAVATION FLOOR SLOPED TO SUMP PIT

END OF P.T. WOOD FOOTING PLATE MAY EXTEND BEYOND WALL TO MINIMIZE CUTTING AND FIELD TREATING OF ENDS

P.T. WOOD FOOTING LINE

SUMP PIT

EXCAVATION LINE

Figure 6-40   Setting the footing plate

as shown in Figure 6-40. If it is necessary to cut the plate, the cut end must be field coated as previously mentioned.

Pressure-Treated Walls

Walls for PWF are framed in a similar manner to walls used for above-grade work. Consisting of a single bottom plate and either a single or double top plate, studs are placed at 300 or 400 mm on centre. Sheathings are applied either vertically or horizontally to the frame (Chapter 8—Plate Layout and Wall Assembly). Fabrication of the wall can take place in a shop, with the sections being

TOP PLATE EXTENDED TO THE STUD THICKNESS, ALLOWING FOR A LOCKED CORNER WHEN THE TWO WALLS ARE JOINED

CORNER LAP EQUAL TO STUD THICKNESS PLUS PLYWOOD

P.T. STUDDING

FACE GRAIN

BLOCKING AT PLYWOOD JOINT

FACE GRAIN

P.T. SHEATHING APPLIED VERTICALLY

CAULKING APPLIED BETWEEN ALL VERTICAL AND HORIZONTAL JOINTS AND ON SURFACE

P.T. SHEATHING APPLIED HORIZONTALLY

P.T. WOOD FOOTING PLATE

GRAVEL BED

Figure 6-41(a)   Preserved wood foundation wall

shipped out to the job site, or the walls can be stick-framed at the site. Sections should be constructed to sizes that can be managed easily with the help available. Any lumber 200 mm **above** the finished grade line can be untreated, with the exception of the wall studs. **Sizes** and **spacing** of studs along with the **thickness** of the plywood is dependent on the *height* of the backfill, *grade* and *species* of lumber used, and the *type* of foundation floor (Fig. 6-41).

Where doors or windows occur in the foundation wall **special considerations** must be taken into account. **Doors** are framed-in according to the building code requirements for doors in wood-frame construction. **Window openings** must conform to the building code and also require *extra nails* and *joist hangers* to transfer loads to the regular wall studs (Fig. 6-42) where backfill heights *exceed* 2400 mm. Openings for split-entries are treated in the same manner as window openings.

When applied, sheathing must be caulked between the joints. Any joint not occurring on a plate or stud below grade must be backed with a 38 × 89 mm block (Fig. 6-41(a)). The *length, spacing* and *type* of nail allowed must conform to the job specs or local requirements.

Sealing the Preserved Wood Foundation

Sealing the outside of the PWF is necessary to keep the *moisture content* in the plywood and lumber below 20%. This reduces the possibility of nail corrosion. It

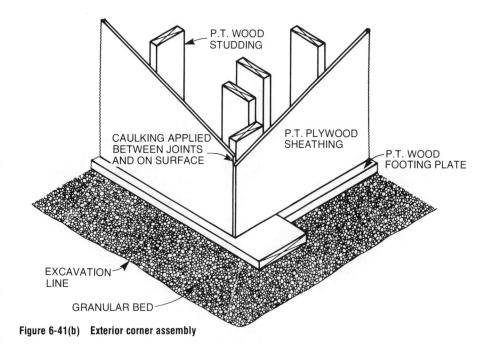

Figure 6-41(b)   Exterior corner assembly

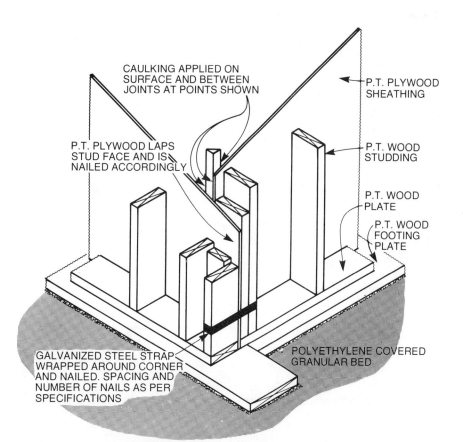

Figure 6-41(c)   Interior corner assembly

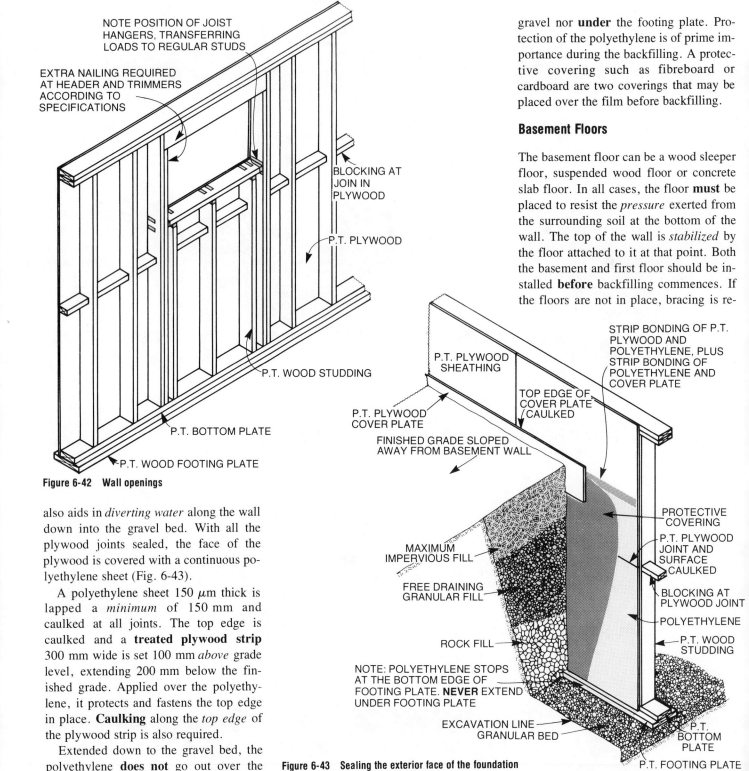

NOTE POSITION OF JOIST HANGERS, TRANSFERRING LOADS TO REGULAR STUDS

EXTRA NAILING REQUIRED AT HEADER AND TRIMMERS ACCORDING TO SPECIFICATIONS

BLOCKING AT JOIN IN PLYWOOD

P.T. PLYWOOD

P.T. WOOD STUDDING

P.T. BOTTOM PLATE

P.T. WOOD FOOTING PLATE

Figure 6-42   Wall openings

also aids in *diverting water* along the wall down into the gravel bed. With all the plywood joints sealed, the face of the plywood is covered with a continuous polyethylene sheet (Fig. 6-43).

A polyethylene sheet 150 μm thick is lapped a *minimum* of 150 mm and caulked at all joints. The top edge is caulked and a **treated plywood strip** 300 mm wide is set 100 mm *above* grade level, extending 200 mm below the finished grade. Applied over the polyethylene, it protects and fastens the top edge in place. **Caulking** along the *top edge* of the plywood strip is also required.

Extended down to the gravel bed, the polyethylene **does not** go out over the

gravel nor **under** the footing plate. Protection of the polyethylene is of prime importance during the backfilling. A protective covering such as fibreboard or cardboard are two coverings that may be placed over the film before backfilling.

Basement Floors

The basement floor can be a wood sleeper floor, suspended wood floor or concrete slab floor. In all cases, the floor **must** be placed to resist the *pressure* exerted from the surrounding soil at the bottom of the wall. The top of the wall is *stabilized* by the floor attached to it at that point. Both the basement and first floor should be installed **before** backfilling commences. If the floors are not in place, bracing is re-

STRIP BONDING OF P.T. PLYWOOD AND POLYETHYLENE, PLUS STRIP BONDING OF POLYETHYLENE AND COVER PLATE

P.T. PLYWOOD SHEATHING

TOP EDGE OF COVER PLATE CAULKED

P.T. PLYWOOD COVER PLATE

FINISHED GRADE SLOPED AWAY FROM BASEMENT WALL

MAXIMUM IMPERVIOUS FILL

FREE DRAINING GRANULAR FILL

ROCK FILL

NOTE: POLYETHYLENE STOPS AT THE BOTTOM EDGE OF FOOTING PLATE. **NEVER** EXTEND UNDER FOOTING PLATE

EXCAVATION LINE GRANULAR BED

PROTECTIVE COVERING

P.T. PLYWOOD JOINT AND SURFACE CAULKED

BLOCKING AT PLYWOOD JOINT

POLYETHYLENE

P.T. WOOD STUDDING

P.T. BOTTOM PLATE

P.T. FOOTING PLATE

Figure 6-43   Sealing the exterior face of the foundation

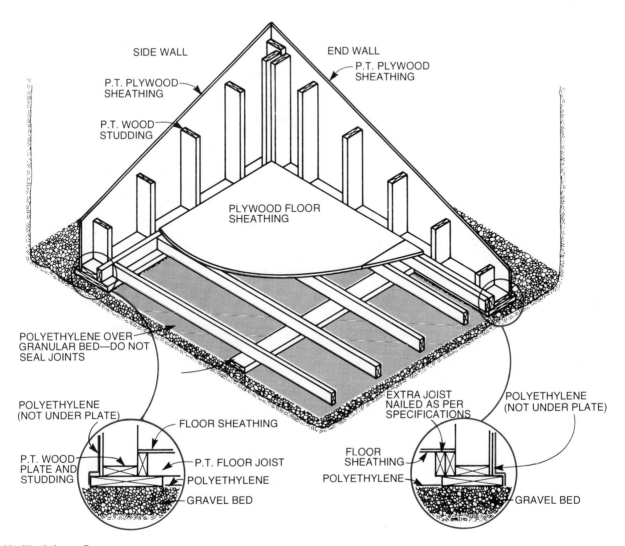

Figure 6-44   Wood sleeper floor

Wood Sleeper Floor

quired to prevent any movement of the walls.

Before beginning to place the wood floor a 100 μm polyethylene membrane cover is laid over the gravel bed. Individual sheets should be no more than 1200 mm wide with **no more** than a 100 mm lap at the joins. Allowance has to be made so that if water is spilled and seeps through the floor, it can work down into the gravel. In no way do you want the polyethylene to form a continuous moisture barrier. **Wood sleepers** (minimum 38 × 89 mm) are laid *flat* over the gravel bed. The **spacing** of the sleeper is dependent on the *size* of floor joist used and the *maximum span* accepted by the code under Residential Standards. The top face of the sleeper must be *level* with top of the footing plate.

Floor joists are cut and end-treated where required to fit between the opposing walls. The ends of the joist butt *against* the header joist, which is used to distribute the load and act as a firestop and subfloor support. A *minimum* bearing of 38 mm is required for the joist ends. If the individual floor joists are joined, the joint *must* occur over a sleeper and the ends *butted* together to resist imposed loads transferred from the outside walls into the floor joist (Fig. 6-44).

Special framing to resist the *lateral* pressures must also be provided for walls running *parallel* to the floor joist. This is accomplished by doubling the joist next to the wall, and installing blocking at right angles to the floor joist *directly* opposite the wall stud. The doubled joist and blocking at this point provide the increased nailing area. Extra nailing of the subfloor along this area completes the requirements (Fig. 6-44).

Plywood used on the floor acts like a *diaphragm* to resist lateral pressures exerted from the soil. Plywood thicknesses must conform to the building code requirements for subflooring under wood-frame construction. To prevent the floor of the building from buckling due to the lateral pressure exerted, a **centre beam** carrying the inner ends of the first floor joist is substituted for a **bearing wall** which sits on *top* of the sleeper floor. When this occurs, the P.T. wood sleeper, which is placed under the inner ends of the floor joist of the basement floor, becomes a footing plate. It must therefore be sized accordingly. **All** materials, including the plywood used for the subfloor in a sleeper floor must be *preserved wood*.

Suspended Wood Floor

Preparatory work for a suspended wood floor is *similar* to that for a sleeper floor in regard to polyethylene application. With the height of the floor established over the crawl space, a 38 × 89 mm **ledger** is nailed to the perimeter wall at the correct height. Support for the *inner ends* of the floor joist is provided by a **dwarf wall**, built to come flush with the *top* of the ledger. As with sleeper floors, the floor joist size must conform to the building code and must butt against the wall studs, while interior ends sitting on the dwarf wall must butt against one another. End walls running parallel to the floor joist must have extra joists, be blocked at right angles to the studding, and require additional nailing as do sleeper floors.

When a suspended floor system is used, the **exterior walls** on a full basement are often 600 mm *longer* than on a

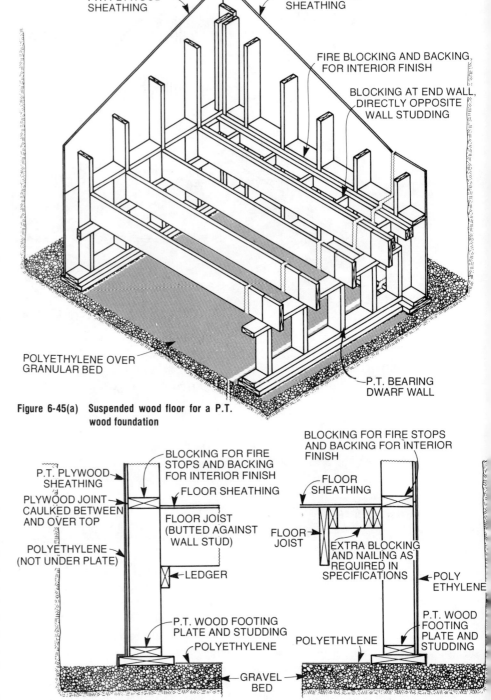

Figure 6-45(a)   Suspended wood floor for a P.T. wood foundation

Figure 6-45(b)   Floor joist at right angles to the wall studs

Figure 6-45(c)   Floor joist parallel to the wall studs

normal basement. This requires an *extra* row of plywood sheathing on the frame. If the **joint** is made to coincide with the *top* of the floor joist when it is installed, the full width **block** at this point provides *fire blocking* and blocking for both the *exterior* sheathing and the *interior* finish when applied (Fig. 6-45).

As with the sleeper floor system, a bearing wall is recommended instead of the conventional beam to carry the first floor joist. Since the crawl space is vented, the only materials that need to be constructed of preserved wood are the members which support the floor; all others (floor joist, plywood) can be built using regular grade construction material.

Concrete Slab Floor

When placing a concrete slab in conjunction with a wood foundation, the pouring, finishing and curing is carried out using the same procedure as discussed previously. Before pouring can take place, however, preparation for the concrete slab (which differs slightly from a concrete slab poured in conjunction with a concrete or masonry wall) must occur.

The gravel bed is first covered with 150 μm polyethylene. To transmit the lateral force of the soil into the concrete slab, it is necessary to pour the slab *above* the lower ends of the wall studding. This distance above the ends of the wall studs is governed by the depth of backfill and the type of wood used. A 19 mm thick treated board is nailed to the perimeter wall at the height of the finished concrete slab. This board becomes the outer screed while intermediate screeds must be set as required (Fig. 6-46).

Main Floor-Foundation Wall Connection

Just as the basement floor requires certain precautionary measures to prevent move-

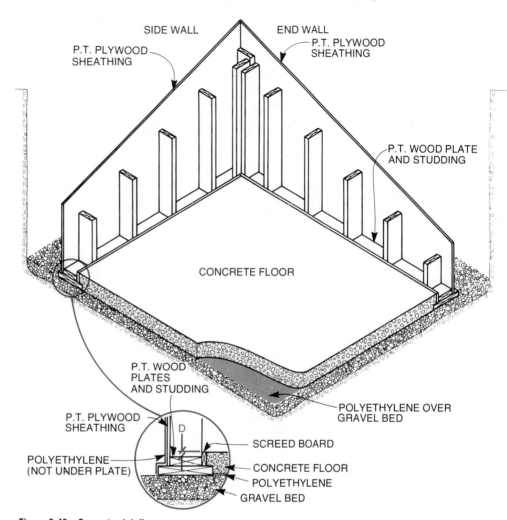

Figure 6-46   Concrete slab floor

ment at the bottom of the wall, the top of the foundation must also be *anchored* to the first floor joists to resist lateral soil pressure.

With the floor joist resting on the *top* of the wall plate as shown in Figure 6-47, the header and floor joist are nailed into the top plate. Some builders feel that these nails alone are not sufficient to resist lateral pressure and are using a galvanized steel strap to give additional support. This strap is fastened to the inside face of the wall stud, bent up over the top plate and back up onto the exterior face of the header. This same method is also used on the end wall running parallel to the floor joist, where blocking and extra nailing, similar to that used on sleeper or suspended floor systems, is required.

An alternate method of fastening the joist to the wall (Fig. 6-48), occurs when the floor joist is set *inside* the foundation wall. If the floor joists are butted against the wall stud, the header may be omitted.

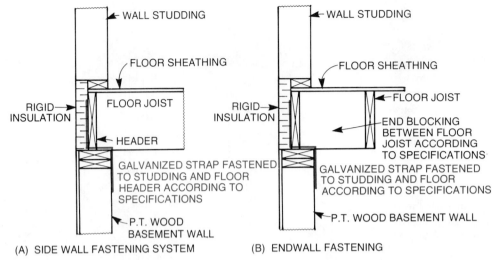

Figure 6-47 Floor joist fastened to the top of the foundation wall

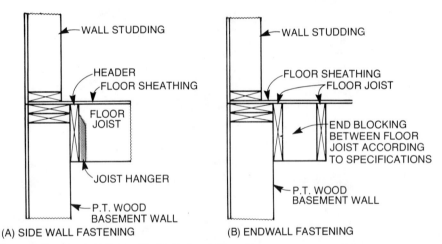

Figure 6-48 Floor joist fastened inside the foundation wall

Although normal nailing is accepted for fastening the ends of the floor joist, it is recommended that joist hangers be used for added support. As in the other systems, end walls must be blocked at the studs and extra nailing is required.

A **stairwell** framed alongside or within 1200 mm of the side wall and 1800 mm of the end wall requires special consideration to provide the required lateral support. Special framing requirements for the stairwell, along with any additional support required for a split entry, must be designed to provide lateral resistance to soil loads at the top of the foundation wall. Close adherence to the job specs is required.

DRAINAGE AND BACKFILL

Once the foundation for the building has been completed, it is usual practice to fin-ish the outside of the wall and footing area so that the backfilling of the excavation can be completed. The one *prime* objective when working on the exterior of the wall is to provide a basement area that will remain dry all year.

A dry basement is possible as long as the water movement is controlled. **Water control** is necessary for both *surface* and *sub-surface* water control. Surface water exists because of the (a) rainfall (b) melt-water from snowfall and (c) watering of the property, especially in urban areas.

Proper grading of the lot around the building is the easiest way to control surface water. The minimum gradient away from the building should be 1 mm in 12 mm or more. The finished grade on site should be such that all water will flow outward and around the building toward the front or back of the lot, depending on the local drainage pattern. Other ways of dispersing surface water away from the building include:

(a) The use of eavetroughing, down-spouts, extensions to downspouts or splashblocks.

(b) Making sure that any driveways, sidewalks, patios, etc. do not impede the flow of water and that they slope away from the foundation wall.

(c) Making allowance for settlement of the backfill.

(d) Finishing the backfilled area next to the foundation with soils which are impervious (i.e.: clay soils).

The other form of water to be controlled is *sub-surface* water. If it is allowed to build up outside the foundation wall, it will pressurize enough to be forced through the wall area or up through the basement floor. Referred to as *hydrostatic pressure*, it must be relieved to maintain a dry basement. To accomplish this, a drainage system should be installed.

Drainage

Unless proven unnecessary, drainage around all residential buildings must be provided by installing either *drain tile* around the outside perimeter of the foundation wall, or by a *gravel* or *crushed rock* collector bed. Although considered today by many as an inferior drainage system, drain tile is still widely used for control of sub-surface water. It is available in many forms, from individual **clay tile** (Fig. 6-49), to the latest corrugated **plastic drain tile**, and is placed around the outside perimeter of the footing at the bottom of the excavation.

When drainage tiles have open joints, as is the case with clay tile, the joints must be left from 5 to 10 mm wide, allowing the water to enter the system. The top part of the joint is covered with polyethylene or No. 15 felt paper strips to prevent sand or silt from entering and plugging the system (Fig. 6-49). With composition or corrugated plastic tile, the holes must be placed *downward* to allow the water to enter. All drain tile must have a minimum inside diameter of 100 mm.

The perimeter of the building surround-

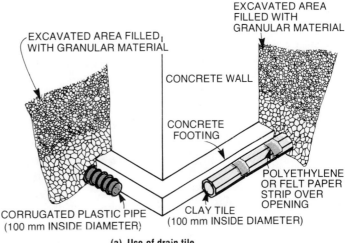

(a) Use of drain tile

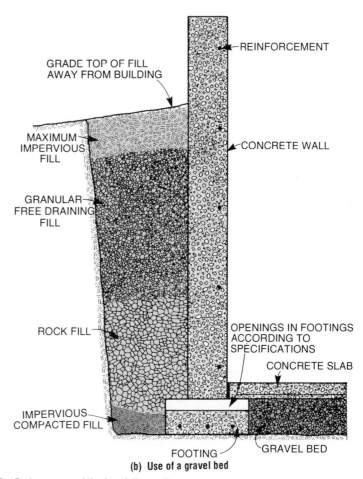

(b) Use of a gravel bed

Figure 6-49   Drainage around the foundation wall

ing the drain tile can be connected to the sewage disposal system. This can be either a direct connection into the sewage line (usually via the floor drain, a sump pit or a drywell). The alternative is to connect the drain tile to the storm sewer system. Check with local authorities regarding code requirements to determine which system is allowed. The entire drainage system around the perimeter wall is then covered with a minimum of 150 mm of crushed rock. This coarse material near the system assures quick disposal of the water before pressure can build up.

The **gravel bed**, which is a requirement of the preserved wood foundation system, is considered by many a superior system to drainage tile. Since the introduction of the 1980 National Building Code, it is now accepted as an alternate system—eliminating the use of drain tile system. When using a concrete footing, it is necessary to leave *weep holes* at intervals to allow the water from the exterior of the wall to flow into the gravel bed under the concrete slab or floor. Gravity causes the water to flow into the sump pit (Figs. 6-39 and 6-49). Ideally, the gravel bed should extend under the concrete footing just as it does under the preserved wood footing plate.

Waterproofing and Damp-proofing

Where hydrostatic water pressure in the subsoil exists, the exterior face of the wall must be *waterproofed*. This is accomplished by applying two coats of bituminous (tar) material from the footing to just below grade level. Brushed, mopped or sprayed on, it must form a complete seal over the concrete or masonry wall face. *Damp-proofing* consists of the same material, applied the same way, but with only one coat.

In the preserved wood foundation system, the polyethylene barrier applied to the exterior wall face performs the same purpose as the bitumen coating on the concrete wall. However, it must be stressed again that the polyethylene does not return under the footing plate. This is so in the event that moisture forms between the polyethylene barrier and the pressure-treated plywood, it can get away into the gravel bed.

Backfill

With the exterior wall protected, backfilling can be completed. This can take place prior to installing the basement and first floors when using concrete foundations. In masonry or preserved wood founda-

tions however, it is strongly recommended that backfilling occur after the two floors are in place to provide adequate support for the walls.

If the area around the foundation has poor drainage quality, the backfill used should be a *free draining* type (preferably coarse sand or gravelly soil brought to within 300 mm of the final grade. The top layer of soil should be a clay soil similar to that described under the control of surface water so that the water drains away (Fig. 6-50). In no case should harmful debris, boulders over 250 mm, or frozen earth be used as backfill material.

Any backfill used around foundations should be placed *uniformly;* this is especially critical with masonry or preserved wood foundations. Equipment should work at right angles to the foundation, pushing the fill into the excavated area rather than dumping it in. In no case should equipment be used parallel to the wall while backfilling or to compact the fill.

No matter what material is chosen, or which system is used, properly constructed footings and foundation walls which are adequately protected and drained will provide the owner with a dry, usable basement area.

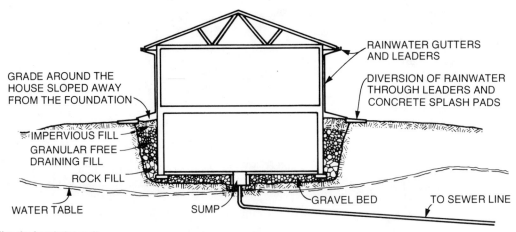

Figure 6-50   Backfilling the foundation wall

REVIEW QUESTIONS

6—FOOTINGS AND FOUNDATIONS

Answer all questions on a separate sheet

A. Replace the Xs with the correct word.

1. That portion of the foundation which distributes the load of the house is called the XXXXXX.
2. Foundation work must be designed to meet XXXXXX conditions and ensure that they are below XXXXXX for the area in which the house is being built.
3. Three types of footings commonly found under house foundations are XXXXXXX, XXXXXX and XXXXXX.
4. Enlargements in footings occur where XXXXXX, XXXXX and/or XXXXXX are required.
5. Footing positions can be easily transferred from the batter boards through the use of builder's lines and a XXXXX.
6. XXXXX are formed into the footing to provide a means of locking the concrete footing and wall together.
7. Stakes driven into the excavation floor support and position XXXXX.
8. Footings for bearing walls are normally constructed with a XXXXXX section, on which the wall sits.
9. Foundation wall height, thickness and construction depend upon (a) XXXXXXX, (b) XXXXXXX, (c) XXXXXX and (d) XXXXXX.
10. XXXXXX are vertical footings designed to transmit loads from the house.
11. Vertical footings are usually constructed of XXXXXX, XXXXXXX or XXXXXX.
12. The vertical footing which is easily adapted to house construction is the XXXXXX pile.
13. Wood pilings are normally cut off and capped to prevent XXXXXX from occurring.
14. In areas where soil must remain frozen to retain its stability, upper portions of pilings become XXXXXX. Beams or girders are then fastened to the extended pile and the building is suspended above the frozen ground.

B.

15. List the different types of foundation walls used in house construction. Sketch each type and indicate the important features of each one.

C. Select the correct answer

16. A major advantage of mechanical form ties is their ability to
    (a) Add reinforcement to the concrete
    (b) Keep the form vertical
    (c) Maintain a predetermined wall width
    (d) Supply a means of fastening wall reinforcement
17. Concrete forms are designed to
    (a) Be reused a number of times
    (b) Be easily transported from job to job
    (c) Support the floor joist in joist filled construction
    (d) Sit on the footing while the wall is being poured
18. Members used to stiffen and align the formwork are called
    (a) Sheathing
    (b) Tie bars
    (c) Walers
    (d) Bracing
19. Of the following, which factor produces the greatest pressure on forms while pouring?
    (a) Wall thickness
    (b) Wall height
    (c) Workability of the concrete
    (d) Rate of pour
20. Concrete pressure exerted on forms is greatest at
    (a) The bottom of the form
    (b) The centre of the form
    (c) The top of the form
    (d) All points throughout the form
21. Which of the following statements is true of unframed panel forms?
    (a) Framing material is not required with this system
    (b) Short-end form ties are used
    (c) Bracing is used on both inner and outer forms
    (d) Not normally used on house construction
22. "Roughbucks" are placed in the wall
    (a) While concrete is being poured
    (b) After the concrete is poured
    (c) Before the forms are constructed
    (d) While forms are being constructed
23. While pouring the concrete, screeds
    (a) Prevent movement of the joist
    (b) Provide a form for the concrete
    (c) Give a level point which concrete can be levelled to
    (d) Keep the reinforcement in position
24. Using the following information, find the number of cubic metres of concrete used in the foundation footing, wall and floor:
(a) Building size—10 × 13 m
Wall thickness—225 mm
Wall height—2.4 m
Footing size—600 × 250 mm
Slab thickness—100 mm
Find
    (i) Centre line length (C.L.L.)
    (ii) Concrete required for the footing
    (iii) Concrete required for the wall
    (iv) Concrete required for the floor

(b) Building size—8.9 × 11.8 m
Wall thickness—200 mm
Wall height—1800 mm
Footing size—450 × 200 mm
Slab thickness—75 mm
Find
   (i) Centre line length (C.L.L.)
   (ii) Concrete required for the footing
   (iii) Concrete required for the wall
   (iv) Concrete required for the floor

E. Mark as either TRUE or FALSE

25. When using wood for constructing a basement, all materials must be pressure-treated.

26. All pressure-treated material cut on-site must have the cut surface field-treated.

27. On a P.W.F., it is recommended that all footings be constructed on a gravel bed.

28. Sealing materials (polyethylene) extending down the outside wall, return under the bottom of the wall.

29. Caulking must be applied at all exterior joints on a wood foundation.

F. Replace the Xs with the correct word(s)

30. With P.W. foundations, the basement floor may be constructed in three different ways:
   (a) XXXXXXX, (b) XXXXXX and (c) XXXXXX.

31. Floor joists are anchored to the top of the P.W. foundation to prevent XXXXX pressure.

32. Along with nails, pressure from the foundation wall is taken up at the floor joist by a XXXXX strip.

33. Drainage around a foundation can be accomplished by using a XXXXXXX or XXXXXXX as a drainage system.

34. Exterior walls of a foundation must be XXXXXX to resist hydrostatic water pressure.

35. Backfill around a foundation must be placed XXXXXX.

7 FLOOR FRAMING

In wood frame house construction the floor joists are the main framing members. They support the flooring materials as well as the exterior walls and interior partitions. The function of the other members of the floor frame is to support the floor joists. The outer ends of the floor joists rest on a sill plate which in turn rests on the foundation wall and is fastened to it. Sometimes the outer ends of the floor joists are embedded in the foundation wall; this is called *joist filled construction*. The inner ends of the floor joists are supported by a centre beam or girder which runs the length of the building. The beam is supported at the ends by the foundation wall, but must be supported with columns along its length. The centre beam and columns are sometimes replaced by a load-bearing wall.

BEAMS OR GIRDERS

Floor joists usually are placed across the width of the building. The width of most houses is such that it is not possible or practical to use floor joists of sufficient length to span the entire distance without support. An intermediate support therefore must be used. The most common method is to use a beam or girder.

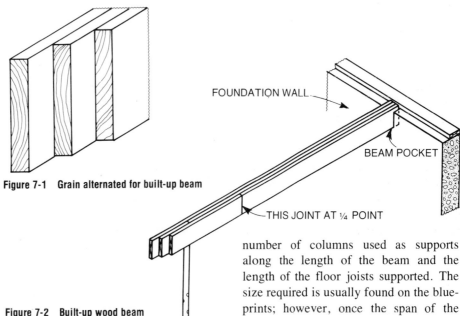

Figure 7-1   Grain alternated for built-up beam

Figure 7-2   Built-up wood beam

Types of Beams

There are several types of beams used in construction today; perhaps the most common is the *built-up beam* because it can be fabricated on the job with readily available material. It consists of three or more pieces of 38 mm stock set on edge and spiked together from each side with 89 mm nails. The size and number of pieces of stock used will depend on the number of columns used as supports along the length of the beam and the length of the floor joists supported. The size required is usually found on the blueprints; however, once the span of the floor joists and the spacing of the columns is known, the size of the beam may be determined from the residential standards.

When placing the pieces of stock together before nailing, the grain of the wood should be alternated to prevent the beam from twisting after being placed (Fig. 7-1).

The joints should occur over columns or within 150 mm of the quarter points in the span (Fig. 7-2).

Another type of wood beam is the **solid type**. This is not used very often because

127

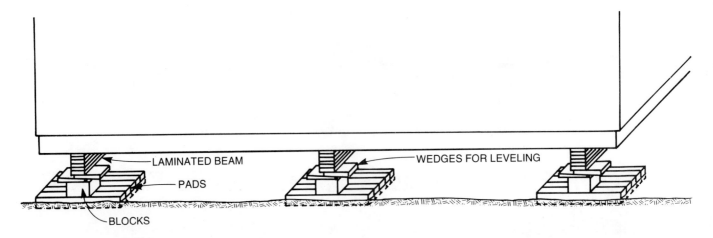

LAMINATED BEAM

WEDGES FOR LEVELING

PADS

BLOCKS

Figure 7-3   Laminated beams used under a prefabricated house

of inavailability and it also has a tendency to twist easily.

Laminated wood beams are also used; they consist of layers of 19 or 38 mm stock placed on the flat and factory glued to form a beam. The number of laminations depend on the size of the beam required. Laminated wood beams are often used with prefabricated buildings where there are no basements. They are usually pressure-treated to prevent decay from contact with the soil (Fig. 7-3).

Steel beams are becoming quite common in house construction because they can span longer distances, eliminating the need for many support columns. There is also no shrinkage or warping. The standard ''I'' beam and wide flange beam are the most commonly used styles.

COLUMNS

The beams or girders are supported at the ends by the foundation wall. Intermediate supports may be required between the end supports and are usually wood or steel. They are set under the beam, and spaced according to the size of the beam and the load it is to support.

The most common type of column in use today is a round **steel pipe** fitted with steel plates at both ends. The column is made of two lengths of structural steel, one telescoping into the other and adjustable for length by means of holes and a large pin. The top plate should be as wide as the beam it supports and is situated on top of a large adjusting screw at the top of the pipe. After the approximate length of column is arrived at by means of the pin on the telescoping pipes, the final position is achieved by adjusting the top screw. One advantage of this adjusting screw is that it can be moved any time during or after construction to compensate for shifting due to lumber shrinkage or soil movement (Fig. 7-4).

If **wood columns** are used they should be at least 140×140 mm and may be either solid, glue-laminated, or built-up of 38 mm lumber. Wood columns should be the same width as the thickness of the beam they support. Frequently, there is a *bolster* at the top of the wood column which helps to distribute the load (Fig. 7-5). Some method of providing damp-proofing at the bottom of the column is

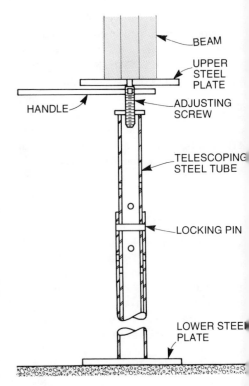

BEAM

UPPER STEEL PLATE

HANDLE

ADJUSTING SCREW

TELESCOPING STEEL TUBE

LOCKING PIN

LOWER STEEL PLATE

Figure 7-4   Steel column

required; this can be 50 μm polyethylene or number 50 roll roofing placed between the column and the concrete base (Fig. 6-4).

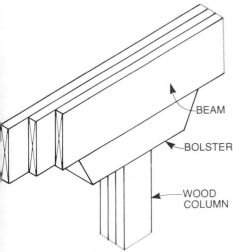

Figure 7-5   Wood column and bolster

Figure 7-6   Truss joists

FLOOR JOISTS

Floor joists are the main members of the floor frame and must be of sufficient strength to provide support for the floor and the load imposed upon them, both live and dead. The floor joists must also give rigidity to the floor frame to prevent cracking of wall and ceiling finishes due to vibration from people walking or moving furniture and equipment.

Types of Floor Joists

Wood joists are usually 38 mm thick and come in widths of 140, 184, 235 or 286 mm. The width will depend on the length between supports (known as span), the joist spacing, species and grade of lumber used and the load imposed. The size required is usually given on the blueprints; however, if all the information regarding span, spacing, load, etc. is known, the size can be determined from tables found in the residential standards. Joist spacings are usually 300, 400, or 600 mm on centre so that plywood panels may be used for floor sheathing. Joist spacings of 400 mm are the most common.

Truss joists are becoming increasingly

popular for house construction because they are able to span the complete width of the building, thus eliminating the need for a centre beam. The open webs allow for the installation of plumbing, heating and electrical services within the floor system. This gives more headroom in the basement. Truss joists are usually spaced at 600 mm on centre (Fig. 7-6).

FLOOR JOIST INSTALLATION

There are several acceptable methods of attaching floor joists, both at the foundation wall and at the beam or load bearing wall.

Foundation Wall and Floor Joist Connection

One acceptable method of attaching floor joists to the foundation wall is to rest the outer ends of the floor joists on a sill plate that is anchored to the foundation wall. Another method is to embed the ends of the joists in the concrete foundation wall.

Box-Sill Method

With this method the ends of the floor joists are kept in alignment with a framing member (header joist) the same dimension as the floor joists. The header joist and the joist ends are toe-nailed to a sill plate which is anchored to the foundation wall. Various methods may be used to frame the joists into the foundation wall as shown in Figure 6-26, 6-33, 7-7.

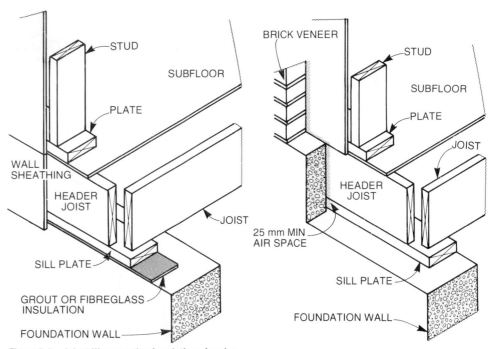

Figure 7-7   Joist/sill connection for platform framing

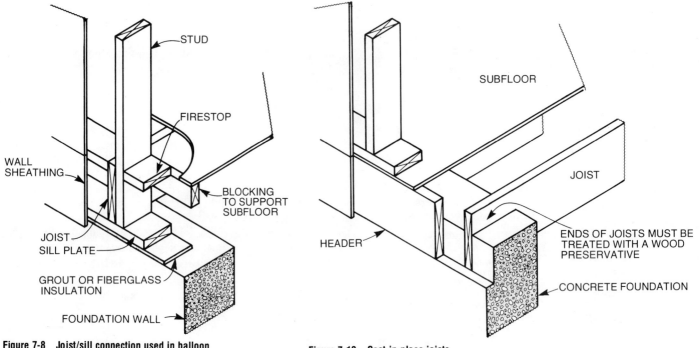

Figure 7-8   Joist/sill connection used in balloon frame construction

Figure 7-10   Cast-in-place joists

When a building is framed using the balloon type of framing, the joist-foundation connection is slightly different. A header joist is not used and the floor joists are fastened to the studs and toe-nailed to a sill plate anchored to the foundation wall (Fig. 7-8).

With the increased emphasis on fuel conservation every effort is being made to improve construction methods to help prevent the loss of heat in buildings. One of the areas most frequently neglected in the past was the connection of the floor frame to the foundation wall. One method of improving this situation is illustrated in Figure 7-9; this can be used with either a concrete foundation or a pressure-treated wood foundation.

Joist Embedded Method

This method is used with concrete foundations. The beams, joists and headers are positioned on top of the forms before

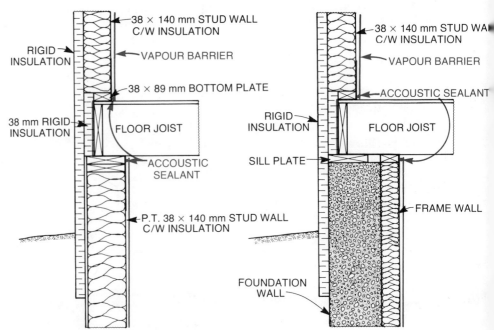

Figure 7-9   Joist/sill connection designed to conserve heat

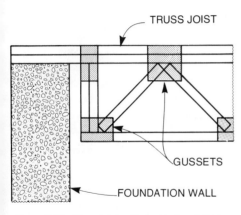

Figure 7-11   Truss joists attached to foundation wall

the concrete is placed. The ends of the beam and joists must be treated to prevent decay (Fig. 6-27, 7-10).

Truss Joists

Truss joists are attached in the same manner as regular wood joists (Fig. 7-11).

Joists Fastened to Centre Beam

The easiest method of fastening joists to a centre beam is to have the joist ends rest on top of the beam and toe-nailed. The joists may be lapped, or butted and scabbed as in Figure 7-12.

When more clearance is desired in the basement, the floor joists may be framed into the side of the beam. They can be supported on a *ledger strip* which has been securely fastened to the beam. When this is done, the joists should be tied together over the beam with a 38 ×38 mm splice at least 600 mm long. Care must be taken to allow at least a 12 mm space be-

tween the splicing scab and the top of the beam to allow for shrinkage (Fig. 7-13). Joists may also be supported by mechanical *joist hangers* attached to the beam (Fig. 6-48).

Wood joists connected to a steel beam requires the fastening of a wood plate to the beam first. The joists may be placed on top of the beam as in Figure 7-14 or framed into the side of the beam as in Figure 7-15, 7-16.

Layout and Placement of Floor Joists

Floor joists are usually laid out 400 mm on centre. When the span changes in the floor system, centres are sometimes changed to 300 or 600 mm. For example, if 235 mm deep floor joists are used for a set span at 400 mm o.c. and there is a projection in the building causing the span to be increased by 600 mm; rather than increase all the floor joists to 286 mm deep because of the increased span, the joist spacing could be decreased to 300 mm. Header joists are marked at

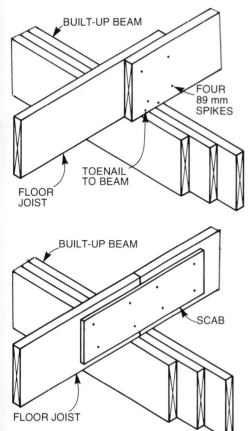

Figure 7-12   Joists on top of beam

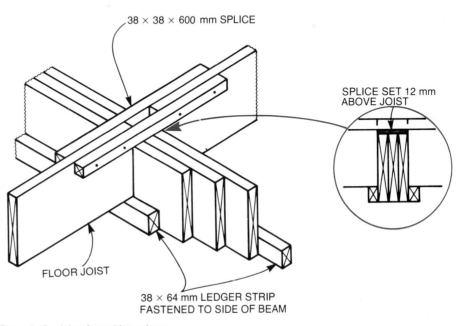

Figure 7-13   Joists framed into a beam

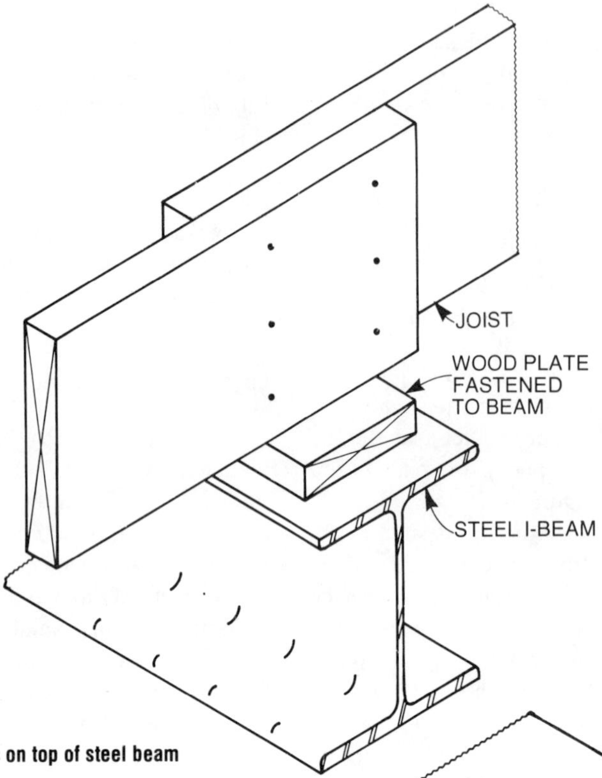

to the regular joists, called *headers*. The regular joists that have been cut are called *tail joists*. Joists on either side of the opening are called *trimmer joists*. When the header is longer than 1200 mm, it should be doubled. When the header joist is longer than 800 mm, the trimmer joists should be doubled. Figure 7-18 shows a floor frame with a typical opening for a stairwell.

Joist hangers are available for single and double framing members and are used to support headers and tail joists. Joists that have a slight bow edgewise should be placed with the crown up, causing them to straighten when the subfloor and normal floor loads are applied. When joists are not the same depth they must be either notched or shimmed at the ends to ensure a level floor; this is called *sizing*.

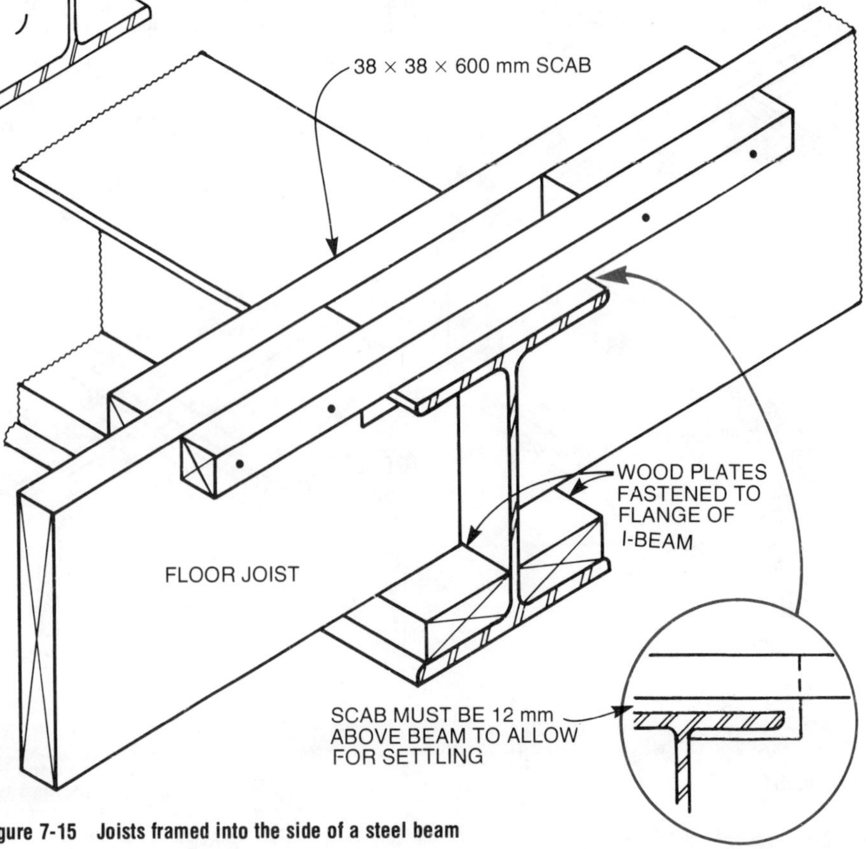

Figure 7-14   Joists on top of steel beam

the required centres, but remember that the first spacing will be 20 mm less than the required spacing (Fig. 7-17).

Joists should be doubled under *non-loadbearing partitions* that are over 1800 mm in length and which contain openings that are not full ceiling height. Where such partitions contain no openings or have openings that are full ceiling height the joists need not be doubled. Doubled joists may be separated up to a maximum of 200 mm by blocking if the blocking is no less than 38 × 89 mm lumber spaced not more than 1200 mm apart.

When it is necessary to cut some of the regular joists to provide openings for stairwells, chimneys or fireplaces, etc., special framing procedures are required. Regular joists are cut and framed into framing members running perpendicular

Figure 7-15   Joists framed into the side of a steel beam

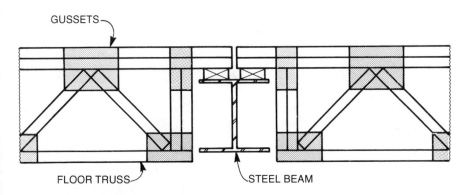

Figure 7-16   Truss joists and steel beam

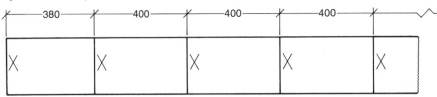

Figure 7-17   Header layout

Joist Restraint

Joists must be prevented from twisting. This is prevented at the foundation wall by the header joist, and at the beam by toe-nailing to the beam. To prevent twisting between these points, bridging is installed at intervals of not more than 2100 mm between supports. Intermediate support may be provided by a 19 × 89 mm continuous wood strap nailed to the underside of each joist and attached to the end joists by 19 × 64 mm or 38 × 38 mm cross-bridging cut to an angle to fit diagonally between the joists, by 3 × 25 mm continuous steel

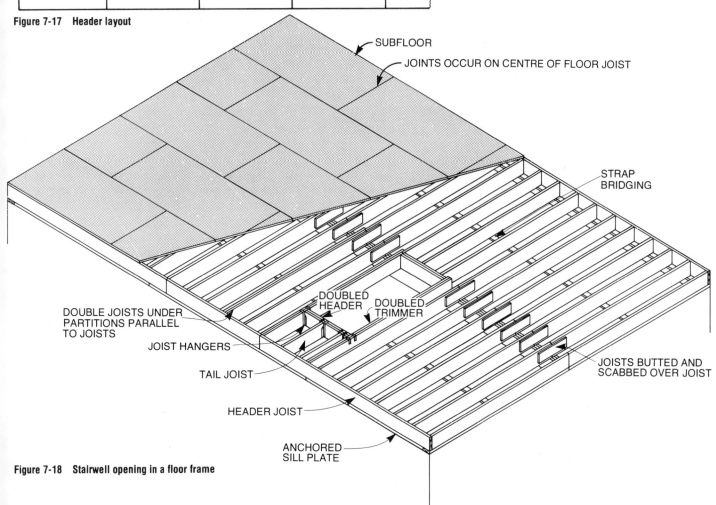

Figure 7-18   Stairwell opening in a floor frame

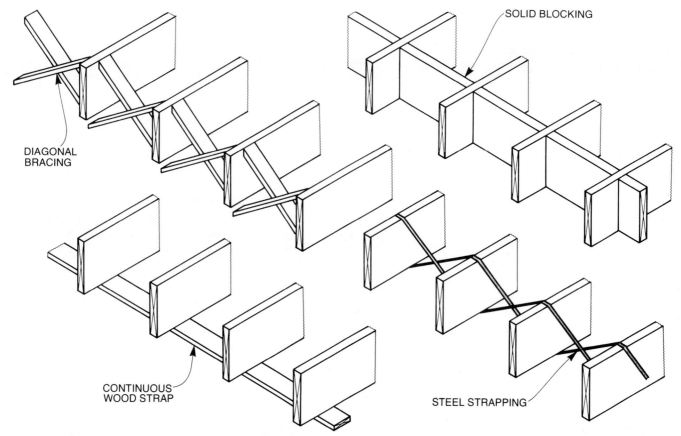

Figure 7-19   Joist restraint

Layout for Cross-Bridging

To lay out and cut diagonal bracing two measurements must be known—the distance between joists and the depth (235 mm) of the joists. When joists are spaced 400 mm o.c., the space between will be 362 mm. Place the framing square on the bridging stock as shown in Figure 7-20, and mark along the tongue. Also mark along the body to indicate the length. Reverse the square, and using the same figures, lay out the second cut, marking along the tongue (Fig. 7-21). The layout lines marked along the tongue

of the square should be parallel. Cut the piece of bridging and try it between the floor joist; if the bridging is the correct size, use this as a pattern and cut as many as required.

The top of the cross-bridging is securely nailed in place in the predetermined location. A chalk line snapped across the top of the joists will aid in keeping the bridging straight. The bottom of the bridging is nailed in place when the house is nearly finished, allowing the floor joists that were placed with the crown up to level off.

Cantilevered Floor Framing

Floor joists sometimes project beyond the foundation wall to provide additional living space without the added work and expense of forming the foundation walls to

match. If the cantilever is parallel to the regular floor joists, the joists simply extend past the foundation wall. If the cantilever is at 90° to the regular floor joists, then the floor framing is more complicated, as illustrated in Figure 7-22.

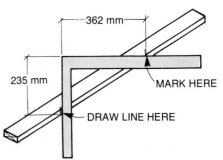

Figure 7-20   Layout for bridging using 235 mm joist at 400 mm o.c. (Step 1).

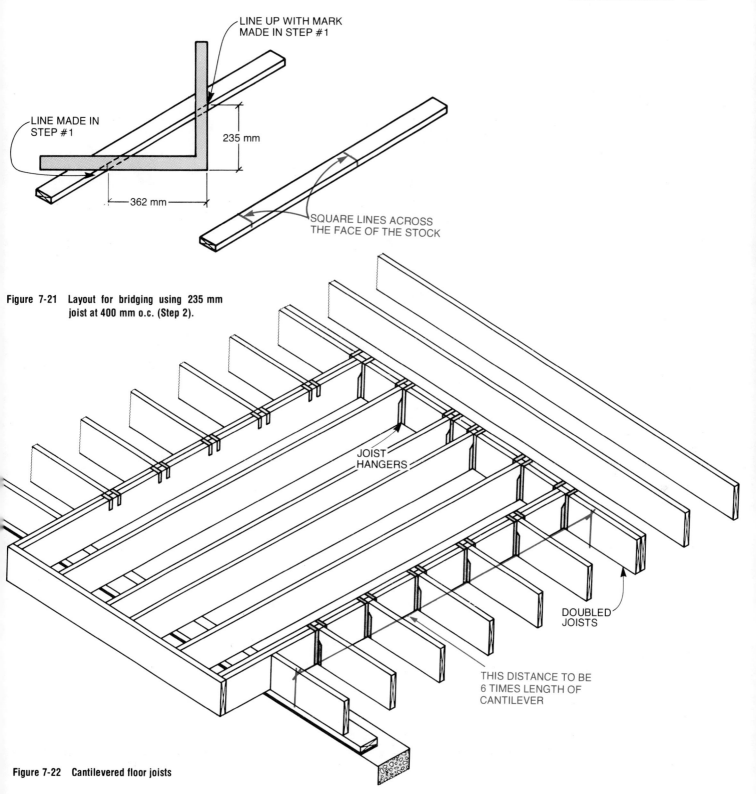

LINE UP WITH MARK
MADE IN STEP #1

LINE MADE IN
STEP #1

235 mm

362 mm

SQUARE LINES ACROSS
THE FACE OF THE STOCK

Figure 7-21   Layout for bridging using 235 mm joist at 400 mm o.c. (Step 2).

JOIST
HANGERS

DOUBLED
JOISTS

THIS DISTANCE TO BE
6 TIMES LENGTH OF
CANTILEVER

Figure 7-22   Cantilevered floor joists

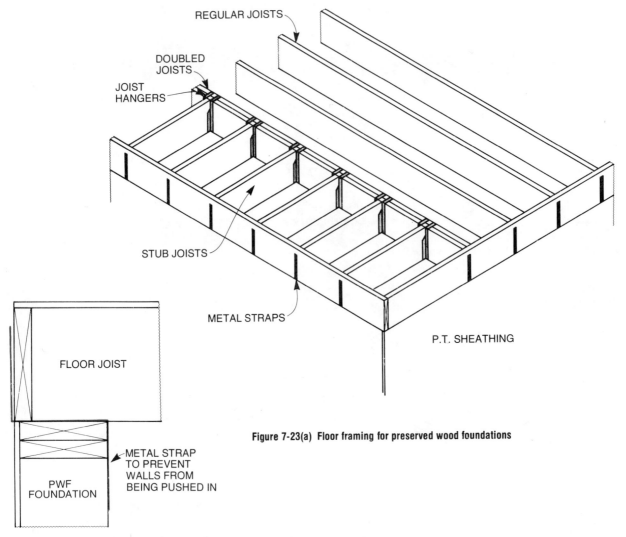

REGULAR JOISTS

DOUBLED JOISTS

JOIST HANGERS

STUB JOISTS

METAL STRAPS

P.T. SHEATHING

Figure 7-23(a) Floor framing for preserved wood foundations

FLOOR JOIST

METAL STRAP TO PREVENT WALLS FROM BEING PUSHED IN

PWF FOUNDATION

Figure 7-23(b) metal strap for lateral restraint

Preserved Wood Foundations

Preserved wood foundations are becoming more popular because of the ease of insulating, thus obtaining a higher RSI rating than conventional concrete foundations. The end wall of a preserved wood foundation, unlike the end wall of a concrete foundation, is very unstable and therefore must be securely fastened to prevent movement in the floor frame. This is achieved by using stub joists at 90° to the regular studs as illustrated in Figure 7-23. Walls must be securely fastened to floor joists to prevent them from being pushed in when backfilling. This is done by the use of 3 × 25 × 450 mm metal strapping nailed to the joists and the foundation wall.

Squaring Floor Frame

Before anchoring the floor frame to the foundation (or forms as in the case of joist filled construction), make certain the floor frame is square. Use a string and blocks to insure the header joists are straight, then adjust the floor frame until the diagonals are equal (Fig. 7-24).

SUBFLOOR

The subfloor provides a solid base for the installation of finish flooring. The subfloor may be solid lumber; either shiplap, tongue and groove, or common boards. When **lumber sheathing** is used the

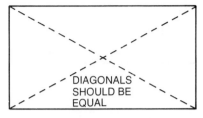

DIAGONALS
SHOULD BE
EQUAL

RECTANGULAR HOUSE

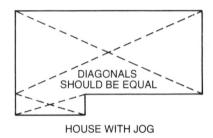

DIAGONALS
SHOULD BE EQUAL

HOUSE WITH JOG

Figure 7-24   Check for squareness

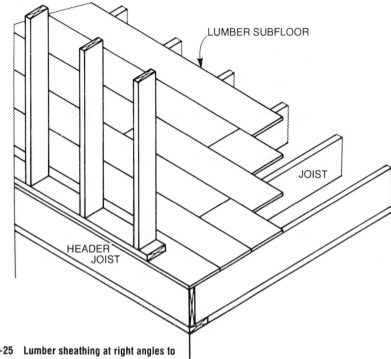

LUMBER SUBFLOOR

JOIST

HEADER
JOIST

**Figure 7-25   Lumber sheathing at right angles to
floor joists**

joints should occur over joists; end joints should be staggered throughout the floor. Boards may be applied at right angles to the joists (Fig. 7-25) or diagonally (Fig. 7-26).

Lumber laid diagonally results in increased waste material because of the angled cuts. It also takes longer to install. When lumber sheathing is used, a panel type of underlay is used to provide a smooth surface for the finish floor. Underlay is usually smooth particleboard or plywood (Fig. 7-27).

Plywood is used extensively for subflooring; it should be installed with the surface grain perpendicular to the joists. Joints should be staggered parallel to the floor joists (Fig. 7-18).

Sometimes only one layer of plywood is used, and acts as both subfloor and underlay. When this method is used the edges should be tongue and groove (Fig.

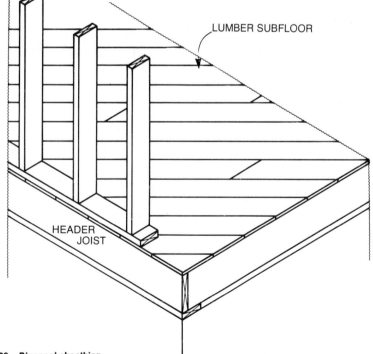

LUMBER SUBFLOOR

HEADER
JOIST

Figure 7-26   Diagonal sheathing

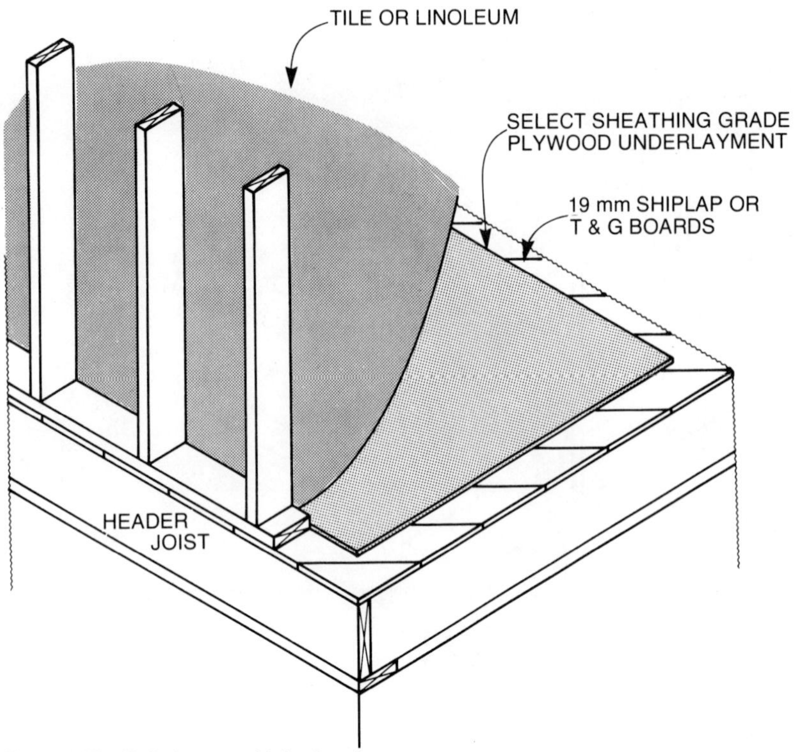

TILE OR LINOLEUM

SELECT SHEATHING GRADE
PLYWOOD UNDERLAYMENT

19 mm SHIPLAP OR
T & G BOARDS

HEADER
JOINT

Figure 7-27   Underlay over shiplap boards

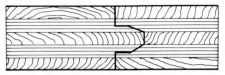

Figure 7-28   Tongue and groove plywood

PLYWOOD
SUBFLOOR

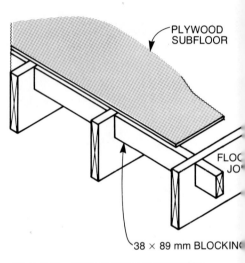

FLOO
JO

38 × 89 mm BLOCKING

**Figure 7-29   Edges of plywood supported by
38 mm blocking**

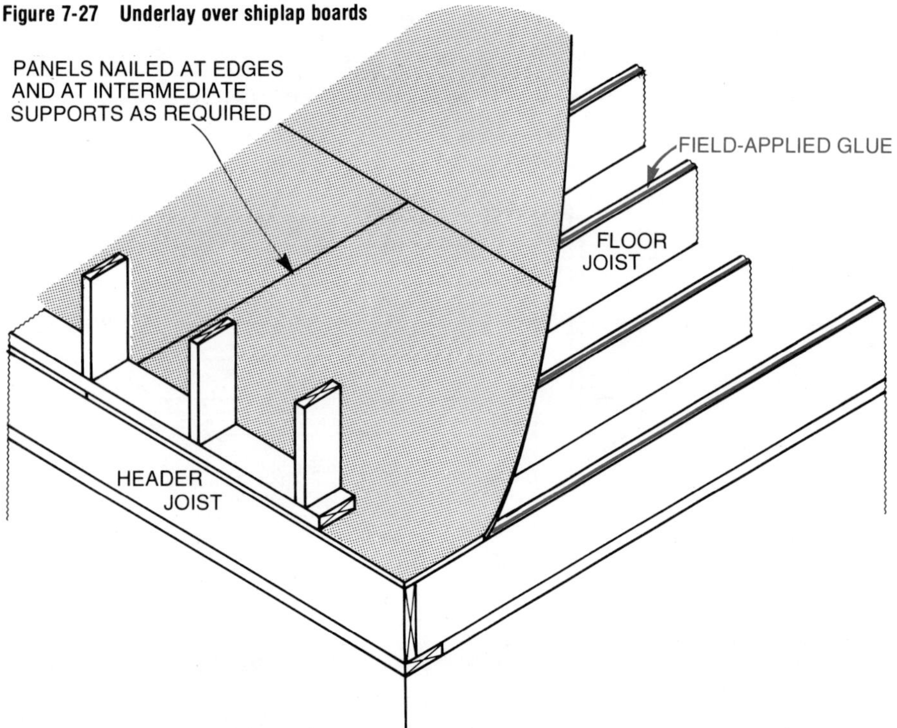

PANELS NAILED AT EDGES
AND AT INTERMEDIATE
SUPPORTS AS REQUIRED

FIELD-APPLIED GLUE

FLOOR
JOIST

HEADER
JOINT

Figure 7-30   Plywood glued to floor joists

7-28), or supported by 38 mm blocking installed between the joists (Fig. 7-29).

Another method of providing a solid floor system is to install a plywood subfloor, with a panel underlay applied at a later date. Annular grooved nails should be used instead of common nails to fasten underlay.

Particleboard is also used as subfloor sheathing, and the installation is similar to plywood.

Another method used for securing plywood subfloor sheathing to floor joists is to glue them together. Several *elastromeric adhesives* suitable for field application are available from building supply dealers. These special construction adhesives are not to be confused with ordinary drywall glues. Most are conveniently packaged in spouted, ready-to-use caulking guns. Some brands are available in bulk for use with portable pneumatic sys-tems. Use only those glues specified for field-glued subfloors and apply according to the manufacturer's instructions (Fig. 7-30). In field-glued plywood subfloors the glue firmly bonds the plywood to the joists, eliminating squeaks due to movement. This method produces a stiffer, better quality floor system.

REVIEW QUESTIONS

7—FLOOR FRAMING

Answer all questions on a separate sheet

A. Select the correct answer

1. What is meant by the term "clear span" of a girder?
   (a) The distance between the opening
   (b) The distance between the joists
   (c) The length of the building
   (d) The distance between supports

2. Trimmer joists are used:
   (a) Under the load bearing partitions
   (b) At the ends of the building
   (c) When bridging is not required
   (d) At the sides of floor openings
   (e) Only when plywood sheathing is to be used

3. When joists are installed it is important to:
   (a) Have the crown up
   (b) Overlap them
   (c) Pick only straight ones
   (d) Put in extra joist

4. Besides carrying the floor load, joists serve what other purpose?
   (a) Holds the bridging
   (b) A more economical method
   (c) Acts as a lintel
   (d) Gives lateral support to foundation walls

5. Floor joists are sitting on, and nailed to, the:
   (a) Sill plate
   (b) Foundation wall
   (c) Sole plate
   (d) Strapping

6. What is the purpose of bridging?
   (a) To keep joists at their proper spacings
   (b) To distribute imposed loads on the floor
   (c) To decrease the number of joists required
   (d) To permit the use of smaller joists

7. When should cross-bridging be nailed at the bottom?
   (a) At the same time as the top
   (b) When the building is otherwise complete
   (c) After the sub-floor is nailed in place
   (d) Before the top is nailed

B. Write full answers

8. Name two types of sill construction and explain each.

9. Joist sizes are determined by three things. Name them.

10. What is a "Ledger Strip" and where is it used?

11. What is meant by the terms:
    (a) Crowning
    (b) Sizing

12. List three types of bearing posts:

13. How are floor joists restrained from twisting?

14. What 4 factors should you consider when laying out flooring members?

C. Mark as either TRUE or FALSE

15. When framing an opening in a floor, the short joists running parallel with the full length joists are called tail joists.

16. Trimmer joists are used only when plywood subfloor is to be used.

17. In residential or small house construction a built-up girder is usually made of two planks in width.

18. A horizontal member that supports the joists is called a bearing post.

19. Floor joists rest on and are nailed to a sill plate.

20. A girder is a horizontal member that supports the joists.

8 WALL AND CEILING FRAMING

WALL FRAMING

The vertical and horizontal members which make up exterior walls and interior partitions are called *wall frames*. A wall frame consists primarily of vertical members called *studs* which are kept in alignment by horizontal members called *plates*. Wall frames serve as a nailing base for interior and exterior wall finishes and support the upper floors, ceiling and roof. When the studs are cut to allow for doors and windows, horizontal members called *lintels* or *headers* are installed to transfer the overhead load to the remaining studs. Interior wall frames are used to divide the area enclosed by the exterior walls into rooms; these are called *partitions*. There are two types of partitions, *loadbearing* and *non-loadbearing*. Loadbearing partitions, as well as serving as room dividers, support floor, ceiling or roof loads. They are framed in the same manner as exterior walls. Non-loadbearing partitions are simply room dividers; they must be strong enough to provide adequate support for wall finishes, doors, and general living accommodation abuse.

There are two general types of wall framing used in construction today; *platform or western* framing and *balloon* framing. (Figs. 8-1 and 8-2)

PLATFORM CONSTRUCTION

In platform construction, the wall frame is constructed independently of the floor. The floor framework is constructed first and covered with the subfloor; this gives a large flat surface on which to assemble and nail the wall frame. The bottom plate in platform construction is commonly called a sole plate. It is used to support and align the lower ends of the studs. The supporting studs extend one storey from the sole plate to a top plate which, like the sole plate, keeps the studs in alignment. To add strength in supporting the load of the next floor or the ceiling and roof, a second plate is used. This also aids in tying wall sections together at corners and interior partitions. Wall sections that are assembled on the subfloor are usually sheathed and sometimes even the windows and siding are installed, eliminating the need for scaffolding for these operations.

After the wall sections are assembled they are raised in place and securely nailed, using the second top plate to tie together the corners and partitions. In the case of a multi-storey building, a floor frame is built on top of the wall frame, and the subfloor is installed, providing a working platform for the next wall section. In single storey houses the ceiling and roof members are installed directly on top of the wall sections.

BALLOON FRAMING

Balloon framing is used in two-storey buildings. The wall studs extend in one piece from the sill, which is situated on the foundation wall, to the roof. The second floor joists are nailed to the studs and supported by a *ribbon* that has been let into the studs. This ribbon is usually 19×89 mm stock. The top of the studs are aligned with the top plates, which provide support for the ceiling joists and rafters. The advantage of this type of construction is that there will be less shrinkage, which is desirable when stucco or brick veneer is used as an exterior finish. The long exterior wall studs make it necessary to install firestops.

FRAMING MEMBERS

The most common type of framing used in construction today is platform construction. The principal members in this type of construction are studs and plates with special framing used when there are openings for doors and windows.

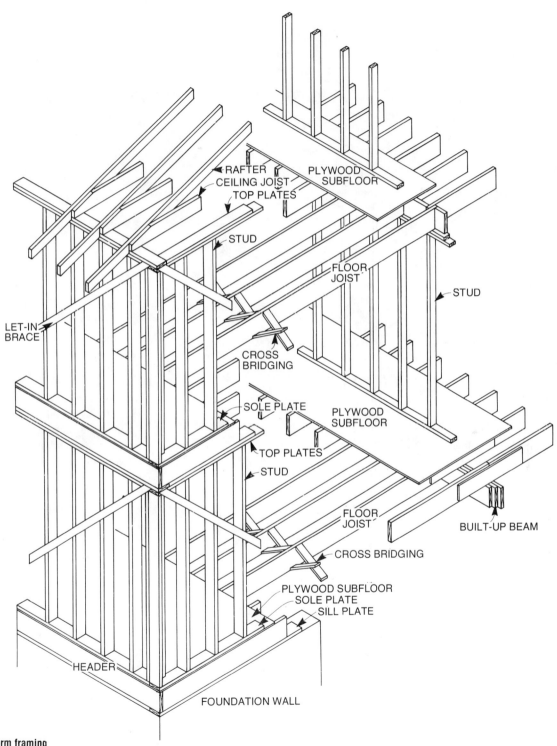

RAFTER
CEILING JOIST
TOP PLATES
PLYWOOD SUBFLOOR
STUD
FLOOR JOIST
STUD
LET-IN BRACE
CROSS BRIDGING
SOLE PLATE
PLYWOOD SUBFLOOR
TOP PLATES
STUD
FLOOR JOIST
BUILT-UP BEAM
CROSS BRIDGING
PLYWOOD SUBFLOOR
SOLE PLATE
SILL PLATE
HEADER
FOUNDATION WALL

Figure 8-1   Platform framing

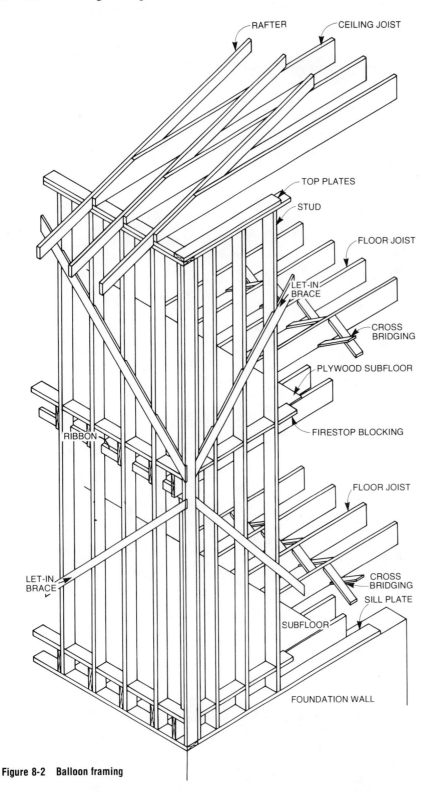

Figure 8-2 Balloon framing

Wall Studs

Exterior studs should be long enough so that the distance from the finish floor to the finish ceiling, called *house headroom,* is at least 2400 mm high. This facilitates the application of wallboard, as most wallboard is 2400 mm, whether put on vertically or horizontally in two rows. In calculating the length of studs required, add the thickness of the ceiling finish and the thickness of the floor finish to the desired headroom. From this figure subtract the thickness of three plates.

Example:

House headroom required

	2400 mm
Finish ceiling	12 mm
Finish floor—underlay	7 mm
—tile	3 mm
TOTAL	2422 mm
Subtract the thickness of three plates	114 mm
	2308 mm

For this house, the length of studs should be cut to 2308 mm (Fig. 8-3).

Lumber dealers sell studs that are precut to a standard length and sold individually or in lift lots. These are called *precision end trimmed* (PET) studs. The length allows for a double top plate and a single bottom plate with a ceiling height of 2400 mm. The length of the **PET** studs is 2310 mm.

Wall studs should be at least 38 ×89 mm, but with the emphasis today on fuel conservation, studs 38 ×140 mm are becoming more popular because more insulation can be used. Wall studs may be spaced 400 or 600 mm on centre in one-storey buildings, 400 mm o.c. being the most common spacing in house construction.

Wall Plates

Wall studs are kept in alignment by top and bottom plates. Wall covering,

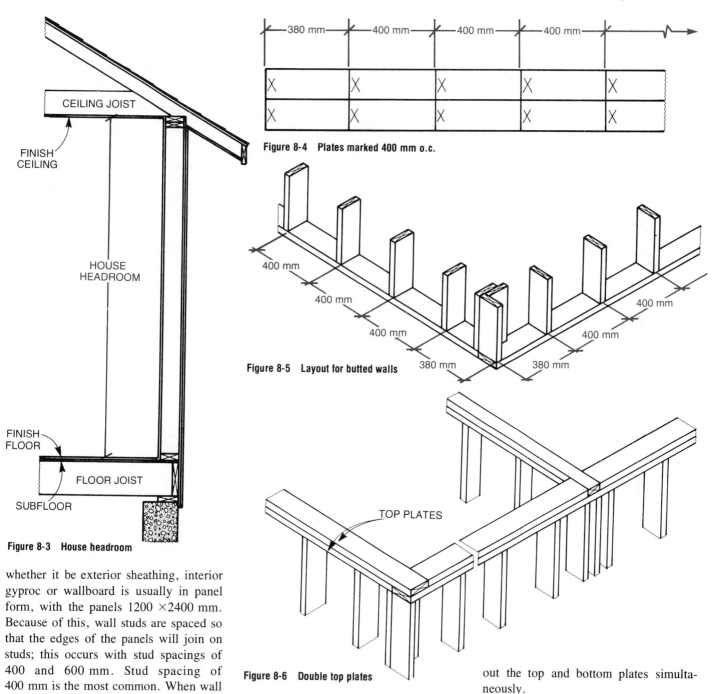

FINISH
CEILING

HOUSE
HEADROOM

FINISH
FLOOR

FLOOR JOIST

SUBFLOOR

Figure 8-3   House headroom

CEILING JOIST

380 mm — 400 mm — 400 mm — 400 mm

Figure 8-4   Plates marked 400 mm o.c.

400 mm
400 mm
400 mm
380 mm
400 mm
400 mm
380 mm

Figure 8-5   Layout for butted walls

TOP PLATES

Figure 8-6   Double top plates

whether it be exterior sheathing, interior gyproc or wallboard is usually in panel form, with the panels 1200 ×2400 mm. Because of this, wall studs are spaced so that the edges of the panels will join on studs; this occurs with stud spacings of 400 and 600 mm. Stud spacing of 400 mm is the most common. When wall plates are marked out for stud spacings of 400 mm o.c., the first dimension, from the end of the plate to the edge of the second stud will be 380 mm. A line is drawn and an "X" is marked to signify on which side of the line the stud should be placed (Fig. 8-4). From the edge of that stud to the edge of the next will be 400 mm. This dimension is continued down the length of the wall plate, laying out the top and bottom plates simultaneously.

When plates are laid out and butt against one another, allowance must be made for the width of studs used. (Fig. 8-5).

A double top plate is used to tie walls

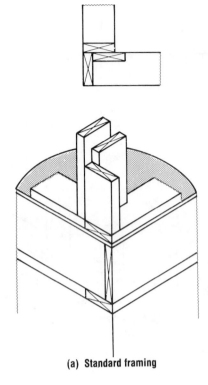

(a) Standard framing

Figure 8-7   Three stud corner

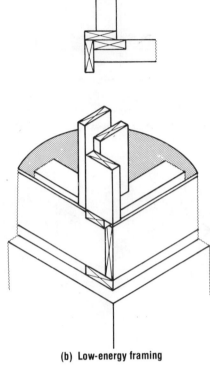

(b) Low-energy framing

Figure 8-9 shows a four-stud assembly which can be used only with 38 × 89 mm studs. With this type of construction it is not possible to insulate. Even though wood provides some insulation, in extremely cold weather there will be a build-up of frost in the corners due to the transfer of cold by conduction. Because of this the four-stud assembly is not recommended.

Interior Wall Intersections (Partitions)

To provide a nailing surface for interior finish, it is necessary to install backing at interior-exterior wall intersections. One of the most popular means of providing this backing is to make up a three-stud assembly in the shape of a "U" (Fig. 8-10). With this type of assembly, insulation must be installed before exterior sheathing is applied.

and partitions together as well as give added support for overhead loads. (Fig. 8-6). Joints in top plates should lap at least one stud spacing.

Exterior Corners

To provide a strong tie between adjoining walls and to provide nailing support for interior and exterior finish, a corner-stud assembly of at least three studs is used. There are several stud arrangements which can be used; Figure 8-7 shows a stud arrangement where insulation can be installed after the exterior sheathing is applied.

Figure 8-8 shows a three stud assembly where insulation must be installed before the sheathing is applied.

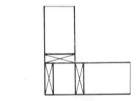

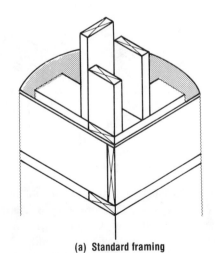

(a) Standard framing

Figure 8-8   Three-stud corner

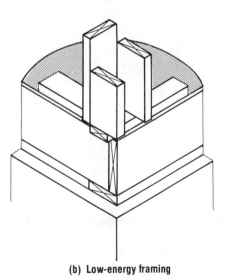

(b) Low-energy framing

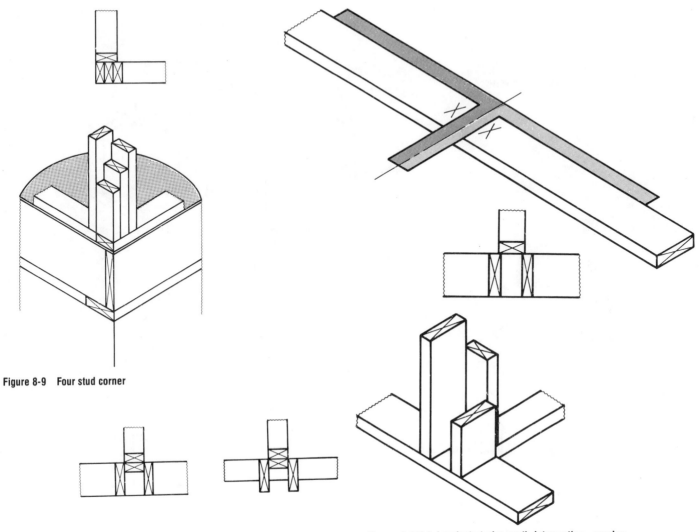

Figure 8-9   Four stud corner

Figure 8-11(a)   Interior/exterior wall intersection—overlapping studs (standard framing)

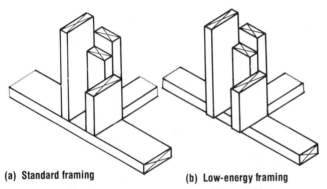

(a) Standard framing   **(b) Low-energy framing**

Figure 8-10   Interior/exterior wall intersection "U" assembly

Backing can be provided by installing studs which are the thickness of one stud on either side of the partition centre line (Fig. 8-11). Insulation must be installed before exterior sheathing is applied.

Backing can be provided by nailing 38 ×89 mm members between the regular studs, not more than 600 mm o.c. (Fig. 8-12). Insulation may be installed after exterior sheathing is applied.

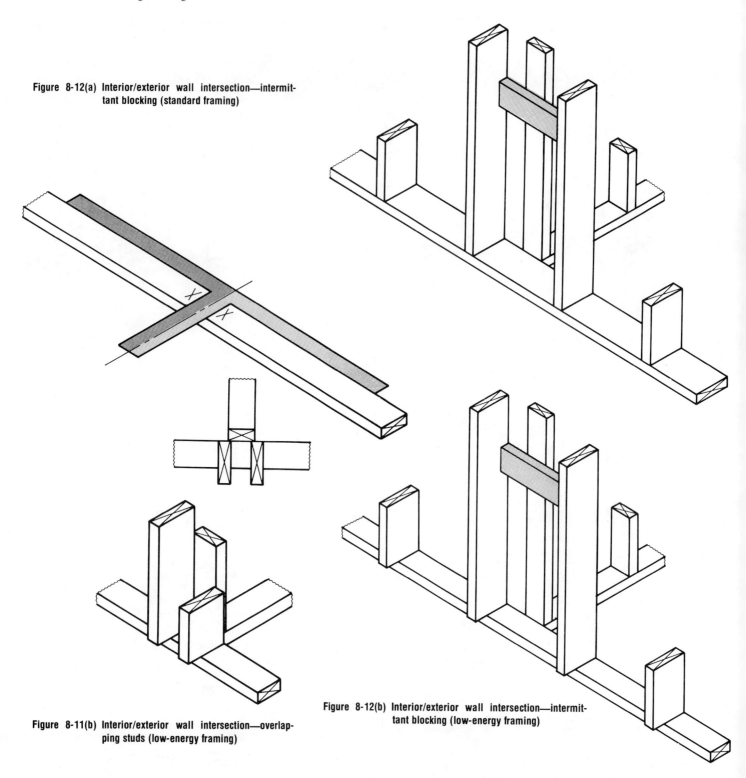

Figure 8-12(a) Interior/exterior wall intersection—intermittant blocking (standard framing)

Figure 8-11(b) Interior/exterior wall intersection—overlapping studs (low-energy framing)

Figure 8-12(b) Interior/exterior wall intersection—intermittant blocking (low-energy framing)

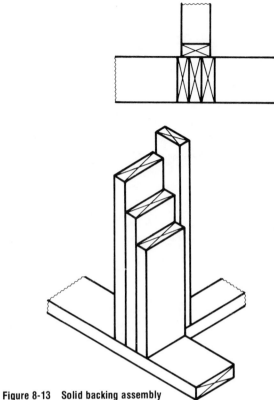

Figure 8-13   Solid backing assembly

Parts of a Rough Opening

Lintel—A lintel is a horizontal member installed over door and window openings to carry the overhead loads to the adjoining studs. Lintels, sometimes called headers, are usually made of two pieces of 38 mm lumber separated with spacers to make them the same width as the studs used for the wall. These 38 mm members, along with the blocking, are nailed together as a single unit. The depth of the lintel is determined by the width of the opening and the overhead loads to be supported. There are tables found in the residential standards giving minimum requirements for lintels.

Trimmers—Trimmers are vertical members extending from the underside of the lintel to the sole plate and are used to support the lintel. They are nailed to full-length regular studs for added support.

Upper and Lower Cripples—Cripples are regular studs that have been cut to

A solid type of backing can be provided by the use of three studs. (Fig. 8-13). However, it is not possible to insulate this and, as in the solid corner assembly, there may be frost build-up in extremely cold weather. This method is therefore not recommended.

OPENINGS IN EXTERIOR WALLS

Where doors or windows occur in exterior walls, regular studs must be cut and provisions made to support the overhead load of ceiling and roof. These openings are referred to as *rough openings* (R.O.) as shown in Figure 8-14.

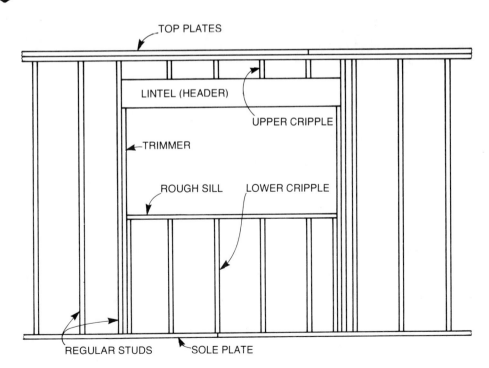

Figure 8-14   Parts of a rough opening

allow for door or window openings. There are no lower cripples in rough openings for doors. Frequently, to speed up the construction of wall sections, lintels are made full depth between the top plate and the trimmers, thus eliminating the need for upper cripples. When 38 × 140 mm wall studs are used, the space between the 38 mm supporting members will be sufficient to warrant insulating to decrease the possibility of heat loss.

Rough Sill—A rough sill is a horizontal member at the bottom of the rough openings for windows, and is used to support the window frame and keep the lower cripples in alignment.

Sole Plate—A sole plate is a horizontal member at the bottom of the wall frame used to keep the regular studs and lower cripples in a rough opening in alignment, as well as giving support to the wall frame. In rough openings for doors the sole plate must be cut out.

Size of Rough Openings

When determining the size of rough openings, the *outside measurement* (O.S.M.) of the door or window frame must be known. Most door and window frames are factory-built and are usually not on site when the wall sections are built, therefore the O.S.M. must be determined from the blueprints or the manufacturer's specification sheet. After the O.S.M. is determined, the rough opening must be enlarged enough for insulation and alignment between the window and door frame and the rough opening members. Usually a 10 mm allowance on each side of the frame is sufficient. Since most door and window frames have slanted sills it is only necessary to allow for insulation on the top; 10 mm is allowed for this. Insulation will be installed under the slanted sill. If the window frames have flat sills,

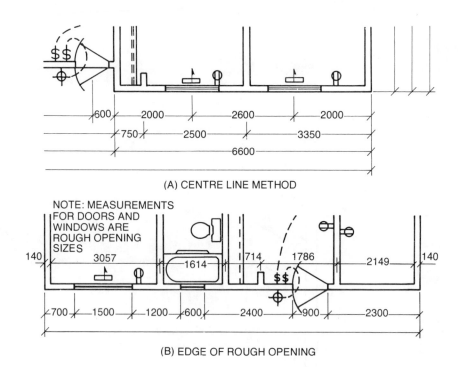

(A) CENTRE LINE METHOD

Figure 8-15   Dimensioning for doors and windows

10 mm must be allowed at both the top and bottom for insulation.

The height of the front door usually determines the height of the windows. When the length of trimmer studs are determined for the front door, this measurement will be used for the window trimmer length. When the size of door and window frames are given, the first figure is always the width and the second figure the height.

Architects use two methods of dimensioning for doors and windows. One method is to dimension to the centre line of the window; when this method is used, the O.S.M. must be found elsewhere, either on another part of the plans or from specs. The other method used is to dimension to the edge of windows. Some architects will give the measurement as a rough opening. The measurement used must be signified on the blueprint (Fig. 8-15).

Sometimes only the size of the door is given and the O.S.M. of the frame must be determined before the rough opening for the wall section can be laid out. To find the R.O. width for exterior doors, add to the door width the thickness of the jambs at the rabbet, plus 20 mm for plumbing and insulation. To find the R.O. height, add to the door height 60 mm for the sill, plus the thickness of the header jamb at the rabbet, plus 10 mm for insulation. To find the R.O. width for interior doors, add to the door width the thickness of both side jambs, plus 20 mm for plumbing clearance. To find the R.O. height, add to the height of the door the thickness of the finish floor, plus the necessary clearance beneath the door (15 mm), plus the thickness of the header jamb, plus 10 mm for alignment. If forced-air heating is used and the cold air returns are located in the hallway, the clearance under the interior doors must be

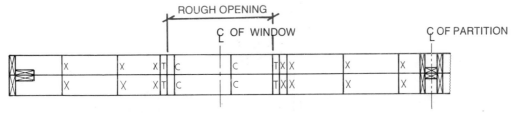

Figure 8-16   Plate layout

approximately 19 mm in order for the system to operate properly.

Layout of Plates for Rough Openings

The top and bottom plates are laid out at the same time, with the main side walls usually laid out first. The plates should be positioned side by side along the length of the wall section. If the plates must be joined, the joints must occur on the centre of the studs. Regular studs should be laid out from the same end of the building on both sides of the building. The regular studs are marked out at the desired centres with the first stud from the end being 20 mm less than the stud centres. Lines are drawn across the plates at these points. An ''X'' beside the line will signify on which side of the line the stud should be nailed. If a blueprint is used where the windows and doors are dimensioned to a *centre line* ($\mathcal{C}$L), then all the $\mathcal{C}$L of the doors and windows should be marked on the plates. Also, all the $\mathcal{C}$L for partitions should be marked. Once the width of the rough opening for doors and windows is determined, measure half that distance either way from the centre line. There will be trimmer studs at the ends of this measurement which should be marked with a ''T''. Next to the trimmer studs there will be a regular stud to be marked with an ''X''. Between the trimmer studs there should be several markings for regular studs; because these are now in a rough opening they will become cripple studs. To be sure regular studs will not be nailed there, a line should be drawn from trimmer stud to trimmer stud.

This will signify that they are no longer regular studs, but are now cripple studs. Indicate the positions of the studs which will be used at corners and partitions. This depends on the assembly used (Fig. 8-16). If special backing is required in the wall section, write that information on the wall plates.

When a blueprint is used, the dimensioning is to the edge of the door or window, and the O.S.M. is given on the floor plan, remember that a 10 mm clearance must be allowed on either side of the opening without moving the position of the opening.

All markings for regular studs, centres for openings, and centres for partitions should be laid out in one operation as you move down the plates. This not only speeds up the operation, but there is less possibility of damage to the steel tape used for the layout.

ASSEMBLY OF WALL FRAMING MEMBERS

In platform framing the wall sections are usually assembled and nailed on the subfloor. After the top and bottom plates are marked out they are positioned on the subfloor the length of the studs away from each other. Pre-cut studs are laid on the floor where regular studs will occur and then nailed in place. The studs are placed with the bow up. This is done to prevent a *zig-zag* effect in the wall which would result if no attention were paid to the direction of the bows. Frequently, corner and partition assemblies are made up sepa-

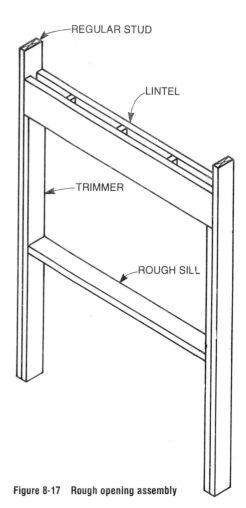

Figure 8-17   Rough opening assembly

rately and dropped in place as they are needed in the wall section. Rough opening assemblies are often made up as units and dropped into position (Fig. 8-17). Because of the varying sizes of rough openings, upper and lower cripples are cut and nailed in place after the rough assembly

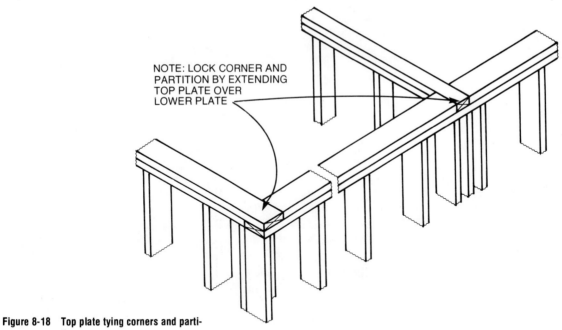

NOTE: LOCK CORNER AND
PARTITION BY EXTENDING
TOP PLATE OVER
LOWER PLATE

Figure 8-18   Top plate tying corners and partitions

(R.O.) is positioned, and are nailed secure.

After the wall section is framed with the extra studs for corners and partitions and the rough opening members are nailed in place, the wall section can be erected. However, it is often sheathed and further finish such as windows, cornice, siding, etc. can be installed while the frame is on the subfloor. When wall sheathing is applied before the wall is erected, the wall should be straightened and squared. Wall sections are straightened by lining up either the top or bottom plate with a straight line achieved by snapping a chalkline on the floor. After the plates are straightened, the wall section may be squared by measuring diagonals. Prior to erecting the wall frame a chalkline should be snapped on the subfloor to show where the inside edge of the exterior wall sole plate will be positioned. Chalklines are also snapped to align the interior partitions. It is only necessary to snap lines on one side of the partition.

After the chalklines are snapped, the wall sections are raised into position and temporarily braced. Then the bottom plates are nailed through the subfloor to the floor framing members. The wall sections are then plumbed and nailed together at the corners and intersections. A second top plate is added which laps the corners and partitions, providing an additional tie to the walls. (Fig. 8-18).

Backing

When the exterior walls and interior partitions are being framed, plan for backing, which will be needed for the installation of various fixtures and appliances. This will prevent loss of time and needless frustration later. Backing may be required to support drapery rods, clothes closet rods and shelves, shower curtain rods, toilet paper holders, cold air return openings, recessed cabinets, lavatory fixtures, and bathtubs. Figure 8-19 shows a few of the backings required in an average

house. With experience you will find other places where backing will make installation or application of various accessories easier.

Bracing

Bracing is not necessary if wall sections are sheathed with an acceptable thickness of plywood. Bracing is required when fibreboard paneling or horizontal boards are used as sheathing. Since the trend today is to build houses that are more energy-efficient, some contractors are installing 50 mm rigid insulation on the exterior wall frame. Since rigid insulation is quite expensive, these contractors are eliminating plywood as a method of adding rigidity to the wall frame and are using diagonal bracing instead, thus keeping construction costs down and at the same time increasing the RSI rating of the wall.

The most common method of providing diagonal bracing is to notch a

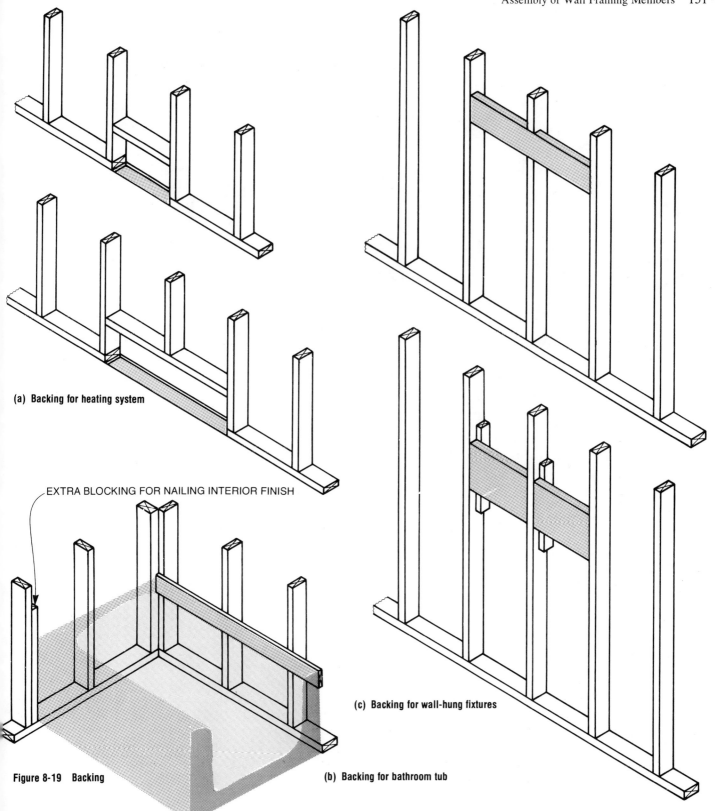

(a) Backing for heating system

EXTRA BLOCKING FOR NAILING INTERIOR FINISH

(c) Backing for wall-hung fixtures

Figure 8-19   Backing

(b) Backing for bathroom tub

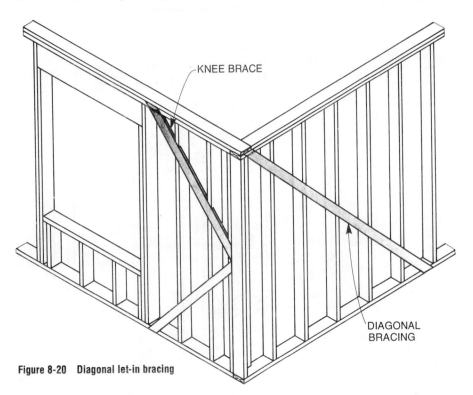

Figure 8-20   Diagonal let-in bracing

loads by means of *dwarf walls*. In multistorey buildings, the joists supporting the ceiling finish of the first storey are also the floor joists of the storey above. Their size can be determined by treating them as floor joists. Ceiling joists are similar to floor joists except that they are of a smaller dimension and no header joists are used. The ends are kept in alignment by nailing them to the rafters and top plate (Fig. 8-22).

The size of ceiling joists are determined by the length of span and the joist spacing used. When ceiling joists are used as roof support, they should be increased by one dimension greater than otherwise required. Provisions for openings for chimneys, attic access, etc. are framed in the same manner as the floor framing.

When low slope hip roof construction is used, it is not possible to install all the joists parallel to one another. The joists

19 × 89 mm member into the studs at approximately a 45° angle. This is known as *let-in bracing*. When a rough opening does not allow this type of bracing a knee brace may be installed (Fig. 8-20).

Diagonal *cut-in bracing* may be used; with this method 38 mm thick stock the same width as the studs is cut and installed between the studs. Braces extend from the bottom plate to the top plate crossing at least three stud spaces (Fig. 8-21). Galvanized *metal strapping* is also used as diagonal bracing.

CEILING FRAMING

Ceiling joists are horizontal members situated on top of the wall plates. They are used to tie the exterior walls together and also to provide nailing support for ceiling finishes. As discussed in Chapter 9, they sometimes provide added support for roof

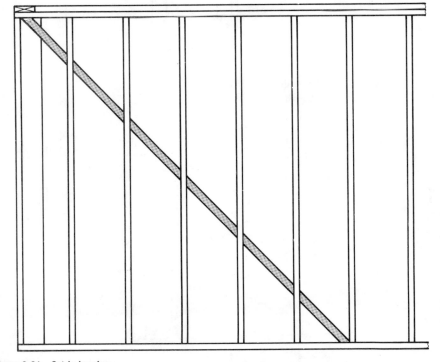

Figure 8-21   Cut-in bracing

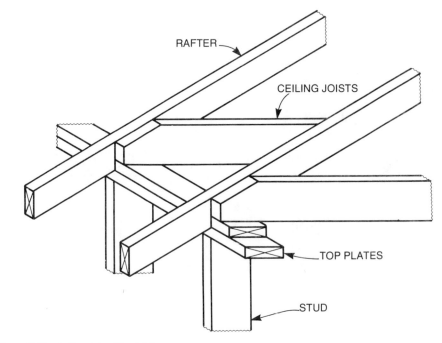

Figure 8-22   Rafter and ceiling joist

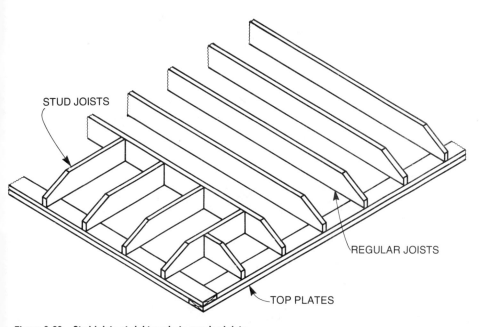

Figure 8-23   Stud joists at right angle to regular joists

near the end wall must be framed at right angles to the regular joists (Fig. 8-23).

In houses with a centre bearing wall, if a room requires that part of the bearing wall be eliminated, such as an L-shaped living-dining room, the inner ends of the ceiling joists may be supported on a beam. This beam would be framed into the ends of the bearing or outside wall with adequate vertical members to support it. If it is desirable for the ceiling to be continuous on one plane a flush beam may be used. This beam would be supported on top of the top plates and the ceiling joists framed into it. These joists would be attached using metal *joist hangers*. (Fig. 8-24).

Backing at Ceiling and Walls

Nailing support for interior finish is required at the meeting of wall and ceiling when the partitions run parallel to the ceiling joists. An extra joist need only be installed when a regular joist is within 38 mm of one side of the partition. Another method is to nail to the top plate a 38 mm member which is one dimension larger than the top plate. If this method is used, a cross-member should be nailed between the joists at 1200 mm o.c. to prevent the backing from moving when the interior finish is attached to it (Fig. 8-25).

FRAMING FOR LOW ENERGY HOUSES

With the increased emphasis on energy conservation, efforts are being made by contractors to provide a practical means of attaining low energy houses and yet keep the cost competitive in the housing market. One practical method is to increase the amount of insulation in walls and ceilings. To provide for the added insulation, new methods of wall framing have been devised.

One means of increasing the insulation

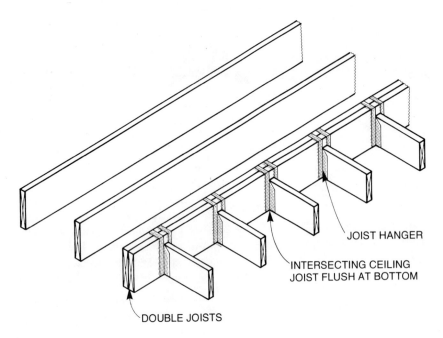

JOIST HANGER

INTERSECTING CEILING
JOIST FLUSH AT BOTTOM

DOUBLE JOISTS

Figure 8-24   Flush beam

and still stay with conventional framing methods is to increase the dimension of the wall studs from 89 to 140 mm minimum. By eliminating exterior plywood sheathing the cost is reduced. If this is done, diagonal bracing must be installed to add rigidity. The wall frame is then sheathed with 50 mm rigid insulation and covered with wood siding. This gives a RSI rating of slightly over 5. This method of construction is frequently used with pressure-treated wood foundations. (Fig. 8-26).

Other methods of framing have been tried using a *double wall* to gain space for added insulation. These are used with both concrete foundation and pressure-treated wood foundations. Figures 8-27, 8-28, and 8-29 show various framing methods using the double wall construction.

The double wall method allows for the use of large amounts of wall insulation using lower cost batt insulation while minimizing conduction through the studs.

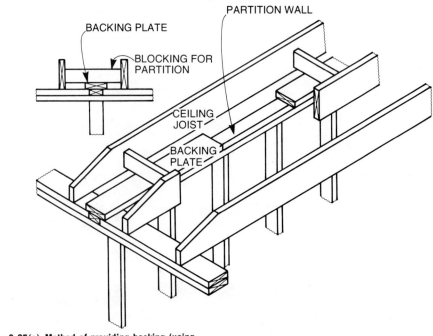

PARTITION WALL

BACKING PLATE

BLOCKING FOR
PARTITION

CEILING
JOIST

BACKING
PLATE

**Figure 8-25(a) Method of providing backing (using
a backing plate)**

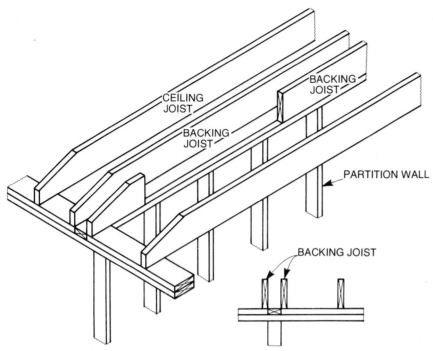

Figure 8-25(b) Method of providing backing (using ceiling joist)

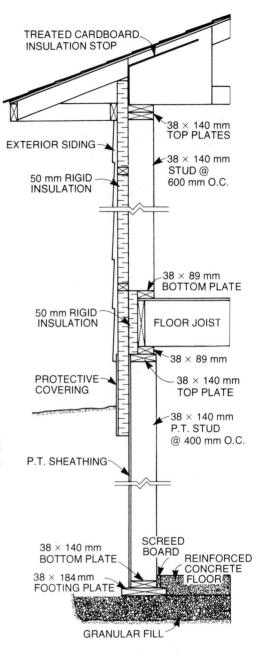

Figure 8-26 Wall section of a low energy house

The RSI rating for this wall is 7. With this method the vapour barrier is placed *away* from the inside surface of the walls to protect it during construction and to minimize penetration by electrical wires and plumbing. A rule of thumb for a climate such as is found on the Canadian prairies, where this method was first used to any extent, is that there should be twice as much insulation outside the vapour barrier as inside in order to prevent condensation occurring in the wall. In areas of high humidity, it is advisable to place the vapour barrier at the conventional location on the inside of the wall.

The construction sequence for a double stud wall is as follows:

1. Build the inner stud wall (this is the structural wall for the house) in the conventional manner. To save time, mark plates for both the inner and outer walls at the same time. Place a 150 μm vapour barrier on top, then nail on sheathing in the conventional manner. Take care not to damage the vapour barrier when cutting the sheathing.
2. Build the outer stud wall on top of the inner wall.
3. Using temporary spacer blocks, nail the two walls together using plywood plates which are nailed to the top and bottom wall plates.
4. Tilt the wall section up and put in place. Staple vapour barrier flaps to the inner face of top and bottom plates to prevent damage. The wall can be wired and insulated when desired.

WALL SHEATHING

Wall sheathing is usually applied to exterior walls to add rigidity to the framework. It is also used to provide insulating

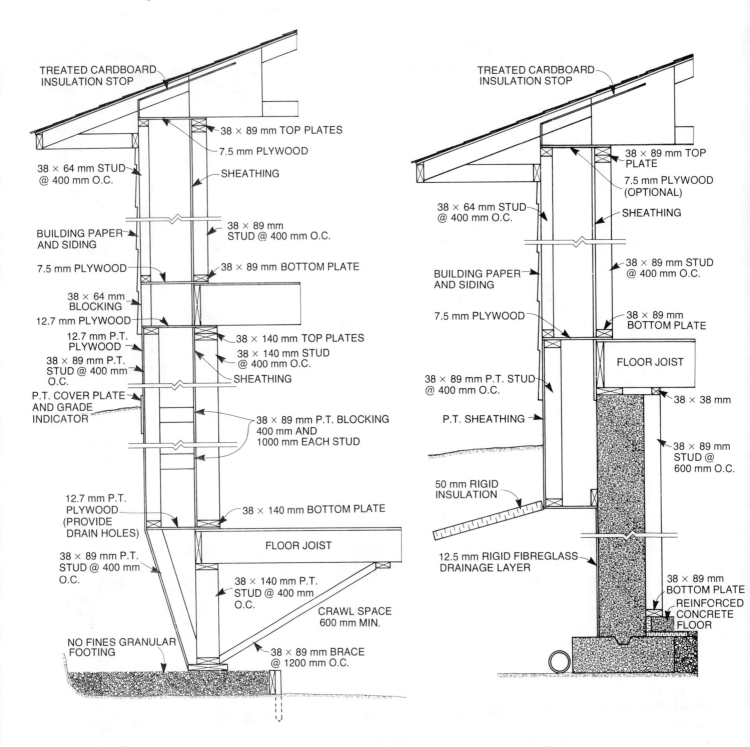

TREATED CARDBOARD
INSULATION STOP

38 × 89 mm TOP PLATES

7.5 mm PLYWOOD

38 × 64 mm STUD
@ 400 mm O.C.

SHEATHING

38 × 89 mm
STUD @ 400 mm O.C.

BUILDING PAPER
AND SIDING

38 × 89 mm BOTTOM PLATE

7.5 mm PLYWOOD

38 × 64 mm
BLOCKING

12.7 mm PLYWOOD

12.7 mm P.T.
PLYWOOD

38 × 140 mm TOP PLATES

38 × 89 mm P.T.
STUD @ 400 mm
O.C.

38 × 140 mm STUD
@ 400 mm O.C.

SHEATHING

P.T. COVER PLATE
AND GRADE
INDICATOR

38 × 89 mm P.T. BLOCKING
400 mm AND
1000 mm EACH STUD

12.7 mm P.T.
PLYWOOD
(PROVIDE
DRAIN HOLES)

38 × 140 mm BOTTOM PLATE

FLOOR JOIST

38 × 89 mm P.T.
STUD @ 400 mm
O.C.

38 × 140 mm P.T.
STUD @ 400 mm
O.C.

CRAWL SPACE
600 mm MIN.

NO FINES GRANULAR
FOOTING

38 × 89 mm BRACE
@ 1200 mm O.C.

Figure 8-27 Double wall—pressure-treated wood foundation

TREATED CARDBOARD
INSULATION STOP

38 × 89 mm TOP
PLATE

7.5 mm PLYWOOD
(OPTIONAL)

38 × 64 mm STUD
@ 400 mm O.C.

SHEATHING

38 × 89 mm STUD
@ 400 mm O.C.

BUILDING PAPER
AND SIDING

38 × 89 mm
BOTTOM PLATE

7.5 mm PLYWOOD

FLOOR JOIST

38 × 89 mm P.T. STUD
@ 400 mm O.C.

38 × 38 mm

P.T. SHEATHING

38 × 89 mm
STUD @
600 mm O.C.

50 mm RIGID
INSULATION

12.5 mm RIGID FIBREGLASS
DRAINAGE LAYER

38 × 89 mm
BOTTOM PLATE

REINFORCED
CONCRETE
FLOOR

Figure 8-28 Double wall—concrete wall and footing foundation

value. Some materials used as wall sheathing are solid lumber, (either shiplap, tongue and groove or common boards). This may be applied either horizontally (Fig. 8-30) or diagonally (Fig. 8-31). When applied horizontally, diagonal bracing must be used to provide rigidity.

Plywood is used extensively as a wall sheathing; while it provides some insulating qualities its main advantage is that it provides rigidity to the wall frame. This is especially important when large wall sections are assembled on the subfloor and tilted into place or when large openings occur in the wall frame. Plywood panels may be applied with the face grain either parallel or at right angles to the wall

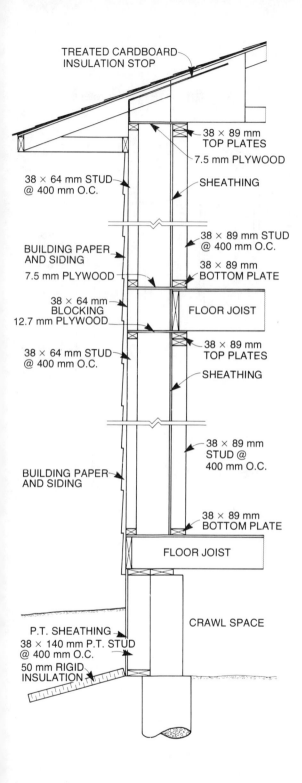

Figure 8-29   Double wall—grade beam and piling foundation

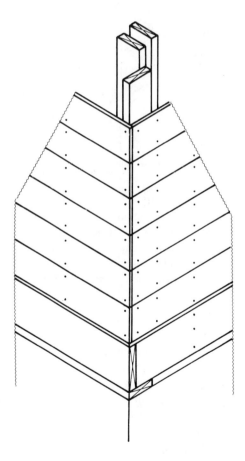

Figure 8-30   Horizontal lumber sheathing

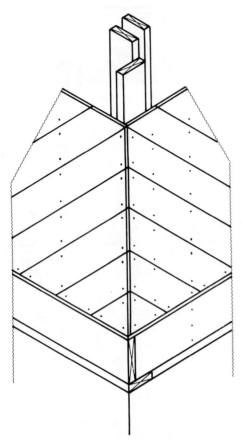

Figure 8-31   Diagonal lumber sheathing

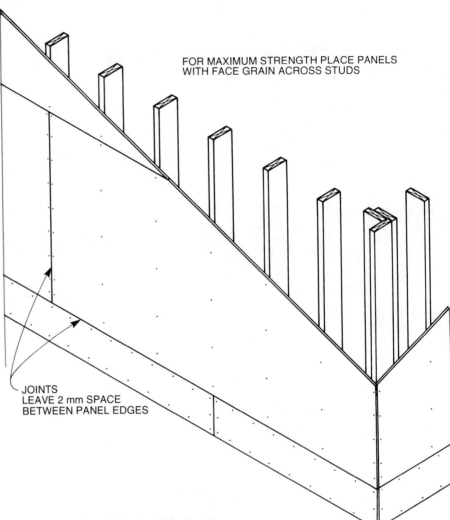

FOR MAXIMUM STRENGTH PLACE PANELS
WITH FACE GRAIN ACROSS STUDS

JOINTS
LEAVE 2 mm SPACE
BETWEEN PANEL EDGES

Figure 8-32   Exterior wall sheathed with plywood

studs. For maximum strength it is recommended that the grain run at right angles to the studs. When installing plywood panels a small gap of approximately 2 mm should be left between the panels to allow for expansion. Figure 8-32 shows plywood being applied to an exterior wall frame.

Particleboard is also used as a wall sheathing, and the application is similar to plywood. No diagonal bracing is required as the panel gives the necessary rigidity to the wall.

Fibreboard sheathing is also used as a wall sheathing, its main value being its insulating qualities. Diagonal bracing should be installed to give rigidity to the wall.

REVIEW QUESTIONS

8—WALL AND CEILING FRAMING

Answer all questions on a separate sheet.

A. Select the correct answer

1. When erecting exterior walls, the bottom plate is usually lined up by:
   (a) Sighting along the bottom plate
   (b) Setting the wall flush with the edge of the floor
   (c) Snapping a chalk line at the edge of the floor and line up the outside of the wall on the chalk line
   (d) Snapping a chalk line at the edge of the floor and line up the inside of the wall on the chalk line
   (e) Using a string line

2. The framing members that stiffen the sides of door openings and support the header are known as:
   (a) Purlins
   (b) Trimmers
   (c) Cripples
   (d) Knee braces

3. In platform construction, the sole plate of the wall rests on:
   (a) The floor joists
   (b) The subfloor
   (c) The underlay
   (d) The header
   (e) The sill

4. The most widely used framing method in modern construction is the:
   (a) Sectional method
   (b) Balloon method
   (c) Post and beam method
   (d) Platform method

5. The two most important factors to be considered in determining the size of headers are:
   (a) The type and size of sheathing used
   (b) The size of lumber and species used
   (c) The width of the opening and the load above
   (d) The amount of nailing required and the spacing of nails

6. The members used to support lintels in door or window openings are called:
   (a) Regular studs
   (b) Trimmer studs
   (c) Cripple studs
   (d) Headers

7. A wall that carries more than its own weight is called a:
   (a) Retaining wall
   (b) Cavity wall
   (c) Bearing wall
   (d) Curtain wall

8. Ribbon boards are most common in:
   (a) Western or platform framing
   (b) Balloon framing
   (c) Braced framing
   (d) All of these

9. The positions of ceiling joists are usually co-ordinated with what other framing members?
   (a) Cripples
   (b) Cap plates
   (c) Studs
   (d) Trimmers

10. The bottom piece in a rough opening for a window, framed in a stud wall is called:
    (a) Rough fascia
    (b) Rough trimmer
    (c) Trimmer
    (d) Rough sill

B. Write full answers

11. Why is it preferable to use a steel tape to mark stud centres rather than stepping out with a framing square?

12. What is the main purpose of a lintel in a wall opening?

13. Explain how to "square" a stud frame wall in preparation for sheathing (2 steps).

14. What are two reasons for allowing clearance around doors and windows when framing up walls?

15. In what type of wall framing are the studs fastened to the sill plate?

C. Mark as either TRUE or FALSE

16. Diagonal braces across the studs are necessary when lumber sheathing is applied horizontally.

17. Trimmer studs are used for supporting the lintel.

18. The first measurement given for a window unit is always the width.

19. If a stud in a frame wall is crowned or warped, one remedy is to make a saw cut, wedge it and then cleat both sides of the stud.

20. It is adequate to nail interior partitions down to the subflooring.

9 ROOF STYLES AND TERMINOLOGY

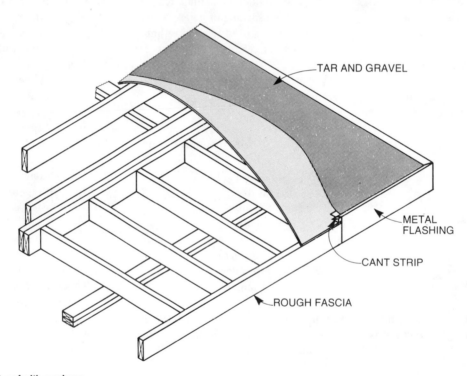

TAR AND GRAVEL

METAL FLASHING

CANT STRIP

ROUGH FASCIA

Figure 9-1   Flat roof with overhang

One of the most essential components of a house is the roof. The roof not only provides protection against the weather but also forms a structural tie between the outside walls. Another purpose of the roof is to enhance the appearance of the building. There are basically two types of roofs—*flat* and *slanted* or *sloped*.

TYPES

A flat roof is constructed with roof joists which support the roof as well as the ceiling. It usually has a slight slope for drainage. The roof may overhang the wall as in Figure 9-1 or may be constructed with a *parapet wall* as in Figure 9-2. Flat roofs

are sometimes used in house construction, but are used mainly on commercial buildings.

The simplest type of sloped roof is the *shed* roof. This roof has only one sloped surface and it normally runs across the width of the building (Fig. 9-3).

The most widely used type of roof is

160

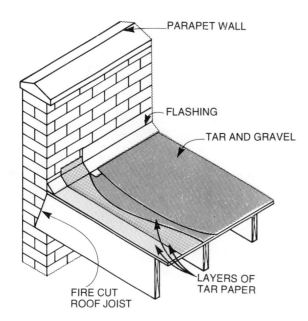

Figure 9-2   Parapet wall

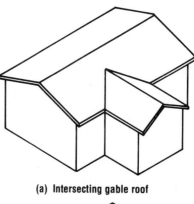

(a) Intersecting gable roof

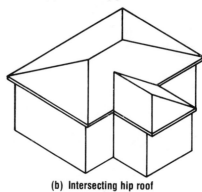

(b) Intersecting hip roof

Figure 9-6   Intersecting roof

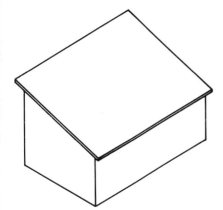

Figure 9-3   Shed roof

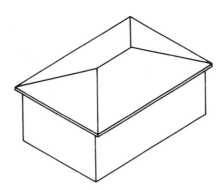

Figure 9-5   Hip roof

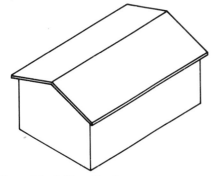

Figure 9-4   Gable end roof

the *gable* roof. In this style there are two sloping surfaces, which meet at an apex known as the ridge (Fig. 9-4).

Another popular type of roof used in house construction is the *hip* roof which has four sloped surfaces meeting at the ridge (Fig. 9-5).

A building which is L-shaped, or which has a projection requires a type of roof referred to as *intersecting*. This type of roof forms a valley at the intersection of the two sloped surfaces. The ridges of the two roof sections usually run at 90° to each other (Fig. 9-6). The roof may be either gable or hip.

By using various combinations of gable and hip roofs, roof types such as *dutch hip, gambrel,* and *mansard* can be built (Fig. 9-7).

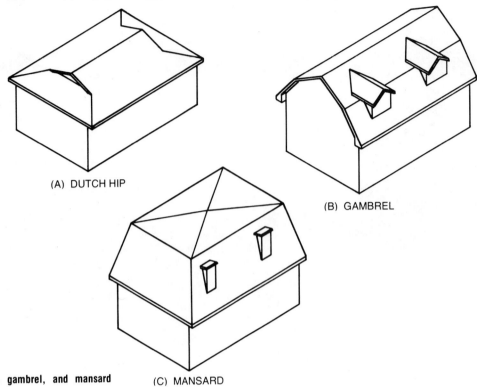

(A) DUTCH HIP

(B) GAMBREL

(C) MANSARD

Figure 9-7 Dutch hip, gambrel, and mansard roofs

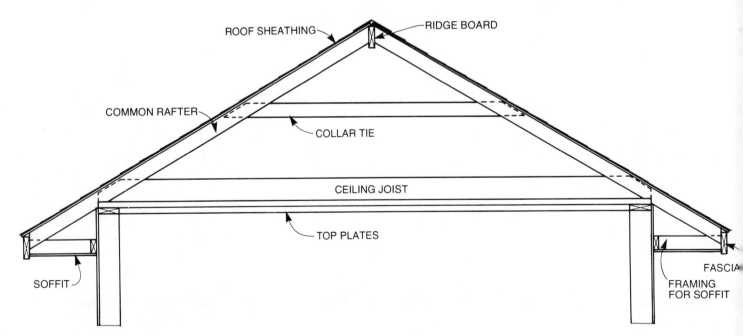

ROOF SHEATHING

RIDGE BOARD

COMMON RAFTER

COLLAR TIE

CEILING JOIST

TOP PLATES

SOFFIT

FASCIA

FRAMING FOR SOFFIT

Figure 9-8   Cross-section of a gable roof

ROOF TERMS AND COMPONENTS

Many of the terms and components found in a gable roof are common to both hip and intersecting roofs. Figure 9-8 shows a cross-section of a gable roof with a number of the following parts indicated.

1. **Ridge Board**—A ridge board is the horizontal member which supports and aligns the upper ends of the roof rafters.
2. **Common Rafter**—A common rafter is a roof member which extends from the top plate of the outside wall to the ridge board. This rafter is used in all types of roofs.
3. **Collar Tie**—A collar tie is a type of roof support which acts as a stiffener by tying two opposite rafters together.
4. **Rafter Tail**—A rafter tail is the portion of the rafter which extends beyond the outside edge of the exterior wall. **Note:** This portion is not considered in the line length of the rafter.
5. **Slope Symbol**—A slope symbol is used on a plan to indicate the amount of slope or inclination of the roof in relation to the horizontal run.
6. **Rise of a Roof** (Total rise)—The rise of a roof is the vertical distance from the top of the wall plate to the intersecting point of the theory lines on a pair of rafters.
7. **Span of a Roof**—The span of a roof is the horizontal distance between the outer faces of the outside wall plates.
8. **Run of a Roof** (Total run)—The run of a roof is the horizontal distance as measured from the outer face of the outside wall plate to the centre of the ridge board. The run is equal to **one-half** the span on equally pitched roofs.

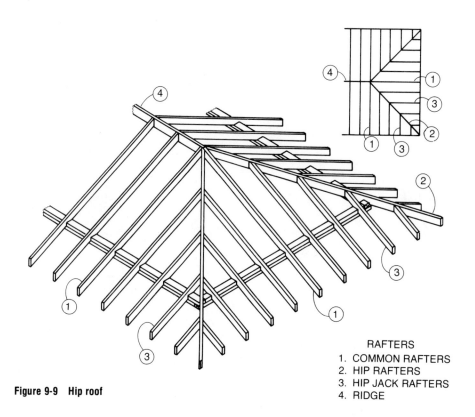

Figure 9-9 Hip roof

RAFTERS
1. COMMON RAFTERS
2. HIP RAFTERS
3. HIP JACK RAFTERS
4. RIDGE

9. **Overhang**—The overhang is the horizontal projection of the rafter extending beyond the exterior wall plate.
10. **Line Length of a Rafter**—The line length of a rafter is the theoretical length from the top outer edge of the wall plate to the centre of the ridge board.
11. **Intersecting Line**—An intersecting line runs parallel to the edge of the rafter at the top of the birdsmouth.

The gable roof is made up of *common rafters* with a ridge board to support and align the upper ends of the rafters. A hip roof has *common rafters* and a ridge board, as well as *hip rafters* formed by the meeting of the two sloping surfaces.

There are also *hip jack rafters* which are basically common rafters with the upper ends cut to frame into the hip rafters (Fig. 9-9).

An intersecting roof has rafters that are also found in a gable roof. If the main roof is a hip roof, there will be hip and hip jack rafters. Where the two roofs intersect there will be *valley rafters*. There will be rafters extending from the ridge board to the valley rafter called *valley jack rafters*. They are common rafters with the lower ends cut to frame into the valley rafter. There may also be *cripple jack rafters* which are common rafters with both ends cut to frame into hip and valley rafters or two valley rafters (Fig. 9-10).

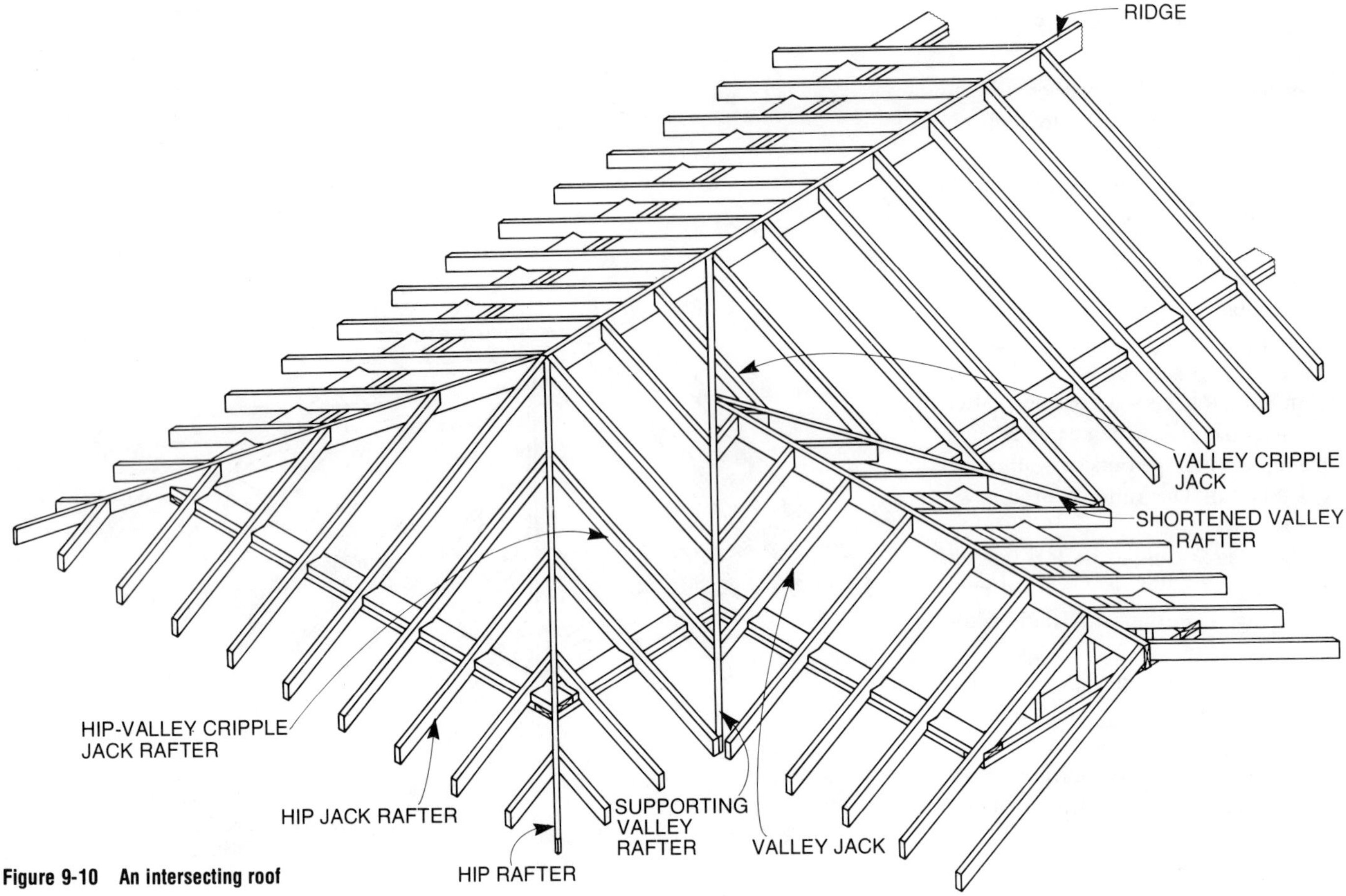

Figure 9-10   An intersecting roof

Labels on figure: RIDGE, VALLEY CRIPPLE JACK, SHORTENED VALLEY RAFTER, HIP-VALLEY CRIPPLE JACK RAFTER, HIP JACK RAFTER, SUPPORTING VALLEY RAFTER, VALLEY JACK, HIP RAFTER

ROOF SLOPE

Roof slope is expressed as a ratio, with the vertical component of the roof, which represents the rise, given first. For slopes less than 45°, the first number should always be unity. For example, a ratio of 1:4 indicates that for every unit of measurement of rise, there will be four units of measurement of horizontal run. In a ratio, no unit of measurement is given; so 1:4 can represent one inch of rise to four inches of run or it can represent one millimetre of rise to four millimetres of run. For slopes steeper than 45°, the horizontal component becomes unity. The rise however, is still given first. Therefore, a ratio of 4:1 means for every four units of rise there will be one unit of run.

RATIO AND PROPORTION

Since ratio and proportion play a very important part in the calculation of roof members, a brief review of ratio and proportion will be given. A ratio is a relation between two like numbers or values. The ratio may be written as a fraction (1/4), as a division (1 ÷ 4) or with a colon or ratio sign (1:4). When the last of these forms is used, it is read "1 is to 4". Ratios may be expressed by the word *per* as in kilometres *per* hour, or revolutions *per* minute. In mathematical terms, they are written km/h, or r/min.

A proportion is a method of expressing equality between two ratios. The equal sign between two ratios may be written as the double colon (::) or with an equal sign (=); the equal sign is most often used. Thus, 1:4=2:8 is a proportion that is read 1 is to 4 as 2 is to 8, or 1/4 equals 2/8. In any proportion the first and last terms are called *extremes* and the second and third are called *means*. In 1:4=2:8 the extremes are 1 and 8, the means, 4 and 2. In working with proportions the product of the means is equal to the product of the extremes.

$$1:4 = 2:8$$
$$1 \times 8 = 2 \times 4$$
$$8 = 8$$

No proportion is true unless the two ratios are equal. By using this rule, you can find the missing term of any proportion if the other three terms are given.

Example:

$$2:6=8:?$$

Find the value of the missing term. The letter x is traditionally used to denote a missing term or an unknown quantity.

2:6=8:x

2 times x = 6 times 8 (Product of the extremes equals product of the means)

2x=48

x = 24

The principle of ratio and proportion can be readily applied to roof calculations. If the roof symbol is read as a ratio such as 1:3 the unit of rise will be 83.3 if the unit of run is 250.

$$1:3 = x:250$$
$$3x = 250$$
$$x = 83.3$$

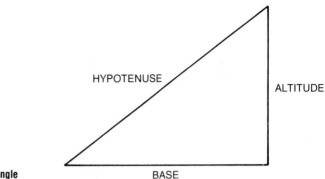

Figure 9-11   Right-angled triangle

ROOF TRIANGLE

The right-angle triangle is the most important geometrical shape used in construction and it definitely is in roof framing. A right-angled triangle is one with an angle of 90°. In Figure 9-11 the side opposite the right angle is called the *hypotenuse*–the other two sides are the *base* and the *altitude*.

The roof triangle corresponds to the right-angled triangle in that the base is the *total run*, the altitude is the *total rise*, and the hypotenuse is the *line length* of the rafter (Fig. 9-12).

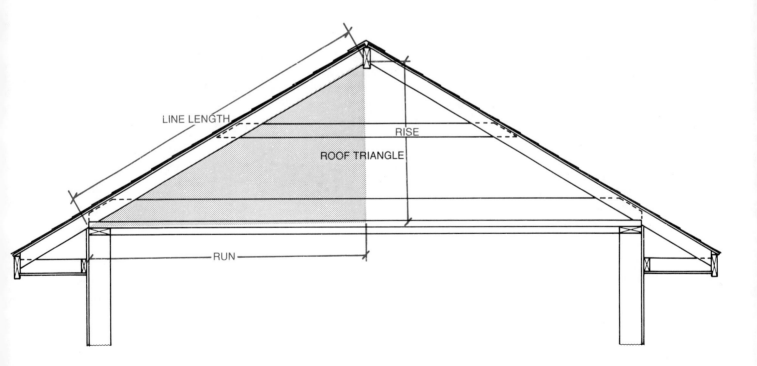

Figure 9-12   Roof triangle

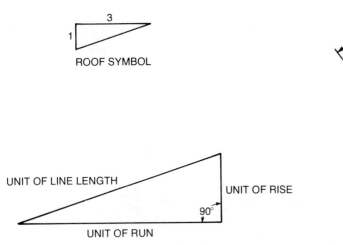

Figure 9-13   Roof symbol and unit roof triangle

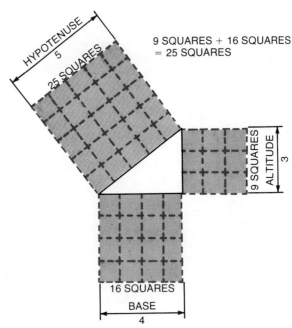

Figure 9-14   Right triangle

The roof slope symbol and the unit roof triangle are also right-angled triangles (Fig. 9-13). The use of the unit roof triangle will be covered in the roof framing section.

Basic Concepts of the Right Triangle

In the 6th century B.C., Pythagoras, a Greek mathematician and philosopher, formulated the Pythagorean Theorem or Right Triangle Law. This law states that the square of the hypotenuse of a right triangle is equal to the sum of the squares of the two sides (Fig. 9-14).

The formula is $c^2 = a^2 + b^2$, where c = length of hypotenuse, b = length of base, a = length of altitude. To find the hypotenuse of a right triangle, square the altitude and square the base, add the results and then find the square root of this sum.

The formula is $c = \sqrt{a^2 + b^2}$

Example:

A right triangle has an altitude of 300 mm and a base of 400 mm. Find the length of the hypotenuse.

1. Formula
   $$c = \sqrt{a^2 + b^2}$$
2. Substitution
   $$c = \sqrt{300^2 + 400^2}$$
3. Evaluation
   $$c = \sqrt{(300 \times 300) + (400 \times 400)}$$
   $$c = \sqrt{90\,000 + 160\,000}$$
   $$c = \sqrt{250\,000}$$
   $$c = 500 \text{ mm}$$

REVIEW QUESTIONS

9—ROOF STYLES AND TERMINOLOGY

Answer all questions on a separate sheet.

A. Write full answers

1. List three functions performed by a well constructed roof.
2. At what angle does the hip rafter meet the ridge board?
3. At what angle does the common rafter meet the ridge board?
4. Do all the jack and cripple jack rafters run parallel to a common rafter?
5. From where is the total rise of a roof measured?
6. What is meant by line length?
7. What is the total rise of a roof for a building with a span of 6600 mm if the slope is 1:3?
8. What is meant by the span of a roof?
9. What is meant by the total run of a building?

B. Replace the Xs with the correct word

10. The basis of all roof framing is a geometric figure called a XXXXXXX.

C. Select the correct answer

11. What is meant by the term "span"?
    (a) The distance a common rafter covers
    (b) The same distance as the length of building
    (c) Twice the distance of the total run
    (d) 4 times the distance of total rise
12. If the span of building is 5 000 mm, and the slope is 1:2.5, the total rise of the roof is:
    (a)   500 mm
    (b) 1 000 mm
    (c) 1 500 mm
    (d) 2 000 mm
13. If the total rise of a roof is 1 500 mm, and the slope is 1:4, the total run is:
    (a) 4 000 mm
    (b) 5 000 mm
    (c) 6 000 mm
    (d) 7 000 mm
14. If the span of a building with a shed roof is 6 000 mm, the low wall is 2 100 mm high, and the slope is 1:5, what is the height of the high wall?
    (a) 2 800 mm
    (b) 3 300 mm
    (c) 3 800 mm
    (d) 4 300 mm
15. A roof has a slope of 1:2. It is desired to put a collar tie down 1 200 mm from the ridge, how long will the collar tie be?
    (a) 2 400 mm
    (b) 3 600 mm
    (c) 4 800 mm
    (d) 6 000 mm
16. If the total rise of a roof is 1 600 mm, and the span is 4 800 mm, the roof slope is:
    (a) 1:1.5
    (b) 1:2
    (c) 1:2.5
    (d) 1:3

D. Mark as either TRUE or FALSE

17. The most common geometric figure used in roof framing is the square.
18. A roof on an L-shaped building is referred to as an intersecting roof.
19. The run of a roof is the same as the total width of the building.
20. Hip jack rafters are common rafters with the upper ends cut to frame into the hip rafters.

10 ROOF FRAMING

FRAMING SQUARE

The framing square is the basic tool used in the layout of roof framing members. The longer and wider leg is called the *body* while the shorter and narrower leg is called the *tongue*. Since the framing square is not big enough to lay out the total run and total rise at one time, it is necessary to take a unit of run and determine a unit of rise. Manufacturers use various measurements as their unit of run for the roof slope. We refer here to the Frederickson Square as the basis for our roof calculations and would at this time like to thank Paul Frederickson for allowing us to benefit from his experience in the manufacture of this framing square. Using this square, the **unit of run** for the common rafter is **250 mm**. The unit of run for the hip rafter is the diagonal measurement of a 250 mm square, which is 353.5 mm. For ease in rafter layout, **354 mm** is used as the **unit of run** (Fig. 10-1). On the framing square, there are marks at 250 and 354 for ease in finding these points.

COMMON RAFTER

The common rafter is the framing member in the roof which extends at a right angle from the top plate to the ridge

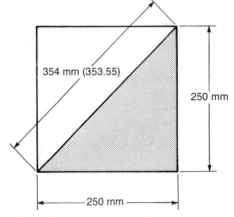

Figure 10-1   Unit of run for a hip rafter

board. It is named *common rafter* because it is used in all types of sloped roofs. There are two basic cuts on a common rafter; a *plumb cut,* which is 90° to the horizontal, and a horizontal cut at the birdsmouth called a *seat cut.* The *birdsmouth* is a notch cut in the rafter which allows for full support on the wall plate. There will be plumb cuts at the ridge board, birdsmouth, and rafter tail (Fig. 10-2). The theoretical line length (commonly called *line length*), is measured from the top outer edge of the wall plate to the centre of the ridge board. The *actual line length* of the common rafter is the length arrived at after deductions have been made for the ridge board.

Calculating Common Rafter Line Length

When the Frederickson Square is used, the unit of run for the common rafter is 250 mm. The first step in finding the line length is to determine the unit of rise and the unit of line length. This is called the *unit roof triangle* (Fig. 9-13). From the roof symbol found in the blueprint, make up a ratio and proportion equation with 250 as the third term. If a ratio of 1:3 is found on the slope symbol, the required equation would read:

$$1:3 = x:250$$
$$3x = 250$$
$$x = 83.3$$

The unit of rise would be 83.3 mm with a unit run of 250 mm. Once the unit of rise is known, look on the rafter table (found on the body of the square), to see if this is a unit of rise for which the line length is given. The first line states the length of the common rafter per unit run (250). Because of the limited space on the square it would have been impossible to give the line length for all measurements of rise. This square gives the line lengths for unit rises from 50 to 500 mm in increments of 50. If the unit of rise is 150 mm, it is a simple matter to look under the number 150 and find the unit of line length, which is 291.5 mm. Since the unit of line length

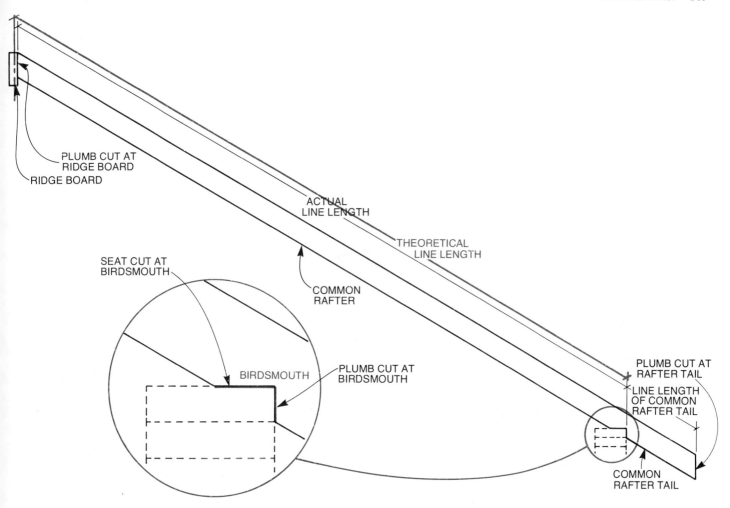

Figure 10-2   Parts and cuts of a common rafter

for 83.3 mm is not given, the Pythagorean Theorem (explained in Chapter 9), must be used. After the unit triangle figures have been found, the next step is to determine the number of units of run there will be in the total run. To find the number of units of run, divide the total run by 250. Calculate the common rafter line length by multiplying the unit of line length by the number of units of run.

Example:
Building Size–7 200 mm × 12 000 mm
Roof Slope–1:3
Overhang–750 mm

From the information:

(a) Span = 7 200 mm

(b) Total run = 3 600 mm

(c) Number of units of run = 14.4

$$3\ 600 \div 250 = 14.4$$

(d) Unit of rise = 83.3 mm

$$1:3 = \times:250$$
$$3x = 250$$
$$x = 83.3$$

(e) Unit of Line Length = 263.5 mm

$$c = \sqrt{a^2 + b^2}$$
$$= \sqrt{250^2 + 83.3^2}$$
$$= \sqrt{62\ 500 + 6\ 938.89}$$

$$= \sqrt{69\ 438.89}$$
$$= \quad 263.5$$

(f) Line Length of
Common Rafter = 3 794.4 mm

$$263.5 \times 14.4 = 3\ 794.4$$

Calculating Line Length of the Common Rafter Tail

When calculating the line length of the common rafter tail, determine the run of the rafter tail, which will be given as *overhang*. Divide the run (overhang) by the unit of run (250) to get the number of units of run. Multiply the number of units of run by the unit of line length of the common rafter; this will give the line length of the common rafter tail.

Example: (from information on common rafter)

a. Run (overhang) = 750 mm
b. Number of Units of Run = 3
   750 ÷ 250 = 3
c. Line Length of Common Rafter Tail
   = 790.6 mm
   263.5 × 3 = 790.6

Common Rafter Deductions

A *deduction* or shortening of the rafter must be made if a ridge board is to be used, as the line length is calculated to the centre of the ridge board. The deduction for a common rafter is one-half the thickness of the ridge board. This deduction must be taken at 90° to the plumb cut and should not be measured along the line length of the rafter.

The other deduction required on the common rafter is at the rafter tail for the fascia board(s) (Fig. 10-5). This deduction is necessary as the overhang is always measured at 90° from the wall to the outer edge of the finished fascia board. The deduction then is the **thickness** of the fascia boards (rough and finished if two are used) and is deducted at right angles to the plumb cut line at the fascia.

LAYING OUT AND CUTTING A COMMON RAFTER

After all the information has been gathered, from the blueprints or by actual measurement, and all the calculations have been made, the actual layout of the rafter can begin. Place the rafter stock selected on two saw horses with the *crown edge* towards you. Grasp the body of the square with your left hand and the tongue with your right hand. Read the unit of run on the body and the unit of rise on the tongue; place these figures on the same edge

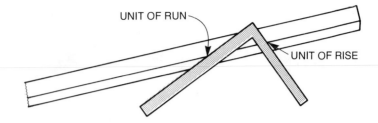

Figure 10-3   Framing square on rafter stock

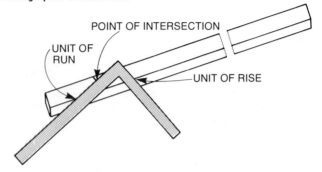

Figure 10-4   Layout for birdsmouth

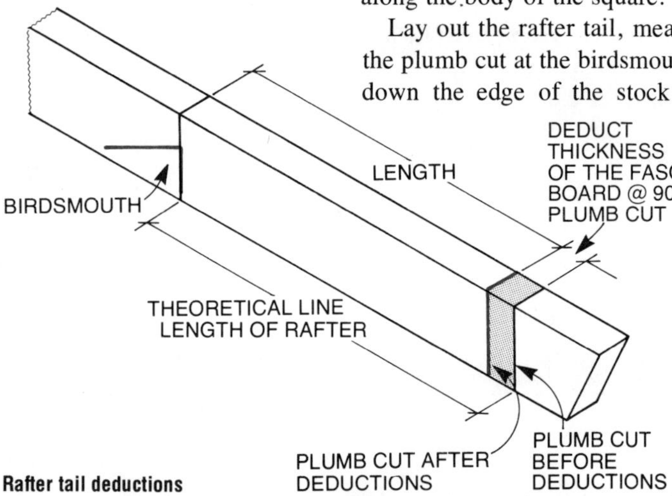

of the rafter stock (Fig. 10-3). Scribe a plumb line along the tongue. The plumb line should be made at the very end of the rafter stock, this will facilitate the measuring of the line length as the tape can be hooked on the end of the stock.

Lay out the line length, measuring from the end of the rafter stock where the plumb cut was scribed. Measure down the edge of the stock the distance determined for the line length and establish a point.

Scribe a plumb line through this point. This will be the plumb cut at the birdsmouth. From the far side of the rafter stock, measure along the plumb cut line to the desired depth of the birdsmouth and establish a point. The depth of the birdsmouth is usually the thickness of the top plate (38 mm). Scribe a level or seat cut line to intersect the point established on the plumb cut line. Place the square as shown in Figure 10-4 and scribe a line along the body of the square.

Lay out the rafter tail, measuring from the plumb cut at the birdsmouth. Measure down the edge of the stock to the line

Figure 10-5   Rafter tail deductions

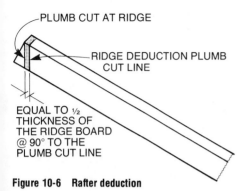

PLUMB CUT AT RIDGE

RIDGE DEDUCTION PLUMB CUT LINE

EQUAL TO ½ THICKNESS OF THE RIDGE BOARD @ 90° TO THE PLUMB CUT LINE

Figure 10-6   Rafter deduction

length of the rafter tail and establish a point. Scribe a plumb cut line through this point. To lay out the deduction at the rafter tail, deduct the thickness of the fascia board(s). This deduction is equal to the **thickness** of the fascia board measured at 90° to the plumb cut at the rafter tail (Fig. 10-5).

Lay out the ridge deductions, which will be equal to **one-half** the thickness of the ridge board measured at 90° to the plumb cut (Fig. 10-6). Cut the rafter on the deduction lines.

Using this rafter as a pattern, cut out a second rafter. At this time it is a good idea to check the layout and cutting by setting the two rafters in place to see if they fit. If they do, continue to cut the remaining rafters required, using the original as a pattern.

With time and experience, you will find there are many jigs and set-ups you can make that will enable you to mass-produce rafters using such power equipment as a radial arm saw or a portable electric handsaw.

RIDGE BOARD

The ridge board is the horizontal member which supports and aligns the upper ends of the roof rafters. The length of the ridge board for a gable roof is determined by the length of the building plus the addition of the overhang at each end. There are two lengths to be considered, the *theoretical length* and the *actual length*. The theoretical length of the ridge board is the same as the length of the building; the actual length is the theoretical length plus the addition for overhang at each end. If the ridge board is not continuous, it must be joined at the centre of two opposing rafters.

COLLAR TIES, DWARF WALLS, AND STRUTS

If intermediate support is deemed necessary to help support the roof load, *collar ties, dwarf walls,* or *struts* may be used. Collar ties are made from 38 × 89 mm, or 19 × 140 mm stock and are used to tie the opposing rafters together. The collar ties should be located in the middle third of the rafter (Fig. 10-7). The line length of the collar tie should be approximately the same length as the total run of the

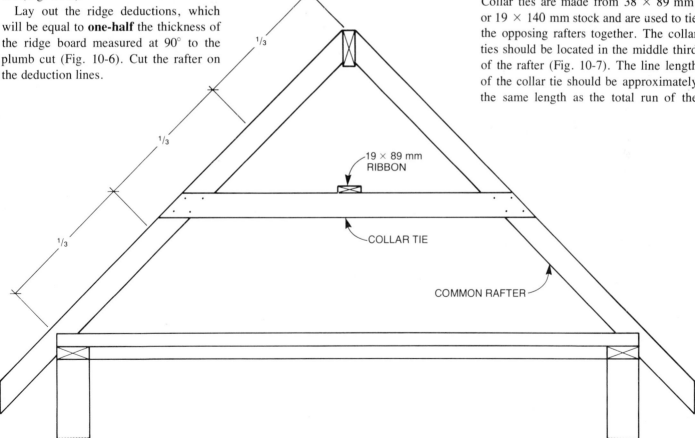

⅓

⅓

⅓

⅓

19 × 89 mm RIBBON

COLLAR TIE

COMMON RAFTER

Figure 10-7   Collar tie and supporting ribbon

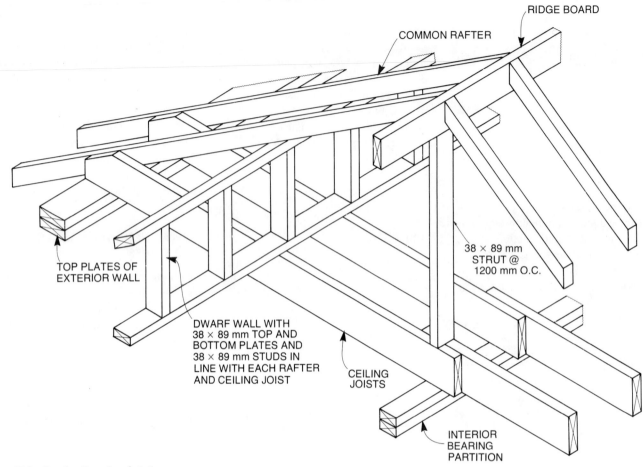

RIDGE BOARD

COMMON RAFTER

38 × 89 mm
STRUT @
1200 mm O.C.

TOP PLATES OF
EXTERIOR WALL

DWARF WALL WITH
38 × 89 mm TOP AND
BOTTOM PLATES AND
38 × 89 mm STUDS IN
LINE WITH EACH RAFTER
AND CEILING JOIST

CEILING
JOISTS

INTERIOR
BEARING
PARTITION

Figure 10-8   Dwarf walls and roof struts

common rafter; this will put the collar tie in the centre of the rafter. All calculations for roof members are taken from centre lines. If a collar tie is more than 2400 mm long, a continuous ribbon of 19 × 89 mm stock must be nailed on top of the collar ties and anchored to the gable end walls. Dwarf walls may also be used as roof support to reduce the span for rafters. Dwarf walls are built of 38 × 89 mm stock with continuous top and bottom plates and with the studs placed on the same centres as the rafters and ceiling joists (Fig. 10-8). When ceiling joists are used to support part of the roof load, they

should be one dimension greater in depth than is required when they are not supporting the roof load.

Struts are made from 38 × 89 mm stock and extend from the ridge board to a loadbearing partition at an angle of 45° or more to the horizontal. Struts are necessary to support the ridge in a roof with a slope of 1:4 or less (Fig. 10-8).

EAVE PROJECTIONS AND GABLE ENDS

The rafter tails are kept in alignment by a framing member called a *rough fascia* which also provides a nailing surface for finishes under the eave called *soffit*.

Gable end projections are framed at the outer edge of the roof with a rafter which is supported at the top by the extended ridge board and at the bottom by the rough fascia. Because it is not supported by the outside wall there is no birds-mouth. This rafter is called either a *rake* rafter or a *barge* rafter. Gable end projections extending more than 400 mm beyond the wall are usually supported by framing members called *lookouts*. One method of framing is to notch the end rafter to hold a 38 × 89 mm member on the flat, which will be end-nailed to the barge rafter and the first rafter from the end (Fig. 10-9). Lookouts are usually placed 600 to 1200 mm o.c. depending on the width of overhang.

Gable end studs provide framing for sheathing to close in the gable end. In calculating the line length of the gable end studs, remember all calculations are taken from centre lines and the intersecting line at the birdsmouth. The line length of a gable end stud is the theoretical perpendicular length from the top wall plates to the intersecting point of the line length of the common rafter (Fig. 10-10). When calculating the line length of a gable end stud, determine the run (which will be the centres indicated on the blueprint). Determine the unit of rise from the roof slope. Use the following equation:

$$\frac{\text{Run of Gable End Stud}}{\text{Unit of Run}} \times \text{Unit of Rise}$$

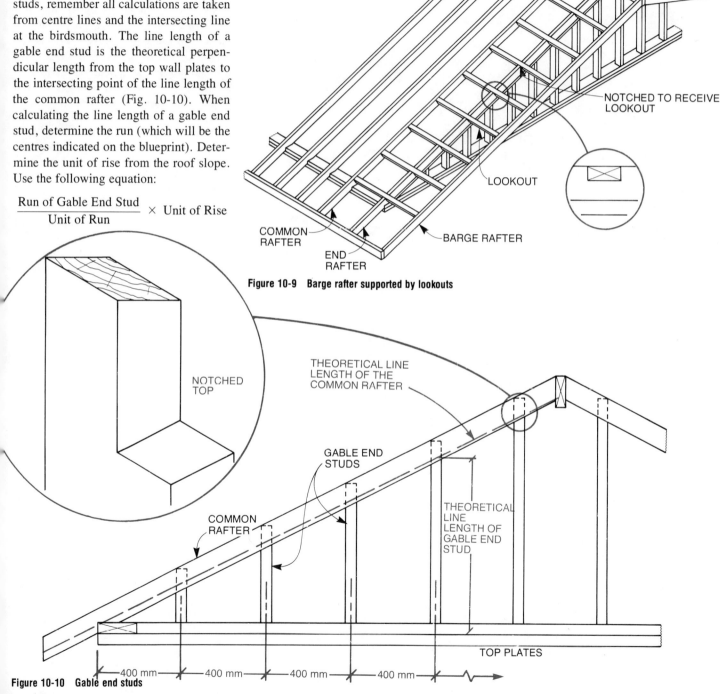

Figure 10-9   Barge rafter supported by lookouts

Figure 10-10   Gable end studs

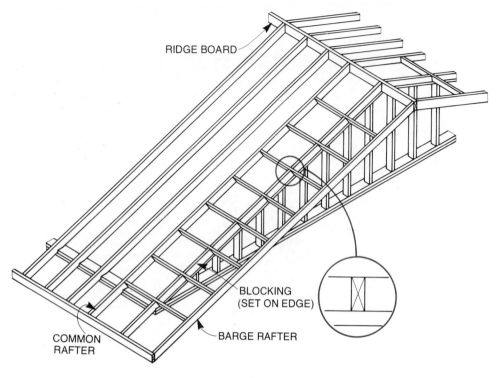

RIDGE BOARD

BLOCKING
(SET ON EDGE)

COMMON
RAFTER

BARGE RAFTER

Figure 10-11   Roof ladder

Example:
The slope of the roof in Figure 10-10 is
1:2, the unit of run is 250 (Frederickson),
the unit of rise is 125, and the stud centres
400 mm.

$$\frac{400}{250} \times 125 = 200$$

The length of the first stud is 200 mm;
each stud after that will increase by
200 mm if they are placed 400 mm o.c.
As can be seen in Figure 10-10, deductions must be made because of centre line
calculations, in order for the gable end
stud to fit.

In construction today, it is common
practice to construct the gable end assembly at the same time as the end wall frame
and to raise both as a single unit (Fig. 10-
11). This is done by laying out the full

size gable end on the subfloor, using a
chalkline and two rafters to give the correct slope. Lay out and cut plates for the
gable end frame, measure lengths for
gable studs, and cut and fasten in place.
Attach these to the wall frame and raise it
into position.

A roof ladder is made by nailing lookouts to the barge rafter and the first rafter
from the end. This is all done on the subfloor and then raised into position. The
upper end of the barge rafter is supported
by the extended ridge board and the lower
end is supported by the rough fascia.

HIP RAFTER

A roof framing member called a hip rafter
is necessary where the two sloping surfaces of a hip roof meet. The hip rafter
meets the ridge board at an angle of 45°.

It should be at least one dimensional size
greater in depth than the common rafter to
provide full bearing for the hip jack
rafters.

Methods of Framing

There are two methods of framing hip
roofs; one is to use a *single check* cut, the
other is to use a *double check* cut. In the
single cut method, the hip rafters are
framed into the ridge board (Fig. 10-12).
In the double cheek cut method the hip
rafters are framed into common rafters
(Fig. 10-13).

Calculating Hip Rafter Line Length

When calculating the line length of a hip
rafter, find the unit of rise from the slope.
Determine the number of units of run for
the hip rafter; these will be the same as
those for a common rafter. Determine the
unit of line length for the hip rafter by
looking on the second line on the rafter
table under the unit of rise. If the unit of
rise is not given on the framing square
the unit of line length must be determined

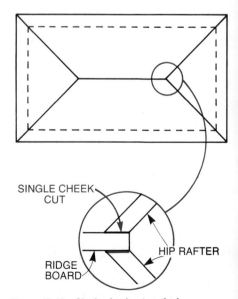

SINGLE CHEEK
CUT

HIP RAFTER

RIDGE
BOARD

Figure 10-12   Single cheek cut method

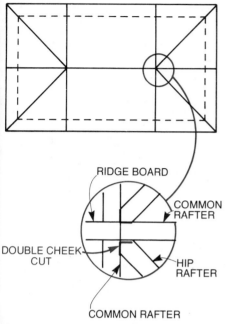

Figure 10-13   Double cheek cut method

by calculation. The formula for finding the unit of line length is:

Unit

of L.L. $= \sqrt{\text{Unit of rise}^2 + \text{unit of run for hip}^2}$

Example:
Slope–1:3
Unit of rise–83.3 mm
Unit of run–353.5 mm
Unit of L.L. $= \sqrt{83.3^2 + 353.5^2}$
$= \sqrt{6\ 938.89 + 124\ 962.25}$
$= \sqrt{131\ 901.14}$
$= 363.2$

After the unit of line length is determined, multiply this figure by the number of units of run. This will give the line length of the hip rafter.

Example:
Building size–7200 × 12 000 mm
Roof slope–1:2.5
Overhang–750 mm
From the information:
(a) Span = 7200 mm

(b) Total run = 3600 mm
(c) Number of units of run = 14.4
(d) Unit of rise = 100
     1:2.5 = x:250
     2.5x = 250
       x = 100
(e) Unit of line length = 367.4
     found on the second line of the rafter table
(f) Line length of hip rafter =
     367.4 × 14.4 = 5290.56

This measurement is from the plumb cut at the birdsmouth to the centre of the ridge board.

Calculating Line Length of Hip Rafter Tail

The hip rafter tail is that portion of the hip rafter which extends beyond the wall plates of a hip roof. When calculating the line length, determine the number of units of run by dividing the overhang by the unit of run for a common rafter. The unit of line length will be the same as for the rest of the hip rafter. Multiply this by the number of units of run.

Example: (from information on L.L. of hip)
(a) Run (overhang) = 750 mm
(b) Number of units of run = 3
       750 ÷ 250 = 3
(c) Unit of Line Length = 367.4 mm
(d) Line Length of hip rafter tail =
       367.4 × 3 = 1102.2

This measurement is made from the plumb cut at the birdsmouth to the plumb cut at the rafter tail.

Deductions on a Hip Rafter

Using the single cheek cut method, the deduction to the upper end of the hip rafter is **one half** the 45° thickness of the ridge board (Fig. 10-14).

Using the double cheek cut method, the deduction to the upper end of the hip rafter is **one half** the 45° thickness of the common rafter (Fig. 10-15). Deductions for both methods are taken at right angles to the plumb cut line.

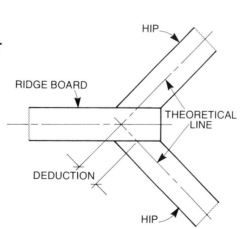

Figure 10-14   Deductions-single cheek cut method

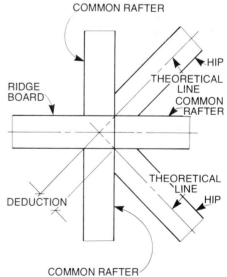

Figure 10-15   Deduction-double cheek cut method

RISE PER UNIT RUN	500	450	400	350	300	250	200	150	100	50
LENGTH OF COM. PER UNIT RUN (250 mm)	559.0	514.8	471.7	430.1	390.5	353.6	320.2	291.5	269.3	255.0
LENGTH OF HIP PER UNIT RUN (353.5 mm)	612.4	572.3	533.9	497.5	463.7	433.0	406.2	384.1	367.4	357.1
DIFF. IN JACKS AT 400 mm O.C.	894	824	755	688	625	566	512	466	431	408
SIDE CUT OF HIP OR VALLEY	144	154	166	178	191	204	218	230	241	248
SIDE CUT OF JACKS	112	121	133	145	160	177	195	214	232	245
ANGLE OF COM. AT PLATE	63.44°	60.95°	58.00°	54.46°	50.20°	45.00°	38.66°	30.96°	21.80°	11.31°

Figure 10-16   Frederickson Framing Square table

Like the common rafter, a deduction must be made on the rafter tail of the hip rafter for the fascia board(s). The deduction is the 45° thickness of the ridge board taken at right angles to the plumb cut line.

Dropping the Hip

The term *dropping the hip* is used for an operation which lowers the hip rafter to allow the top edges of the hip rafter to line up with the top edges of the common and hip jack rafters. This alignment of the rafter edges allows the sheathing to lay flat at the hips. The operation is done when laying out the hip rafters.

Side Cut of a Hip Rafter

The side cut tables on the framing square are used when laying out the *side cut* of a hip rafter. (The side cuts of a hip rafter are on the edge of the rafter stock and not on the side as might be expected). The side cut tables for hip rafters are found on the fourth line of the rafter table. Look under the unit of rise to find the required figures (Fig. 10-16). If the unit of rise cannot be found on the rafter table, use a side cut figure closest to that figure shown on the square and the cut will fit reasonably well. Use the side cut figure (obtained from the framing table) on the tongue of the square and 250 on the body. Mark along the body (Fig. 10-17).

LAYING OUT A HIP RAFTER

After all the information has been gathered and the calculations are complete, the actual layout can begin. Place the rafter stock on two sawhorses with the crown edge towards you. Square the crown edge of the rafter stock at the extreme right hand end. Grasp the square with the body in the left hand and the tongue in the right hand and locate the unit of rise on the tongue and the unit of run (354) on the body. Scribe a plumb line on the side of the rafter stock, along the tongue of the square. The plumb cut line should meet the squared-off line at the edge (Fig. 10-18). Lay out the line length, measuring from the squared-off

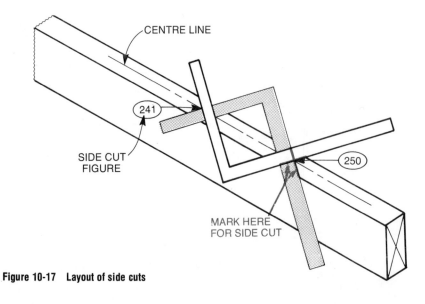

CENTRE LINE

241

SIDE CUT
FIGURE

250

MARK HERE
FOR SIDE CUT

Figure 10-17   Layout of side cuts

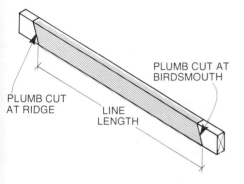

Figure 10-18 Hip rafter plumb cuts and line length

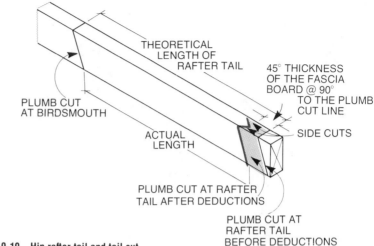

Figure 10-19 Hip rafter tail and tail cut

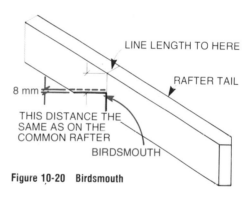

Figure 10-20 Birdsmouth

line on the top of the rafter stock; measure down the edge of the stock the distance determined for the line length and establish a point. Square the crown edge of the stock at this point (Fig. 10-18). Scribe a plumb line through this point. This will be the plumb cut at the birdsmouth.

Starting from the squared-off line at the birdsmouth, lay out the rafter tail. Measure down the edge of the stock the distance determined for the line length of the hip rafter tail and establish a point. Square the edge of the rafter stock at this point. This distance is the theoretical line length of the rafter tail. Scribe a plumb line through this point.

Lay out the deduction at the fascia board. From the plumb cut line measure back a distance equal to the **thickness** of the 45° thickness of the fascia board(s). Establish a point, scribe a plumb cut line through this point, and square to the top edge of the rafter stock through this point. Bisect this squared-off line. With the framing square and the method illustrated in Figure 10-17 mark the side cut lines to intersect the squared-off line where it is bisected (Fig. 10-19). Scribe plumb lines on each side of the rafter stock (Fig. 10-19). Cut on these lines.

Lay out the birdsmouth by measuring a plumb line the same distance as the common rafter and establish a point (Fig. 10-

20). Draw a seat cut line through this point. Drop the hip rafter. The following equation can be used to calculate the amount the hip rafter should be dropped:

Amount of drop =

$$\frac{\text{Unit of rise} \times \frac{1/2 \text{ thickness}}{\text{of rafter stock}}}{\text{Unit of run}}$$

Example:

Where the unit of rise is 150, the unit of run is 354 and the rafter stock is

38 mm, the following equation could be set up:

$$x = \frac{150 \times 19}{354}$$
$$= 8$$

The amount of drop would be 8 mm.

Lay out the deductions at the ridge. From the upper plumb cut line measure at 90° a distance equal to **one half** the 45° thickness of the ridge board and establish a point. Scribe a deduction plumb line through this point. Square the top edge of the rafter stock to meet this deduction plumb line. Bisect this squared-off line.

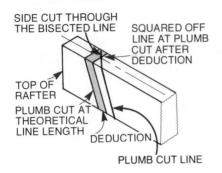

Figure 10-21   Single cheek cut

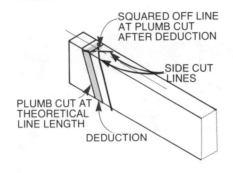

Figure 10-22   Double cheek cut

Draw a side cut line through the point where the squared-off line was bisected. The position of the rafter in the roof will determine the way to draw this line (Fig. 10-21). Draw plumb cut lines from both ends of the side cut line. Cut on these

lines. These are cuts for the single cheek cut method.

If a double cheek cut method is used, make the deductions as for a single cheek cut and then draw a second side cut line (Fig. 10-22). Draw plumb lines from the

ends of the side cut lines. Cut on these lines.

HIP JACK RAFTER

The hip jack rafter is similar to the common rafter in that the birdsmouth and the rafter tail are the same for each. The hip jack rafter is framed into the hip rafter, meeting the hip at an angle of 45°. Hip jacks are cut in pairs, and there are four pairs of hip jacks of each length on a hip roof (Fig. 10-23).

The unit of run of the hip jack rafter is the same as the common rafters, which is 250 mm using the Frederickson square. The total run of the hip jack rafter is equal to the distance it is from the corner of the building (Fig. 10-24).

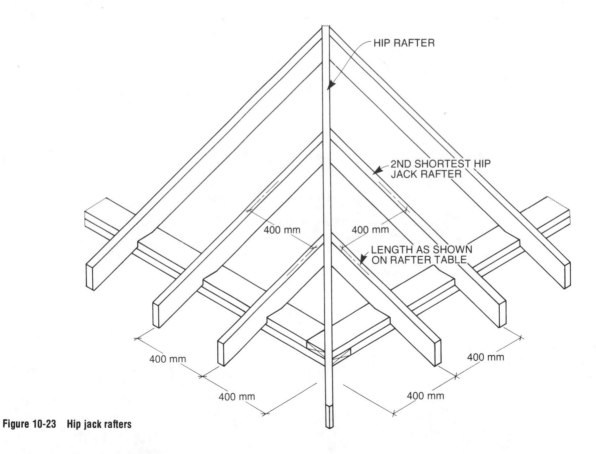

Figure 10-23   Hip jack rafters

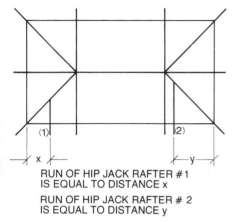

RUN OF HIP JACK RAFTER #1
IS EQUAL TO DISTANCE x

RUN OF HIP JACK RAFTER #2
IS EQUAL TO DISTANCE y

Figure 10-24   Total run for hip jack rafters

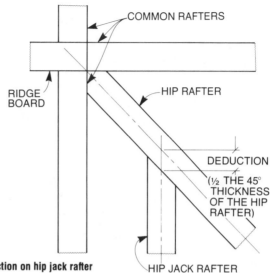

Figure 10-25   Deduction on hip jack rafter

Calculating the Line Length of a Hip Jack Rafter

When calculating the line length of a hip jack rafter, find the unit of rise from the slope. Determine the total run by measuring the distance the rafter is from the corner of the building. Divide the total run by the unit of run (250) to get the number of units of run. Multiply the unit of line length, which is the same as that for the common rafter, by the number of units of run to find the total line length.

Example:
Rafter Layout–400 mm o.c. (starting from the corner)
Roof Slope–1:2.5

Find the line length for the **third hip jack** from the corner:
(a) Unit of rise = 100 mm
(b) Total run = 1200 mm
(c) Number of units of run = 4.8
    1 200 ÷ 250 = 4.8
(d) Unit of line length = 269.3 mm
(e) Line length for hip jack rafter =
    263.5 × 4.8 = 1264.8 mm

This measurement runs from the plumb cut at the birdsmouth to the centre of the hip rafter. The hip jack rafter tail is identical to the common rafter tail; it is cut in the same way, and the line length is the same.

Deduction on Hip Jack Rafters

The deduction made on the upper end of the hip jack rafter is **one half** the 45° thickness of the hip rafter (Fig. 10-25).

Common Difference of Hip Jacks

The *common difference* of hip jacks is a term used to identify the amount by which one hip jack is longer or shorter than the preceding one. There are two methods of determining the common difference of jacks; the calculation method and the difference of length tables. Using the calculation method, the formula is:

$$\frac{\text{Unit of L.L.}}{\text{Unit of run}} \times \frac{\text{Spacing}}{\text{of jacks}} = \frac{\text{common}}{\text{difference}}$$

Example:
Roof Slope–1:3
Rafter spacing–600 mm
Unit of L.L. = 263.5 mm (calculation)
Unit of run = 250

Using the formula: $\dfrac{263.5}{250} \times 600 = 632.4$

The common difference is 632.4 mm, which means that for every rafter placed 600 mm away from the preceding one, it will be 632.4 mm longer in line length. Using the difference in length tables found on the framing square, the third line will be for jacks at 400 mm o.c. Under the unit of rise the difference in length will be given. This table is only of value if the unit of rise can be found on the square and the rafters are placed 400 mm o.c.

Since there are four pairs of hip jack rafters of each length on a hip roof, a hip jack pattern is usually made up. The hip jack pattern starts from the last common rafter. Subtract the common difference of jacks from the common rafter line length to get the line length of the longest hip jack rafter. The second longest hip jack rafter will be the common difference shorter than the preceding hip jack rafter, and so on (Fig. 10-26).

Side Cut of Hip Jacks

The side cut of jacks can be found on the

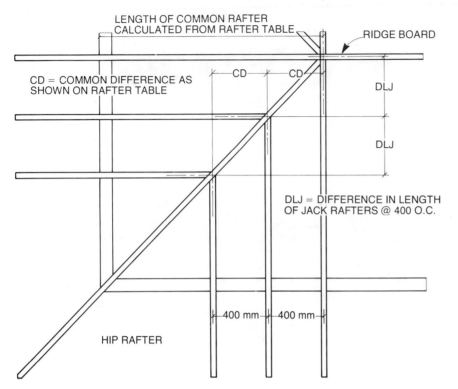

Figure 10-26   Common difference of hip jacks

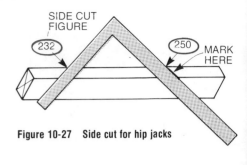

Figure 10-27   Side cut for hip jacks

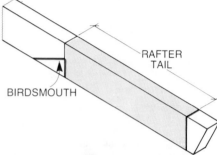

Figure 10-28   Birdsmouth and rafter tail after deductions on the hip jack rafter

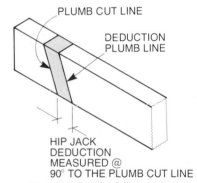

Figure 10-29   Deduction plumb line

framing square on the fifth row of the rafter table. Use the side cut figure on the tongue of the square and 250 on the body. Mark along the body (Fig. 10-27).

LAYING OUT A HIP JACK RAFTER

Place the rafter stock on sawhorses with the crown edge towards you. Square the crown edge of the rafter stock at the extreme right hand end. Scribe a plumb cut line using the unit of run for a common rafter (250) and the unit of rise determined from the slope of the roof. The plumb cut line should meet the squared-off line at the edge. Lay out the line length, beginning at the squared-off line on the top of the rafter stock; measure down the edge of the stock to the distance determined for the line length and establish a point. Square the edge of the stock

at this point. Scribe a plumb cut for the birdsmouth. Scribe the level or seat cut line in the same position as on the common rafter (Fig. 10-4). Starting from the squared-off line at the birdsmouth, lay out the rafter tail by measuring down the edge of the stock to a distance determined for the line length of the common rafter tail and establish a point. Square the crown edge of the rafter at this point and scribe a plumb cut line to meet the squared-off line (Fig. 10-28). From this squared-off line, deduct the thickness of the fascia board for a common rafter (Fig. 10-5).

Lay out the deductions where the hip jack meets the hip rafter by measuring from the upper plumb cut line a distance equal to **one half** the 45° thickness of the hip rafter at 90° to the plumb cut and establish a point. Scribe a plumb cut line through this point. Square the top of the

rafter to meet this plumb line (Fig. 10-29). Bisect the squared-off line.

Scribe a side cut line through the point where the squared-off line was bisected. The position of the rafter in the roof determines which way to draw this line. Scribe plumb cut lines at both ends of the side cut line. Cut on these lines (Fig. 10-30).

LAYING OUT A HIP JACK RAFTER PATTERN

Determine the common difference of hip jack rafters either from the framing square

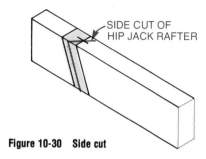

Figure 10-30   Side cut

or by calculation. First lay out a common rafter, then on this same stock lay out the longest jack. On the top edge of the rafter measure down from the squared-off line length the determined common difference and establish a point. Square the top edge of the stock through this point (Fig. 10-31). Scribe a plumb line to meet this line.

Lay out the deduction for the longest hip jack by measuring a distance equal to **one half** the 45° thickness of the hip rafter at 90° to the plumb cut line and establish a point. Scribe a plumb cut line at this point. Square the top edge to meet this line. Bisect this squared-off line. Using the side cut figures for jacks and 250 on the framing square, scribe a side cut line (Fig. 10-27). From the longest point of the side cut line, square across the top edge of the stock (Fig. 10-32). From this squared-off line measure down the top of the rafter the common difference and establish a point. Square the rafter at this point. Repeat this operation as many times as necessary. Number the hip jack rafters 1, 2, 3, etc.

RIDGE BOARD ON A HIP ROOF

Because the hip roof can be framed using either the single cheek cut method or the double cheek cut method, the carpenter must be able to calculate the length of ridge board for either method. There are two lengths to be considered when calculating the length of the ridge board, the theoretical and the actual. The theoretical

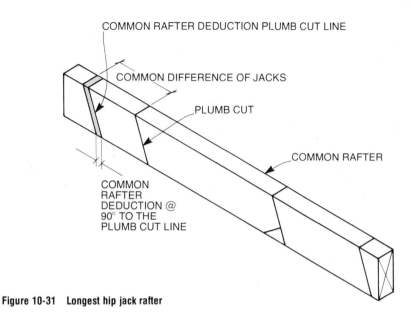

Figure 10-31   Longest hip jack rafter

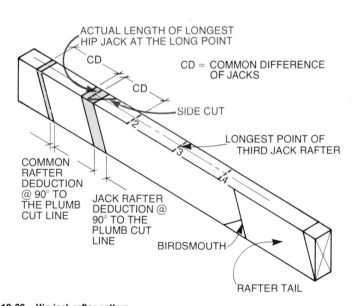

Figure 10-32   Hip jack rafter pattern

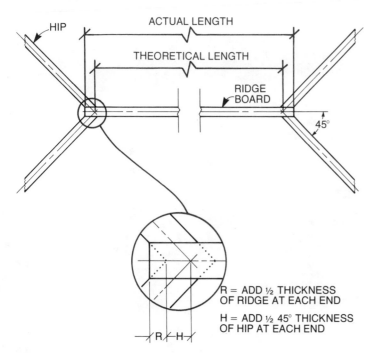

Figure 10-33   Ridge board for single cheek cut method

length for a hip roof is the length of the building minus the width of the building. This will be the same no matter which framing method is used. Using the single cheek cut method, the theoretical length plus the thickness of the ridge board plus the 45° thickness of the hip rafter stock, equals the actual length (Fig. 10-33).

Using the double cheek cut method, the theoretical length plus the thickness of the common rafter stock equals the actual length of the ridge board (Fig. 10-34).

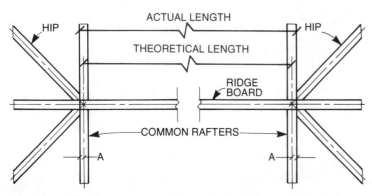

Figure 10-34   Ridge board for double cheek cut method

MATERIAL REQUIREMENTS FOR A HIP ROOF

Listed here are the main components of a hip roof and the methods used to determine the amount of material required:

Ridge Board—Building length minus building width plus additions.

Common Rafters (at 400 mm o.c.)—The number of rafters equals the ridge board length divided by 400. If the single cheek cut method is used, add one, if the double cheek cut method is used, add two. These calculations are for one side of the roof. They must be doubled when calculating the entire roof.

Hip Rafters—Four hips are required for a rectangular shaped roof.

Hip Jack Rafters (at 400 mm o.c.)—The number of pieces required equals the roof width divided by 400. Round-off to the next whole number. This will be the number required for each hip. Use common rafter stock for hip jacks as one common rafter will usually make two hip jack rafters.

Sheathing—The area of a hip roof is the same as the area of a gable end roof. Use the following formula when calculating the area of a hip roof:

$$\frac{\text{Unit of L.L. of common rafter}}{250} \times \frac{\text{width}}{\text{of roof}} \times \frac{\text{length}}{\text{of roof}}$$

THE INTERSECTING ROOF

A building which is L-shaped, or which has a projection, requires a type of roof referred to as *intersecting*. This type of roof forms a *valley* at the intersection of the two sloped surfaces. The intersecting roofs we are concerned with here are of equal pitch. The carpenter must be familiar with all the members in an intersecting roof and of how they relate to one another. Figure 10-35 shows a plan view of

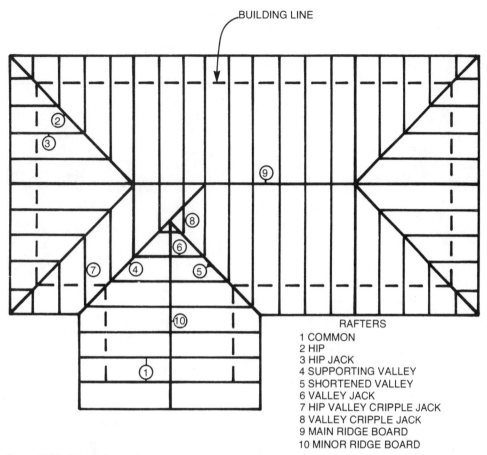

BUILDING LINE

RAFTERS
1 COMMON
2 HIP
3 HIP JACK
4 SUPPORTING VALLEY
5 SHORTENED VALLEY
6 VALLEY JACK
7 HIP VALLEY CRIPPLE JACK
8 VALLEY CRIPPLE JACK
9 MAIN RIDGE BOARD
10 MINOR RIDGE BOARD

Figure 10-35   Intersecting roof

an intersecting roof. A pictorial view is shown on page 164.

Intersecting Roof Members

1. Common rafter
2. Hip rafter
3. Hip jack rafter
4. Supporting valley rafter
5. Shortened valley rafter
6. Valley jack rafter
7. Hip-valley cripple jack
8. Valley cripple jack
9. Main ridge board
10. Minor ridge board

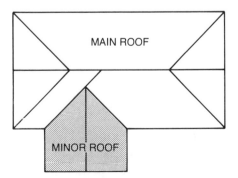

MAIN ROOF

MINOR ROOF

Figure 10-36   Main and minor roofs

MAIN AND MINOR ROOF

The intersecting roof consists of two or more roof sections. These roof sections are called the *main* roof and the *minor* roof. Sometimes called the *major* roof, the main roof is the larger section of an intersecting roof. The minor roof is the smaller section. In Figure 10-36 the main roof is shown in black and the minor roof is shown in colour.

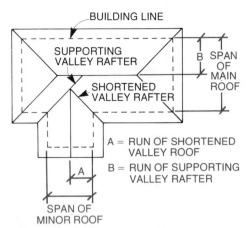

Figure 10-37   Run of valley rafters

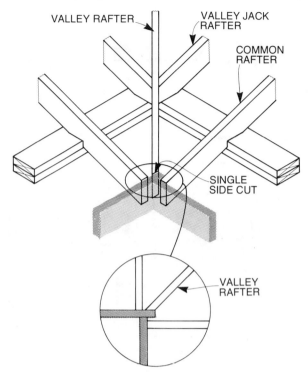

Figure 10-38   Valley rafter tail

VALLEY RAFTER

When an internal corner occurs in a roof, it is necessary to have a rafter at that location, called a *valley rafter*. The valley rafter meets the member to which it is framed at 45°. There are two kinds of valley rafters, the *supporting* valley rafter and the *shortened* valley rafter. When the span of the minor roof is less than the span of the main roof, both kinds of valley rafters are used (Fig. 10-37).

Run of Valley Rafter

The valley rafter is parallel to the hip rafter in the intersecting roof, therefore it also has a unit of run of 353.5 mm. This is true for both the supporting and shortened valley rafters.

For line length calculations, the run of a valley rafter is the same as the run for a hip rafter in that particular section of the intersecting roof. The run of the supporting valley rafter is one half the span of the main roof, while the run of the shortened valley rafter is one half the span of the minor roof (Fig. 10-37).

Calculating Line Length of Valley Rafters

When calculating the line length of a valley rafter, follow the same steps used for

the hip rafter. The line length of the supporting valley will be the same as the line length of the hip rafter on the main roof. The line length of the shortened valley rafter will be the same as the line length of the hip rafter on the minor roof. When calculating the line lengths, remember that the supporting valley rafter is based on the main roof and the shortened valley rafter is based on the minor roof.

Valley Rafter Tail

The line length of the valley rafter tail is the same as the line length of the hip rafter tail. A single side cut is used on the valley rafter tail because the rough fascia will extend past the valley rafter on the main roof. The rough fascia from the minor roof will butt into that of the main roof (Fig. 10-38).

Side Cut on Valley Rafter

The side cut tables on the framing square are the same for the valley as for the hip rafter. Use the same method for finding and laying out the side cut figure as was used for the hip rafter. The supporting valley rafter has a single side cut at its upper end. The side cut is either a left or right hand depending on its location in the roof (Fig. 10-39). The shortened valley rafter has a square side cut at its upper end because it meets the supporting valley rafter at 90° (Fig. 10-39).

When the main and minor spans of a building are equal, the ridge boards will be the same height. When this occurs the valley rafter has a double cheek cut at its upper end (Fig. 10-40).

Deductions on Valley Rafters

The deduction made to the upper end of the supporting valley rafter is **one half** the

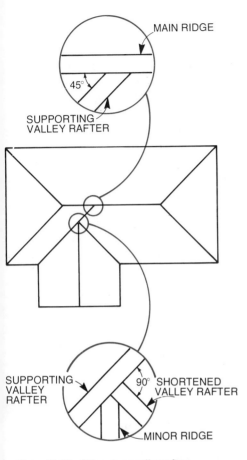

Figure 10-39   Side cuts on valley rafters

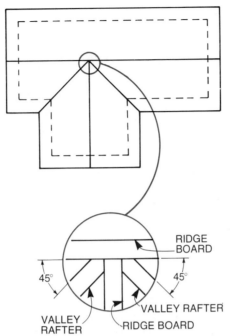

Figure 10-40   Valley rafters when ridges are the same height

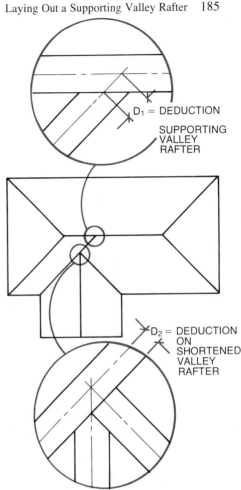

Figure 10-41   Deductions to valley rafters

45° thickness of the main ridge. The deduction made to the upper end of the shortened valley rafter is **one half** the 90° thickness of the supporting valley rafter (Fig. 10-41).

Location of Upper Ends of Valley Rafters

The correct positioning of the upper ends of the valley rafters is very important as an incorrect location will affect all the jacks framed to these rafters. Figure 10-42 shows the correct locations for the valley rafters.

LAYING OUT A SUPPORTING VALLEY RAFTER

Place the rafter stock on the sawhorses with the crown edge towards you. Square the crown edge of the rafter stock on the extreme right hand end. Grasping the square with the body in the left hand and the tongue in the right hand, locate the unit of rise on the tongue and the unit of run (354) on the body. Mark a plumb line on the side of the rafter stock along the tongue of the square to meet the squared-off line. Begin the measurement for the line length at the squared-off line on the top of the rafter stock. Measure down the top of the stock to the distance determined for the line length and establish a point. Square the crown edge of the stock through this point. Scribe a plumb cut line to meet the squared-off line. This will be the plumb cut at the birdsmouth. Lay out

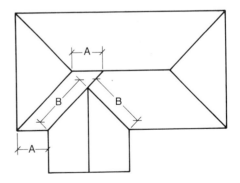

DISTANCES MARKED **A** ARE EQUAL
DISTANCES MARKED **B** ARE EQUAL

Figure 10-42   Location of valley rafters

the supporting valley rafter tail; begin the measurement at the squared-off line at the birdsmouth; measure down the top of the stock to the distance determined for the line length of the rafter tail and establish a point. Scribe a plumb cut line. Make the necessary deductions for the fascia the same as for the hip rafter. Square across the stock at the deduction plumb cut line. Bisect this squared-off line. Lay out the tail cut of the supporting valley rafter by using the side cut line on the crown edge

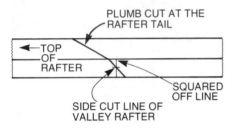

Figure 10-43 Tail cut for valley rafter

of the rafter to intersect the squared-off line where it was bisected (Fig. 10-43). The side cut line may be drawn either way. Scribe plumb lines at both ends of the side cut line. Cut on these marks.

Lay out the birdsmouth on the supporting valley rafter by bisecting the squared-off line on the crown edge of the birdsmouth. Scribe side cut lines from the bisected point back towards the rafter tail (Fig. 10-44). Scribe plumb lines on both sides of the rafter stock to intersect the side cut lines. Transfer the side cut lines

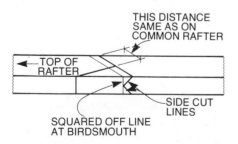

Figure 10-44 Birdsmouth on valley rafter

Figure 10-45 Finished cut on birdsmouth

to the underside of the rafter. The finished birdsmouth will be as shown in Figure 10-45. From the point of intersection of the squared-off line at the birdsmouth and the plumb cut line, measure down to a distance equal to the distance on the common rafter (p. 170). Draw a seat cut line through this point to the plumb line scribed to intersect the side cut line. No drop is required on valley rafters as the sheathing meets the valley rafter on its centre line.

Lay out the deduction for the ridge board; from the upper plumb cut line measure at 90° a distance equal to **one half** the 45° thickness of the ridge board and establish a point. Scribe a plumb line through this point. Square the crown edge of the rafter to intersect this plumb line, and bisect this squared-off line. Draw a side cut line through the point where the squared-off line was bisected. The position of the rafter in the roof determines which way to draw this line. Draw a plumb line to intersect the side cut line. Cut on this line (Fig. 10-46).

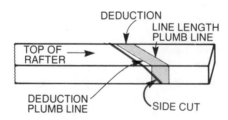

Figure 10-46 Deduction at ridge board

LAYING OUT A SHORTENED VALLEY RAFTER

Follow the same procedure described for a supporting valley rafter. The total run for the shortened valley rafter will be one

half the span of the minor roof. The deduction at the upper end of the shortened valley rafter is the only difference in the layout because the shortened valley meets the supporting valley at 90°. From the upper plumb cut line, measure at 90° a distance equal to **one half** the 90° thickness of the supporting valley rafter and establish a point. Scribe a plumb line through this point. Square the crown edge of the rafter to meet this plumb line. Cut on this mark (Fig. 10-47).

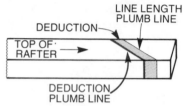

Figure 10-47 Deduction on shortened valley rafter

JACK RAFTERS

After the common rafters, hip rafters, hip jack rafters and valley rafters are installed, the carpenter must install the remaining jack rafters which are: *valley jacks, hip-valley cripple jacks* and *valley cripple jacks*.

VALLEY JACK RAFTER

The valley jack extends from the ridge to the valley rafter. It does not have a birdsmouth or a rafter tail. The upper end of the valley jack is identical to the common rafter (Fig. 10-35). The unit of run for the valley jack rafter is 250 mm. The run of a valley jack rafter is always **equal** to its distance from the intersection of the valley and ridge board (Fig. 10-48).

Calculating Line Length of Valley Jack

To calculate the line length of a valley jack rafter using the rafter tables on the framing square, determine the run of the valley jack. Then determine the number

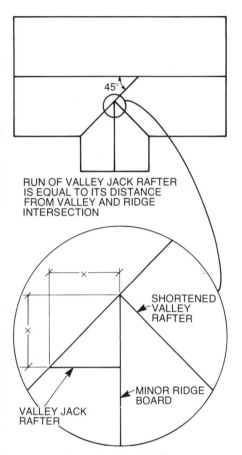

RUN OF VALLEY JACK RAFTER
IS EQUAL TO ITS DISTANCE
FROM VALLEY AND RIDGE
INTERSECTION

SHORTENED
VALLEY
RAFTER

MINOR RIDGE
BOARD

VALLEY JACK
RAFTER

Figure 10-48   Run of valley jack rafter

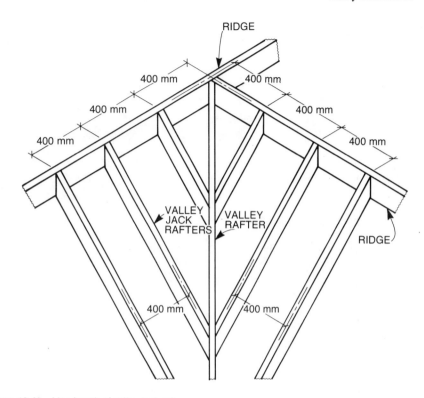

Figure 10-49   Line length of valley jack rafters

of units of run by dividing the total run by 250. From the unit of rise, (which is the same as the common rafter), find the unit of line length from the framing square. The unit of line length for jacks is always the same as those for the common rafter. Multiply the number of units of run by the unit of line length to find the line length of the valley jack rafter.

The line length of the valley jack can be determined by reading the common difference found on the square. When the valley jack rafter is spaced exactly 400 mm from the intersection of the valley rafter and the ridge board, the common difference found on the square is the line length of that rafter. The length of the

next valley rafter is determined by adding the common difference to the length of the preceding rafter (Fig. 10-49).

Deductions on Valley Jack

The deduction made to the valley jack rafter at the upper end is **one half** the 90° thickness of the ridge board. The deduction at the lower end is **one half** the 45° thickness of the valley rafter (Fig. 10-50).

Side Cuts on Valley Jack

To find the side cut figure used for valley jack rafters, look on the fifth row of the framing square, under the unit of rise figure. This figure is used with 250 to lay out the side cut of the valley jack rafter (Fig. 10-27).

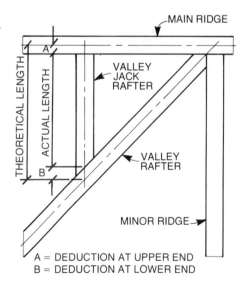

A = DEDUCTION AT UPPER END
B = DEDUCTION AT LOWER END

Figure 10-50   Deductions on valley jack rafters

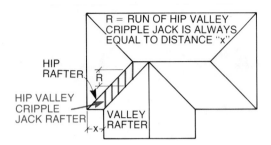

Figure 10-51   Run of hip-valley cripple jack

HIP-VALLEY CRIPPLE JACK RAFTER

The hip-valley jack rafter, sometimes called *cripple jack,* extends from the valley rafter to the hip rafter. It does not have a birdsmouth or rafter tail and its side cuts are parallel to each other (Fig. 10-35).

Run of Hip-Valley Cripple Jack

The unit of run for the hip-valley cripple jack is the same as the common rafter which is 250 mm. The total run of a hip-valley cripple jack is always **equal** to the distance between the centre of the hip rafter and the centre of the valley rafter, as measured along the ridge or plate line (Fig. 10-51).

Calculating Line Length of Hip-Valley Cripple Jack

The method of calculating the line length of the hip-valley jack is the same as the one used for the valley jack rafter. There is no common difference for hip-valley cripple jacks as all the cripple jacks are the **same length**. This is because the hip and valley rafters run parallel to each other.

Deduction on Hip-Valley Cripple Jack

The deduction made on the hip-valley cripple jack at the upper end of the cripple jack is **one half** the 45° thickness of the hip rafter and the deduction at the lower

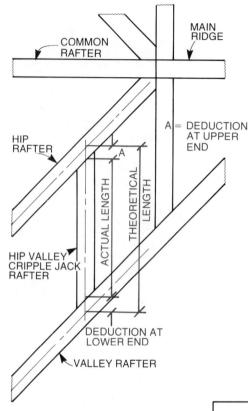

Figure 10-52   Deductions on hip-valley cripple jack

end is **one half** the 45° thickness of the valley rafter (Fig. 10-52).

Side Cuts on Hip-Valley Cripple Jack

The side cuts will be the same as those for the top cuts on hip jacks and the lower cuts on valley jacks.

VALLEY CRIPPLE JACK RAFTER

The valley cripple jack extends from the shortened valley rafter to the supporting valley rafter. It does not have a birdsmouth or a rafter tail. Each end of the valley cripple jack has a side cut, and the angle of the side cuts run in **opposite** directions (Fig. 10-35).

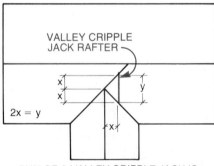

Figure 10-53   Run of a cripple valley rafter

Run of Valley Cripple Jack

The unit of run for a valley cripple jack is 250 mm. The run of a valley cripple jack rafter is found by doubling the distance that the rafter is from the point where the two valley rafters meet (Fig 10-53).

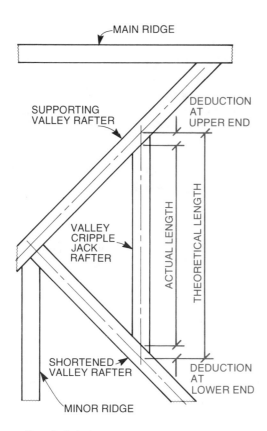

Figure 10-54   Deductions on valley cripple jack

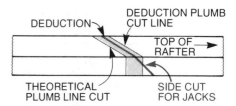

Figure 10-55   Side cut of valley jack

Calculating the Line Length of a Valley Cripple Jack

The method of calculating the line length of the valley cripple jack is the same as the one used for the valley jack.

Deductions on Valley Cripple Jack

The deduction made on the valley cripple jack rafter at the upper end is **one half** the 45° thickness of the supporting valley rafter and the deduction at the lower end is **one half** the 45° thickness of the shortened valley rafter (Fig. 10-54).

Side Cuts on Valley Cripple Jack

To find the side cut figure, use the same method used for valley jacks.

LAYING OUT A VALLEY JACK RAFTER

Place the rafter stock on sawhorses with the crown edge towards you. Square the crown edge of the rafter stock at the extreme right hand end. Scribe a plumb cut line to meet the squared-off line. Remember that the unit of run of all jack rafters is 250 mm. Lay out the line length, measuring from the squared-off line. Measure down the top of the stock to the distance determined for the line length and establish a point. Square the crown edge of the stock at this point. Scribe a plumb line to meet this squared-off line. Make a deduction at right angles to the plumb line; this distance will equal **one half** the 45° thickness of the valley rafter. Establish a point. Scribe a plumb line

through this point. Square the crown edge of the stock to meet this plumb line. Bisect this squared-off line and scribe a side cut line through this point using the side cut line figure for jacks and 250. Scribe a plumb line to meet the side cut line. Cut on this mark (Fig. 10-55). Lay out the deduction at the upper end in the same way as for a common rafter (Fig. 10-6).

LAYING OUT A HIP-VALLEY CRIPPLE JACK RAFTER

Place the rafter stock on sawhorses with the crown edge towards you. Square the crown edge of the rafter stock at the extreme right hand end. Scribe a plumb cut line to meet the squared-off line. Lay out the line length, measuring from the squared-off line. Measure down the top of the stock to the distance determined for the line length. Establish a point here.

Square the crown edge of the stock at this point. Scribe a plumb cut line to meet the squared-off line. Lay out for the deduction at the lower end by measuring back a distance equal to **one half** the 45° thickness of the valley rafters and establish a point. Scribe a plumb line through this point. Square the crown edge of the rafter stock and bisect this line. Scribe a side cut line through the bisected line. Scribe a plumb line to meet the side cut line. Cut on this line (Fig. 10-56).

Lay out the deduction at the upper end by measuring down from the upper plumb

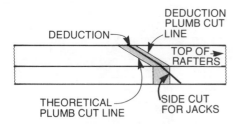

NOTE: LOWER CUTS OF VALLEY JACK AND HIP VALLEY CRIPPLE JACK ARE THE SAME

Figure 10-56   Lower deduction and side cut

cut line to a distance equal to **one half** the 45° thickness of the hip rafter and establish a point. Scribe a plumb line through this point. Square the crown edge of the stock to meet this plumb line. Bisect this squared-off line and scribe a side cut line through this point. Scribe a plumb line to meet the side cut line. Cut on this mark (Fig. 10-57). The top and bottom side cut lines should be parallel to each other.

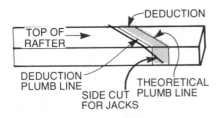

Figure 10-57   Upper deduction and side cut

LAYING OUT A VALLEY CRIPPLE JACK RAFTER

Place the rafter stock on the sawhorses with the crown edge toward you. Square the crown edge of the rafter stock at the extreme right hand end. Scribe a plumb line to meet this squared-off line. Lay out the line length, measuring from the squared-off line. Measure down the top of the stock to the distance determined for the line length and establish a point. Square the crown edge of the stock at this point. Scribe a plumb line to meet this

squared-off line. At right angles to this plumb line, make a deduction equal to **one half** the 45° thickness of the shortened valley rafter and establish a point. Square the crown edge of the stock to meet this plumb line. Bisect this squared-off line and scribe a side cut line through this point. Check the location of the rafter in the roof to determine the direction of the side cut line. Scribe a plumb line to intersect the side cut line. Cut on this mark (Fig. 10-58).

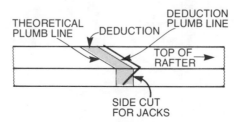

Figure 10-58   Lower deduction and side cut for valley cripple jack

Lay out the deduction at the upper end by measuring at right angles to the plumb cut line a distance equal to **one half** the 45° thickness of the supporting valley rafter and establish a point. Scribe a plumb line through this point. Square the crown edge of the stock to meet this plumb line. Bisect this squared-off line and scribe a side cut line through this point. The side cut line must be **opposite**

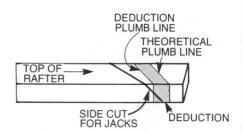

Figure 10-59   Upper deduction and side cut for valley cripple jack

to the lower side cut line. Scribe a plumb line to meet the side cut line. Cut on this mark (Fig. 10-59).

RIDGE BOARDS ON INTERSECTING ROOFS

Because there are many combinations of roofs, both gable and hip, with equal and unequal spans on main and minor roofs, different ridge lengths are required. The carpenter must be able to calculate the correct lengths for these ridges.

Intersecting Roof Using Hips and Unequal Spans on Main and Minor Roofs

The theoretical length of the minor ridge is the **same** as the projection of the minor roof. The theoretical length **minus one half** the 45° thickness of the valley rafter **plus one half** the thickness of the common rafter equals the actual length (Fig. 10-60).

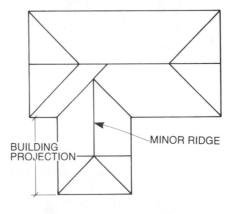

Figure 10-60   Ridge board for intersecting roof with unequal spans

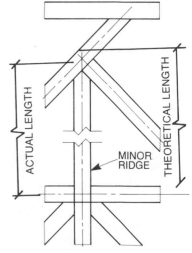

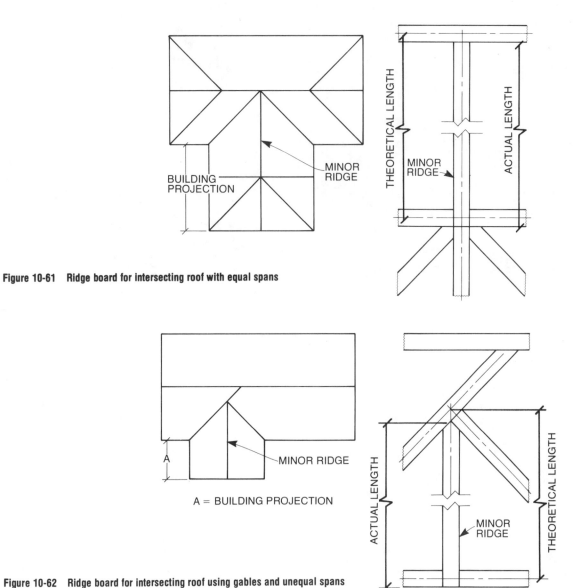

Figure 10-61   Ridge board for intersecting roof with equal spans

Figure 10-62   Ridge board for intersecting roof using gables and unequal spans

Intersecting Roof Using Hips and Equal Spans on Main and Minor Roofs

The theoretical length of the minor ridge is **equal** to the building projection. The theoretical length **minus one half** the thickness of the main ridge **plus one half** the thickness of the common rafter equals the actual length (Fig. 10-61).

Intersecting Roof Using Gables and Unequal Spans on Main and Minor Roofs

The theoretical length of the minor ridge is **equal** to the building projections **plus one half** the minor span. The theoretical length **minus one half** the 45° thickness of the valley rafter **plus the overhang** (if there is any) equals the actual length (Fig. 10-62).

Intersecting Roof Using Gables and Equal Spans on Main and Minor Roofs

The theoretical length of the minor ridge is **equal** to the building projections **plus one half** the main span. The theoretical

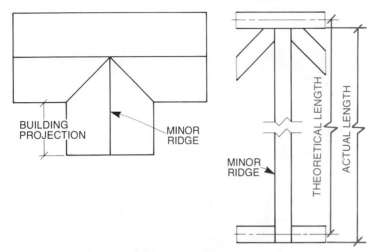

Figure 10-63   Ridge board for intersecting roof using gables and equal spans

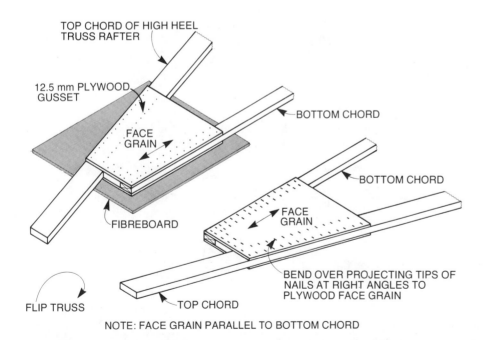

NOTE: FACE GRAIN PARALLEL TO BOTTOM CHORD

Figure 10-64   Nailed plywood gussets

length **minus one half** the thickness of the main ridge **plus the overhang** equals the actual length (Fig. 10-63).

TRUSS RAFTERS

Truss rafter roof construction has become very popular over the past few years. Roof trusses consist of top and bottom *chords* connected by diagonal and vertical members. The top and bottom chords correspond to the rafter and ceiling joists in conventional framing. The diagonal and vertical members are arranged in triangles, which gives the truss its strength. The various members of the truss are connected to each other with either nailed plywood or metal plates called *gussets*. If the trusses are fabricated on-site the gussets used are usually pieces of plywood nailed in place. The number of nails required are predetermined by an architectural engineer (Fig. 10-64).

If the trusses are manufactured in a factory the gussets are usually metal, with teeth formed as a unit with the gusset (integral teeth), and pressed in place (Fig. 10-65).

Advantages of Roof Trusses

There are several advantages to using roof trusses instead of conventional roof framing. Perhaps the greatest advantage of truss rafters is the long distances they can span without intermediate support. It is not uncommon for truss rafters to clear spans of 12 000 mm, which is more than is necessary for the average house. This

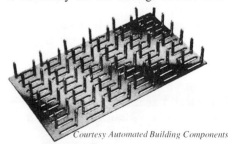

Courtesy Automated Building Components

Figure 10-65   Metal gussets

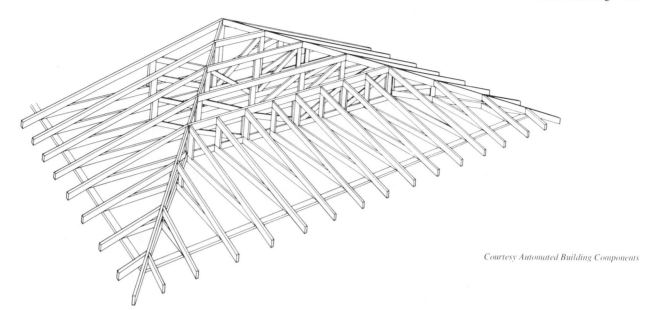

Courtesy Automated Building Components

Figure 10-66   Hip roof using truss rafters

allows flexibility of interior planning as there are no set places for the partitions which must act as bearing walls. Dimensional stock used for truss rafters is usually smaller than that used for conventional framing. This results in a saving in the cost of materials. Roof trusses are usually spaced 600 mm o.c. as opposed to conventional framing where 400 mm o.c. spacing is standard practice. Roof trusses need not be manufactured in a factory, but can be very satisfactorily made on the job site. Most trusses manufactured on-site are used in standard gable end roof construction. Truss rafters manufactured in a factory can be used on any type of roof. Figure 10-66 shows truss rafters being used on a hip roof.

Types of Truss Rafters

There are several types of truss rafters available in standard designs. However, if an unusual roof design is required, factories can custom fabricate trusses to provide the exact span, slope and shape required for virtually any roof line desired.

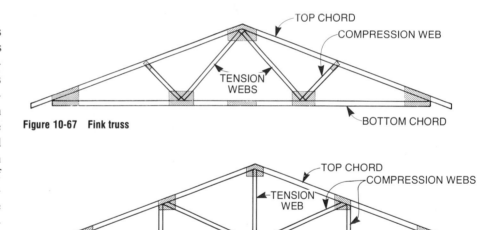

Figure 10-67   Fink truss

Figure 10-68   Howe truss

A type of roof truss very popular for house construction is the *Fink* (W) truss. It is suitable for spans from 6000 to 12 000 mm (Fig. 10-67).

Another popular roof truss is the *Howe* truss. It has the same range span as the Fink truss, but is more efficient for low slopes–below 1:4 (Fig. 10-68).

ROOF SHEATHING

Roof sheathing is applied to the roof frame for the purpose of providing a solid nailing base for the roof covering as well as for adding rigidity to the roof frame. Roof sheathing may be solid lumber, plywood or particleboard. Solid lumber can either be shiplap, tongue and groove or common boards. These boards may be

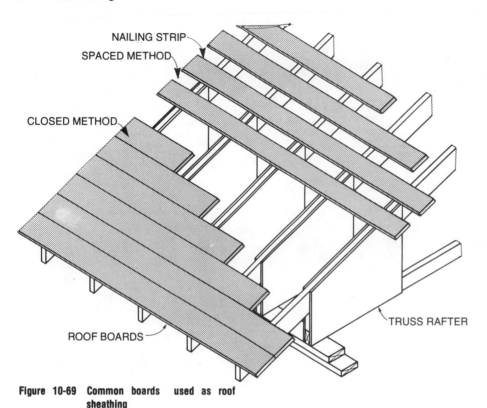

NAILING STRIP

SPACED METHOD

CLOSED METHOD

TRUSS RAFTER

ROOF BOARDS

Figure 10-69   Common boards   used as roof sheathing

applied closed (fitted tight together), as is required when asphalt shingles or a built-up roof is used as a roof covering, or spaced, as when wood shingles and shakes are used (Fig. 10-69).

When plywood is used for roof sheathing the face grain should run at right angles to the roof framing members. The end joints should be staggered. A space of approximately 2 mm should be left between the panels to prevent buckling. The thickness of the plywood used will depend on the spacing of the framing members and also on whether the edges of the plywood will be supported. When the edges are supported, either by 38 mm blocking or metal *H-clips*, the thickness of plywood can often be reduced. Figure 10-70 shows plywood used as roof sheathing when the edges are supported with H-clips.

Particleboard may also be used as roof sheathing. Its application is similar to plywood.

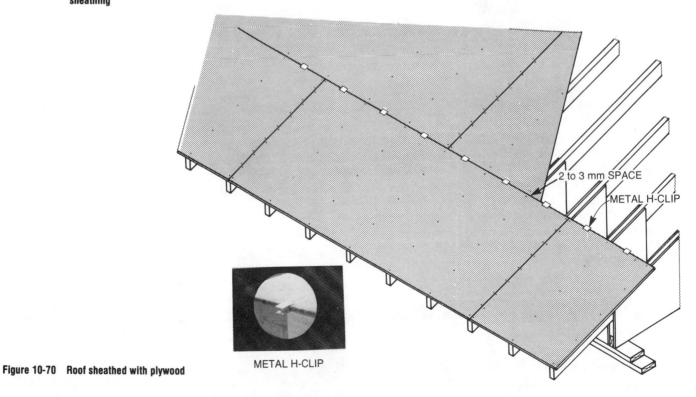

2 to 3 mm SPACE

METAL H-CLIP

METAL H-CLIP

Figure 10-70   Roof sheathed with plywood

REVIEW QUESTIONS

10—ROOF FRAMING

Answer all questions on a separate sheet

A. Select the correct answer

1. The deduction made on the hip jack rafter is:
   (a) The thickness of the ridgeboard
   (b) Half the thickness of the common rafter
   (c) Half the diagonal thickness of the hip rafter
   (d) Half the thickness of the hip jack rafter

2. Shortened measurements at the rafter top should be taken:
   (a) Along the plumb cut
   (b) At right angles to the level cut
   (c) Along the line length
   (d) At right angles to the plumb cut

3. The unit of run for a hip rafter using the Frederickson square is:
   (a) 100
   (b) 250
   (c) 354
   (d) 500

4. The slope of a roof is 1:5 and the span is 6000 mm. What is the length of the third gable end stud from the corner, if the studs are spaced at 400 mm o.c.?
   (a) 220 mm
   (b) 250 mm
   (c) 240 mm
   (d) 250 mm

B. Write a full answer

5. What is the actual length of the ridge board of a hip roof for a building 6000 × 12 000 mm? The hips will be framed using a double cheek cut and will all be cut from 38 mm stock.

6. What unit of run is used when laying out hip jack rafters?

7. In what ways is a hip jack rafter the same as a common rafter?

8. In your own words, write a method for finding the common difference of jacks, without using the difference in length tables on the square.

9. How is the actual length of the ridge board for a hip roof found? Explain both methods.
   (i) Double cheek cut
   (ii) Single cheek cut

10. Is a shortened valley rafter found in all intersecting roofs? Explain.

C. Mark as either TRUE or FALSE

11. The same pattern should always be used when laying out several rafters of the same type.

12. The seat and tail cuts for the **hip jack** are the same as the seat and tail cuts of a common rafter.

13. The total run of a common rafter is always one half of the span in an equal pitch roof.

14. One common rafter will usually make two hip jacks.

15. When hip rafters are framed to the common rafters, the ridge has to be longer than when hips are framed to the ridge.

16. The hip rafter does not need to be dropped at the birdsmouth.

17. A unit of run for a hip rafter is 250 mm.

18. Rafter stock for a hip rafter must be one dimension greater in depth than for a common rafter.

19. Hip rafters can be made out of 19 mm stock.

20. A hip jack rafter pattern is laid out on the hip rafter.

11 CORNICE CONSTRUCTION AND ROOF COVERINGS

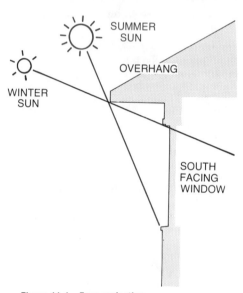

Figure 11-1 Eave projection

The cornice of a building is that part where the side walls meet the roof. There have been many different types of construction used over the years to provide this finish. Years ago this was often a focal point of design and elaborate decorative mouldings were used. Today cornices are of simple design, constructed to be functional as well as decorative.

The horizontal projection of the cornice should be wide enough to provide shade from the high summer sun and yet narrow enough to allow the low winter sun to assist in heating the house during the winter

months (Fig. 11-1). The cornice should blend in with the architectural design of the building, and should provide weatherproof protection to the top of the exterior walls as well as afford a method of ventilating the attic space.

TYPES OF CORNICES

There are basically three types of cornice construction used today: the *open* cornice, where the rafter tails are *exposed,* the *boxed* cornice, where the rafter tails are *enclosed,* and the *snub* cornice, where there are *no* rafter tails.

Open Cornice

In the open cornice the rafter tail is exposed and the underside of the roof sheathing is visible. Since the underside of the roof sheathing is exposed, a better grade of plywood is often used for the first row in order to obtain a better finish when the cornice is painted. Where the sheathing will be exposed, V-joint finished boards or some other type of finished board are often used (Fig. 11-2).

Boxed Cornice

On a boxed cornice, the underside of the rafter tails are covered or closed in with

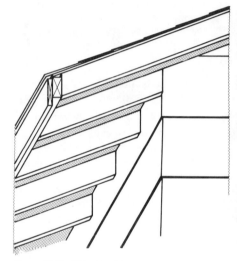

Figure 11-2 Open cornice

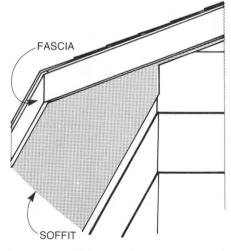

Figure 11-3 Soffit applied directly to underside of rafter tails

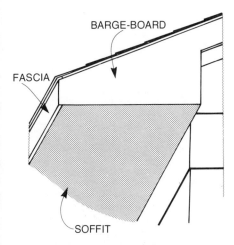

Figure 11-4   Soffit applied horizontally

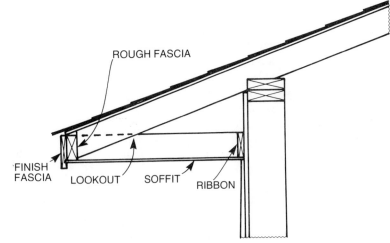

Figure 11-6   Parts of boxed cornice

panels called **soffit**. These panels are available in a variety of materials: plywood, hardboard, metal and vinyl. The soffit can be applied directly to the underside of the rafter tails as in Figure 11-3, or horizontally on a framework as in Figure 11-4.

Snub Cornice

With this type of cornice the rafter tail is omitted. The rafters end flush with the edge of the wall plate. This type of construction is often used in conjunction with boxed cornices, where a portion of the building is cantilevered past the foundation wall, as in the case of bay or bow windows. Construction costs can be kept

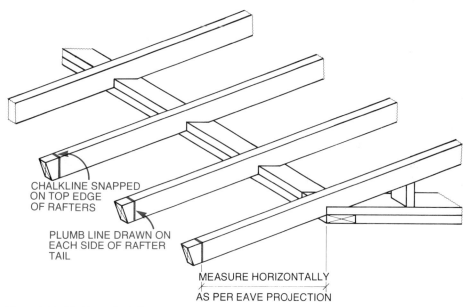

Figure 11-7   Marking rafter tails

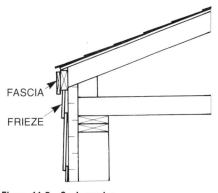

Figure 11-5   Snub cornice

down if the roof line is kept straight, providing an eave projection on the major portion of the building with a snub cornice on the cantilevered part (Fig. 11-5).

Parts of a Boxed Cornice

The boxed cornice is the most popular type. The soffit is installed horizontally

on the framework. The finished parts of the boxed cornice are fascia and soffit; the framework consists of a rough fascia, ribbon and lookouts (Fig. 11-6).

Construction of a Boxed Cornice

The method described incorporates the

use of plywood for the soffit and a combination of plywood and solid lumber for the fascia and barge board.

When the rafter tails are left at random length, (which means they are not cut to the correct length when installed but left slightly longer so they can be trimmed to length), the width of the eave projection is measured horizontally at each end of the building and marked on the end rafters. A plumb line is drawn at these marks and a chalkline snapped on the top edge of the rafter tails (Fig. 11-7). A plumb line is drawn on the face of each rafter tail and then cut on these marks with a portable electric hand saw. This method is far superior to cutting rafter tails to length during rafter installation for achieving a straight line.

The rafter tails are kept in alignment with the rough fascia, and the soffit is then supported on the outer edge by the rough fascia and on the inner edge by a ribbon that has been nailed to the exterior wall. The ribbon is usually a 19 × 64 or 89 mm wide piece of stock placed on top of the wall sheathing and nailed to the studs. A strip of wall sheathing paper should be tacked to the wall before the ribbon is nailed in place. The underside of the ribbon should be level with the underside of the rough fascia. This is achieved by marking each end of the building using a spirit level (Fig. 11-8), then snapping a line between these two points.

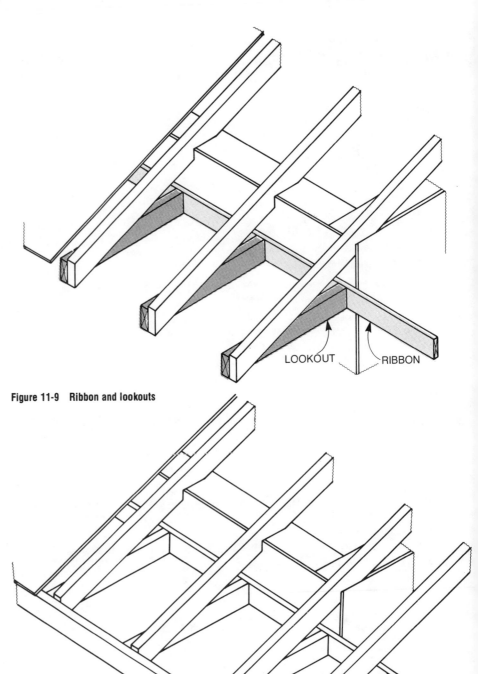

Figure 11-9   Ribbon and lookouts

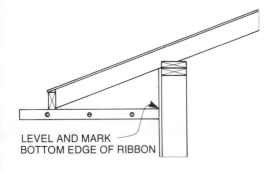

LEVEL AND MARK
BOTTOM EDGE OF RIBBON

Figure 11-8   Leveling ribbon

Figure 11-10   Filler block

Let the ribbon extend past the end wall of the building as far as the outside of the barge rafter (Fig. 11-9). On wide eave projections (400 mm or wider) lookouts are usually installed between the rough fascia and the ribbon to support the soffit. The lookouts may be 19 × 64 mm or 38 × 38 mm, spaced 600 mm o.c. They are toe-nailed to the ribbon and end-nailed to the rough fascia.

A *filler block* of 38 mm stock is required on the underside of the barge rafter between the rough fascia and the ribbon (Fig. 11-10).

The soffit material can now be applied to the framework. The soffit at the eave is referred to as *eave soffit*. The soffit at the gable end is referred to as *rake soffit* and the finish between these two is referred to as *soffit return* (Fig. 11-11).

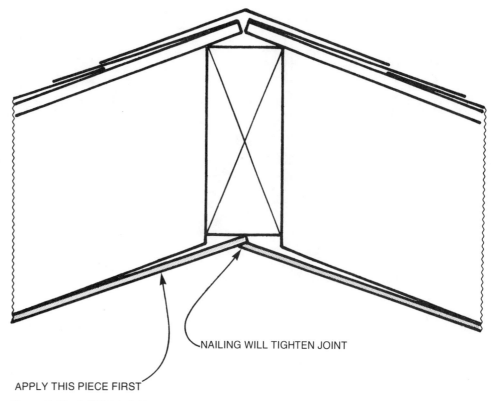

NAILING WILL TIGHTEN JOINT

APPLY THIS PIECE FIRST

Figure 11-12   Soffit joint at ridge

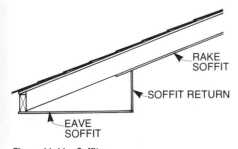

RAKE SOFFIT

SOFFIT RETURN

EAVE SOFFIT

Figure 11-11   Soffits

The **rake soffit** is nailed to the outer edge of the barge rafter. A nailing strip must be nailed on top of the gable end sheathing to correspond with the underside of the barge rafter. This provides a nailing surface for the inner edge of the rake soffit. If the distance between the nailing strip and the barge rafter is more than 400 mm, blocking (usually 38 mm thick), must be installed 600 mm o.c. to support the rake soffit. When applying the rake soffit the joint at the ridgeboard should be made so that when the soffit is nailed, it tightens the joint (Fig. 11-12). The rake soffit is usually allowed to extend past the barge rafter by about 2 mm.

The eave soffit also usually extends 2 mm past the rough fascia. This ensures a tight fit between finish fascia or barge board and the soffit material.

To provide a finish between the eave and rake soffits the soffit return is fitted and nailed in place.

A finish fascia board is nailed to the rough fascia to provide a satisfactory finish between the soffit and the roofing material. The finish fascia usually extends 19 mm below the soffit. The top outside corner should be in line with the roof sheathing. This provides a good painting surface as there are no plywood edges exposed. All joints along the length of the finish fascia should be on a 45° bevel. The finish fascia board should also be cut on a 45° bevel where it fits into the barge board (Fig. 11-13).

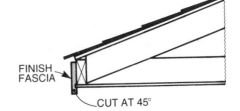

FINISH FASCIA

CUT AT 45°

Figure 11-13   Finish fascia board

The **barge board** is the finish board nailed to the barge rafter. The lower part of the barge board is usually made of 19 mm plywood because it must be wide enough to cover the eave soffit and the soffit return (Fig. 11-14).

To lay out the plywood barge board, select a 19 mm piece of plywood large enough to cover the eave soffit and to extend up the barge rafter to a point a short distance from the soffit return. Cut a 45° bevel on one end and tack in place. Allow the plywood to extend 19 mm below the

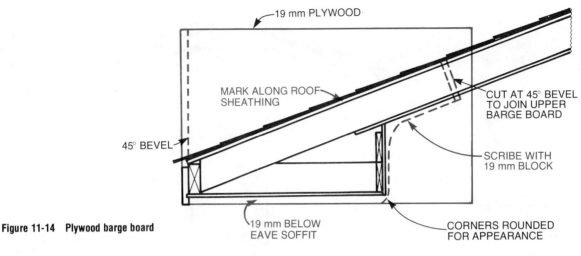

Figure 11-14   Plywood barge board

eave soffit. Mark along the roof sheathing. With a 19 mm block of wood scribe a line along the eave soffit, soffit return and rake soffit. Remove the plywood and cut on the marks. The sharp corners of the plywood are frequently rounded for a better appearance. The top part of the barge board is usually made of common boards. Where the common board joins the plywood at 45° bevel should be used.

Where the barge boards meet at the ridge they will be cut on the same angle as the plumb cut of the rafters. The barge board that is put up first is marked as shown in Figure 11-15. The plumb cut of the rafter is drawn through this mark with a T-bevel square.

After the board is nailed in place, the barge board for the other side is marked on the top edge as shown to ensure a tight fit (Fig. 11-16).

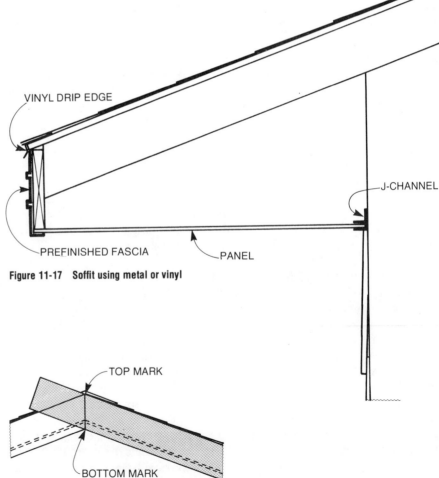

Figure 11-17   Soffit using metal or vinyl

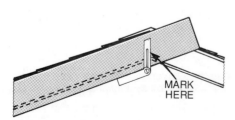

Figure 11-15   Top cut of barge board

Figure 11-16   Second barge board cut

Often, when a boxed cornice is used, the attic space is vented through the eaves. If this is done, the soffit material must have openings installed to allow for the flow of air. If the soffit material is plywood, round holes or a patterned opening is often cut at predetermined intervals. These holes must be properly screened to prevent the entry of birds into the soffit and attic area. Small, manufactured vents may also be installed.

Over the past few years, metal and vinyl products have become increasingly popular as soffit and fascia material. Metal products include aluminum and steel, with aluminum perhaps being the most popular.

Installation is basically the same as the one described for conventional wood materials, but various channels and mouldings simplify the job considerably (Fig. 11-17, 11-18).

Material used for soffit construction is frequently manufactured in 3000 to 3600 mm lengths and therefore must be cut to the required length. The width is usually between 300 and 400 mm with ribs running lengthwise to provide rigidity. The cut pieces lock into one another when installed in the soffit. The material is perforated to provide ventilation into the attic. Cutting the vinyl or metal material is a simple matter if the manufacturer's instructions are followed. After the soffit is secured in place, the prefinished metal fascia is installed to complete the soffit construction (Fig. 11-18).

ROOFING MATERIALS

The purpose of a roof is to protect the building and its contents from the effects of rain and snow. Although this is the main objective of roof coverings, there are other things that should be taken into consideration. The type of roof covering used should enhance the look of the building and should be long-lasting with a minimum of required maintenance. There are several materials which fill these requirements: asphalt shingles, wood shingles and shakes, asbestos shingles, clay tile and slate, roll asphalt roofing, galvanized steel, aluminum, copper and fibreglass. These can all be used on roofs with a slope of 1:3 or greater. On slopes of less than 1:3 a low slope asphalt shingle is available. Another type of roof used for low slopes is a built-up roof, where several layers of roofing felt and tar are applied, topped with a layer of gravel.

Of all the roofing materials mentioned, asphalt shingles are the most commonly used for house construction. They come in a variety of weights, shapes, sizes and colours. The most common is the *strip*

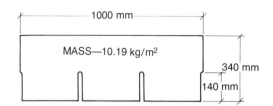

Figure 11-19   Three tab strip shingle

shingle which has three tabs and measures 340×1000 mm and has a mass of 10.19 kg/m² (Fig. 11-19).

Eave Protection

Snow accumulation on a roof will often melt on the building portion of the roof before the snow on the eave portion. This frequently occurs during the freeze-thaw cycle in late fall and early spring. When this happens an ice dam is formed, causing water to back up under the shingles (Fig. 11-20). This results in water damage to the ceiling area. Roofs that are covered with shingles or shakes should therefore have additional protection at the eaves.

This eave protection can be made of polyethylene, asphalt-saturated felt or roll roofing. Eave protection should extend from the edge of the roof a minimum of 914 mm up the roof slope to a line 300 mm or more inside the inner face of the exterior wall (Fig. 11-21).

When the eave protection is polyethylene it should be laid as a continuous sheet and be at least 150 μm thick. If asphalt-saturated felt is used, it should be #15 weight and applied in two layers, with laps of 480 mm cemented together with lap cement. Roll roofing of at least 2.2 kg/m² should be laid with 100 mm or more laps at the ends and head. These laps should be cemented together with lap cement.

An underlay is usually applied under the shingles, which should be an asphalt-saturated sheathing paper of not less than 0.195 kg/m², or #15 plain or perforated asphalt-saturated felt or 50 μm polyethylene.

Figure 11-18   Metal or vinyl soffit

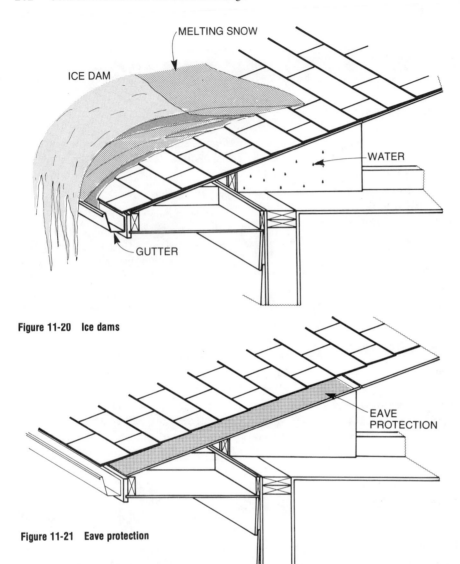

Figure 11-20 Ice dams

Figure 11-21 Eave protection

distance equal to one half the length of the shingle away from the centre line at the eave and ridge, and snap a second line (Fig. 11-22). These lines will be used as a guide in keeping the shingle tabs in a straight line.

Unless the eave protection is made of roll roofing, the first row of asphalt shingles must be double. The shingles for the first row, which is also referred to as the starter strip, should be laid **mineral side up** with the tabs facing up the roof slope. The edge of the shingle should extend approximately 12 mm beyond the eaves and rake of the roof. The edge of this first shingle should be in line with the centre line snapped on the roof. Continue on either side of this first shingle until the row is completed. Shingles 1000 mm long should be fastened with at least four nails or staples located 25 to 40 mm from each end of the strip shingle with the other fasteners equally spaced between them.

Nails used for asphalt shingles should have a head diameter of 9.5 mm or more and a shank thickness of not less than 2.95 mm. They should be long enough to extend 12 mm into the roof sheathing. Some contractors use two fasteners above the cutout, rather than one, to ensure that the shingles are securely fastened to the

INSTALLATION OF ASPHALT SHINGLES ON SLOPES OF 1:3 OR GREATER

After the eave protection is applied, the actual shingling can begin. To achieve the best appearance when applying asphalt shingles, keep the alternate rows of shingle *cutouts* in line. This also keeps the width of tabs on each end of the roof equal. The best method of achieving both these objectives is to find the centre of the roof at the eave and ridge and then snap a line between these two points. Measure a

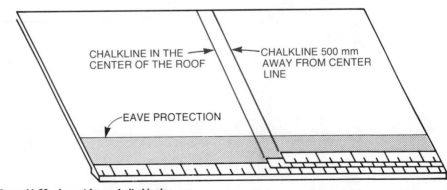

Figure 11-22 Layout for asphalt shingles

roof. Fasteners should be located at least 12 mm above the tops of the cutouts. It is necessary to cut the shingles at the end of each row. This can be done using either a linoleum or utility knife. Use a framing square as a guide. Cut on the backside of the shingle. Care must be taken to ensure the rake tab will extend approximately 12 mm past the finish barge board. The second row of shingles is then applied directly on top of the starter strip row. Line up the edge of the second row shingle along the second chalkline. Complete this row in the same way as the first. The third and subsequent rows are applied with the bottom edge of the shingle in line with the top of the cutouts of the previous row. Line up the edge of the shingles with the appropriate chalkline to make sure that the tabs are in a straight line. To ensure the shingles are in line horizontally, check the exposure every five or six rows by measuring up from the eave at both ends of the roof and snapping a line. Shingles at the ridge are lapped over the ridge and then a ridge cap is installed. Ridge cap shingles are made from regular cap shingles and cut as shown in Figure 11-23.

A chalkline should be used as a guide to keep the ridge cap shingles straight. The distance of this line will be one half the width that the ridge cap shingle is from the ridge. Shingles should start at the opposite end of the building to the direction of the prevailing winds. Use nails

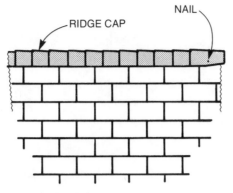

Figure 11-24   Ridge cap

long enough to give good holding power, and allow approximately 150 mm exposure on ridge cap shingles (Fig. 11-24).

Some shingles are manufactured with an adhesive tar strip on the underside of the tabs. This strip melts after the shingles are installed, gluing the tabs securely and preventing damage by wind. If the shingles do not have this adhesive strip, the tabs must be glued individually. A 25 mm patch of plastic roofing cement in the centre of each tab is adequate (Fig. 11-25).

The hips on hip roofs require much the same treatment as the ridge on gable end roofs. The hip cap shingles must begin at the bottom of the roof, and at the top of the hip the ridge cap shingle must lap over the hip cap shingle to ensure a watertight finish.

On intersecting roofs there will be valleys that require special attention. When asphalt shingles are used, the most common method of finishing is a *closed valley*. Flashing must be laid in the valley; it may be either sheet metal (aluminum and galvanized steel being the most popular), 2.7 kg/m² roll roofing. The flashing must be at least 600 mm wide. In a closed valley, alternate rows of shingles coming into the valley are lapped over one another (Fig. 11-26).

INSTALLATION OF LOW SLOPE ASPHALT SHINGLES

Low slope shingles are bigger and heavier than regular asphalt shingles (Fig. 11-27). Because of this there will be three thicknesses of shingle on the roof rather than only two, as there are with regular shingles.

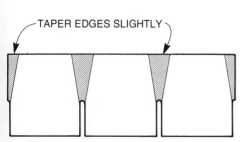

Figure 11-23   Cutting ridge cap shingles

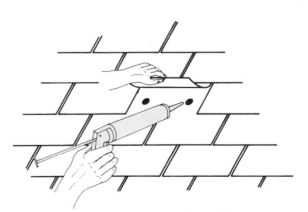

Figure 11-25   Gluing shingle tabs

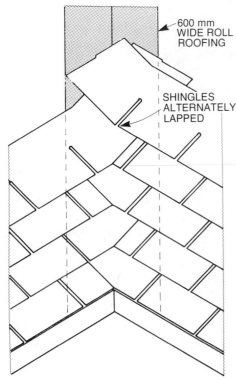

Figure 11-26   Closed valley

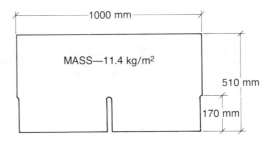

Figure 11-27   Low slope shingle

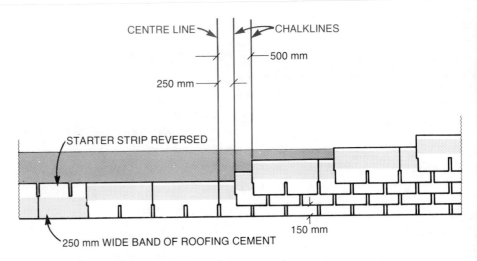

Figure 11-28   Low slope shingle installation

Installation of low slope shingles is basically the same as with regular three tab strip shingles. The main difference is that a **continuous** band of cement is applied under the shingle tabs (Fig. 11-28).

There are several different types of asphalt shingles, each applied in a slightly different manner. Manufacturers always provide detailed instructions, which if followed will ensure a good, watertight roof.

WOOD SHINGLES AND SHAKES

Wood shingles and shakes have been used in house construction for years. When manufacturers started producing asphalt shingles there was a sharp decline in the use of wood as a roof covering. Because of its durability, wood is once again gaining popularity. Shakes in particular are widely used today (Fig. 11-29).

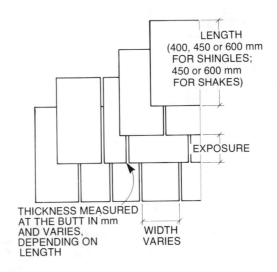

Figure 11-29   Size of shingles and shakes

Shingles are taper sawn with the butts varying in thickness from 10 to 13 mm depending on the length. The tops are approximately 2 mm thick. They have a relatively smooth face and back, are produced in lengths of 400, 450 and 600 mm, and they come in grades ranging from #1 to #3. Shingles and shakes are produced mainly from cedar trees. Shakes come in two lengths, 450 and 600 mm. In producing shakes, cedar logs are first cut into the desired lengths, then planks or boards of the proper thickness are split and run diagonally through a bandsaw to produce two tapered shakes from each plank.

INSTALLATION OF WOOD SHINGLES AND SHAKES

Exposure for wood shingles will vary depending on the length of the shingle, the grade, and the slope of the roof. Table 11-1 shows exposure allowed.

Shingles and shakes may be applied over spaced or solid sheathing. When spaced sheathing is used, the boards should be spaced on centres equal to the weather exposure at which the shingles or shakes are to be laid. Spaced sheathing material should be 19 mm thick and either 89 or 140 mm in width.

Eave protection will be the same as outlined for asphalt shingles. If used, underlay will be the same as that used for as-

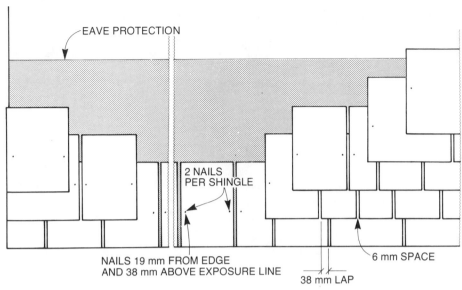

Figure 11-30   Wood shingle installation

phalt shingles, except that it should be of a *breather* type.

Nails used for wood shingles or shakes should have a head diameter of 4.8 mm or more and a shank thickness of not less than 2.0 mm. They should be of sufficient length to penetrate through or extend 12 mm into the roof sheathing, and should be made of hot-dip galvanized or other corrosion-resistant material.

Wood shingles should be 400 mm or more in length and not less than 75 or more than 350 mm wide. The bottom row should be doubled, with the end of the shingles projecting approximately 25 mm

over the edge of the finish fascia board to provide a drip edge. A smaller projection of 19 mm is used at the rake. Shingles should be spaced 6 mm apart to allow for expansion and to prevent possible buckling. Joints in adjacent courses (rows) should be *offset* at least 40 mm. Never have two joints in a line separated by only one course of shingles. Use only two nails per shingle, located approximately 20 mm from either side of the shingle and 40 mm above the exposure line (Fig. 11-30). A chalkline or gauge on a shingling hatchet may be used as a guide to ensure uniform exposure (Fig. 11-31).

TABLE 11-1 SHINGLE EXPOSURE TABLE									
MAXIMUM SHINGLE EXPOSURE FOR ROOFS (mm)									
Slope of roof	Length								
	400 mm			450 mm			600 mm		
	#1	#2	#3	#1	#2	#3	#1	#2	#3
1:4 to 1:3	95	90	75	105	100	90	145	140	125
1:3 and steeper	125	100	90	140	115	100	190	165	140
Do not use shingle if roof slope is less than 1:4.									
Courtesy Council of Forest Industries									

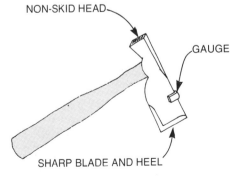

Figure 11-31   Shingling hatchet

TABLE 11-2 EXPOSURE TABLE FOR WOOD SHAKES		
MAXIMUM SHAKE EXPOSURE FOR ROOFS (mm)		
Length of shake (mm)	Minimum butt thickness (mm)	Maximum exposure (mm)
450	9	190
600	9	250
Do not use shakes if roof slope is less than 1:3.		
		Courtesy Council of Forest Industries

Shakes are applied in much the same manner as shingles. Shakes should be 450 mm or more long, and not less than 100 or more than 350 mm wide, with a butt thickness of 32 mm or more. Table 11-2 gives the exposure for wood shakes.

Open Valleys

When wood shingles and shakes are used as roof covering, an open valley is used on intersecting roofs. Open valleys should be flashed with either sheet metal of not less than 600 mm in width, or with 2 layers of roll roofing. Metal flashing is usually used with wood shingles and shakes, galvanized steel or aluminum being the most popular. Copper is not recommended because the natural phenol preservatives in wood have a strong corrosive effect on copper. The flashing will frequently be crimped down the centre to form a water barrier. This will divert the water down the valley rather than allow it to splash up the opposite side and perhaps seep in under the shingles. As the roof shingles are laid in the valley they should be trimmed, giving them the correct mitre to form a shingle line 50 to 75 mm away from the centre of the valley. If the shingles are started from the valley and worked away from it, wider shingles can be used, minimizing the number of nails penetrating the flashing (Fig. 11-32).

Special treatment for the hips and

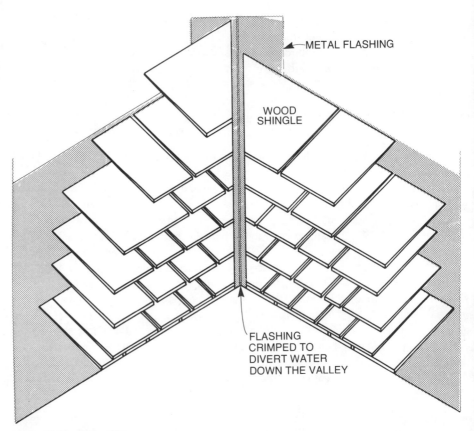

METAL FLASHING

WOOD SHINGLE

FLASHING CRIMPED TO DIVERT WATER DOWN THE VALLEY

Figure 11-32   Open valley

ridge is required with wood shingles and shakes. Either solid wood or a metal cap may be used to finish these areas; however, a hip and ridge finish made from narrow shingles presents a much more pleasing appearance. The capping is made from narrow shingles, 125 to 150 mm in width, which have been sorted

out from the bundles. Two chalklines are snapped on either side of the ridge about 125 mm below the centre line. Starting at one end of the ridge, the first shingle is laid with the edge along the chalkline. The other side extends over the ridge. This protruding edge is trimmed so that the cut surface is parallel to the slope on

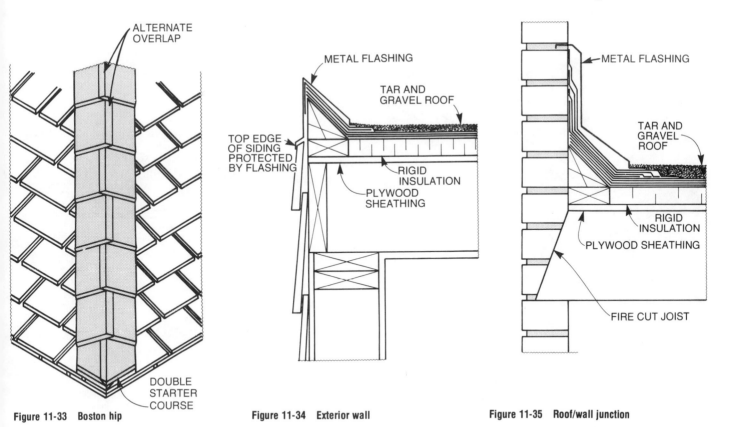

Figure 11-33   Boston hip

Figure 11-34   Exterior wall

Figure 11-35   Roof/wall junction

the other side of the ridge. The shingle on the other side of the ridge should now be put down in the same manner and the projecting edge trimmed so that the cut edge is parallel to the surface of the first shingle. The next two shingles are applied directly on top of the two shingles previously applied. This provides a double starter course. The first shingle is laid on the opposite side of the ridge from the starter shingle on the course below, and the trimmed-off edge slants in the opposite direction. This alternate beveling of the edges to the right and left produces a ridge that is tight and durable. The succeeding courses of shingles are applied at the same exposure as the shingles on the roof. The same procedure is followed in forming hips, but they should be completed before the ridge capping is laid. This permits the first course of ridge shingles to cover the trimmed ends of the last course of hip shingles. This type of finish is often referred to as a *Boston hip* (Fig. 11-33). Be sure to use long enough nails so they will penetrate sufficiently into the roof sheathing.

BUILT-UP ROOFS

Roofs with slopes of 1:6 or less are considered to be flat and therefore not suitable for any type of shingle. Built-up roofs consist of three or more layers of roofing felts, each mopped down with hot tar or asphalt. The final surface is coated with tar or asphalt and covered with an aggregate surface of clean durable gravel, crushed stone, or air-cooled blast furnace slag. Aggregate should be dry and uniformly graded in particles ranging from 6 to 15 mm. Built-up roofs are usually installed by roofing contractors who specialize in this type of roofing system. Figures 11-34 and 11-35 show exterior wall and roof/wall junctions. A *gravel stop* or a *cant strip* must be provided at the edges of the roof. Flashing should extend over the edge of the roof to form a drip. This type of roof does not allow for eaves-troughing. However, excess water may be drained off with either internal drains or exterior down spouts.

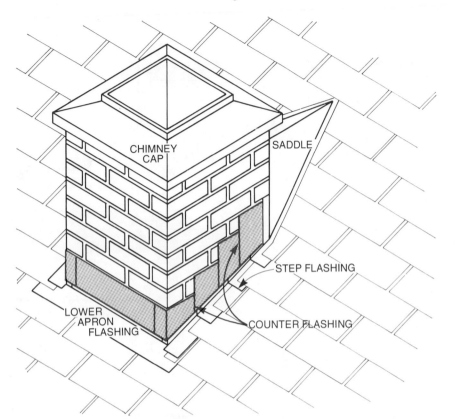

Figure 11-36(a)   Counter flashing. Note how the stepped flashing is tied to the counter flashing.

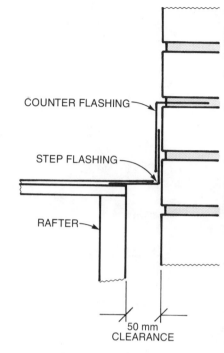

Figure 11-36(b)   Flashing embedded in mortar lines.

Flashings

Flashing is required where chimneys project through the roof. When brick chimneys are built the bricklayer installs *counter flashing* in the mortar lines of the chimney (Fig. 11-36).

When shingling, install the rows of shingles on the roof up to the chimney, then cut and install a lower *apron flashing* at the bottom of the chimney (Fig. 11-37), embedding the apron flashing in roofing cement.

Install the first piece of *step flashing* which will be a piece of metal flashing material at least 150 × 200 mm with a right angle bend in it. It is embedded in roofing cement and bent around the corner of the chimney (Fig. 11-38).

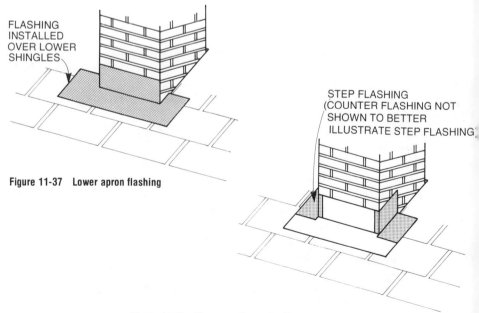

Figure 11-37   Lower apron flashing

Figure 11-38   First row of step flashing

Figure 11-39   Chimney saddle

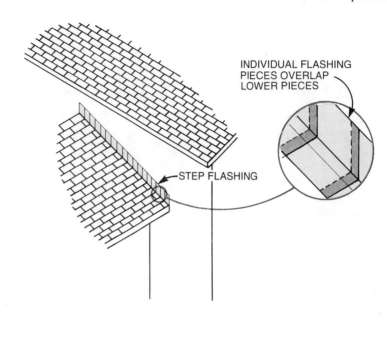

Figure 11-41   Roof/wall junction

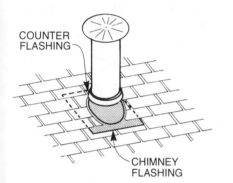

Figure 11-40   Metal chimney flashing

Install the next row of shingles and then another piece of step flashing, and repeat until the top of the chimney is reached. At the top of the chimney an upper apron flashing is installed. If the chimney is more than 750 mm wide the upper apron flashing must be replaced by a chimney saddle. When this is done there will be a change in the counter flashing (Fig. 11-39).

When prefabricated metal chimneys are installed, the roof flashing is supplied with the kit. Although the flashing comes in one piece, the lower portion must be installed on top of the shingles, and the upper portion beneath the shingles (Fig. 11-40).

When a shingled roof butts into a side-wall, as frequently happens with split-level houses, the shingles must be step flashed to prevent leaking behind the individual courses of shingles (Fig. 11-41).

REVIEW QUESTIONS

11—CORNICE CONSTRUCTION AND ROOF COVERINGS

Answer all questions on a separate sheet.

A. Select the correct answer

1. The best type of nail for asphalt shingles are:
   - (a) Hot dipped zinc coated
   - (b) Copper bearing
   - (c) Galvanized wire
   - (d) Finish nails

2. A pre-formed sheet metal strip which is attached to the edge of built-up roofs is a:
   - (a) Wind shield
   - (b) Edge cover
   - (c) Gravel stop
   - (d) Water strip

3. Which of the following is considered a dead load on a roof?
   - (a) Workers
   - (b) Sheathing and common rafters
   - (c) Wind
   - (d) Snow

4. Polyethelene applied under the shingles at the start of shingling is called:
   - (a) Flashing
   - (b) Starter strip
   - (c) Underlay
   - (d) Eave protection

5. The triangular-shaped strip of wood used under the edges of roofing on flat decks is called:
   - (a) Weatherstrip
   - (b) Cant strip
   - (c) Counterflashing
   - (d) Scantling

B. Give the type of roof covering needed for the roofs indicated

6. **Type of roof**
   1. house with a 1:6 slope
   2. flat warehouse roof
   3. rustic roof on motel
   4. temporary construction shack

C. Write full answers

7. Is snow on the roof of a building considered a live or dead load?

8. Are shingles on a roof considered a live or dead load?

9. Draw and name the component parts of a boxed cornice.

10. Name three types of cornice construction used in house construction.

11. Name five materials used for roof covering.

12. When eave soffits are used as part of the ventilation system for the attic, what special care must be taken when insulating?

13. What is the name of the horizontal structural member located between the fascia and the ribbon used to support the soffit?

D. Mark as either TRUE or FALSE

14. Standard 3-tab asphalt shingles may be used on roof slopes of 1:6 or less.

15. The best grade of cedar shingles has a flat grain.

16. A gable roof and a hip roof with the same roof slope and span are equal in area.

17. Metal soffit material is frequently perforated for ventilation purposes.

18. When applying cedar shingles, no attention need be paid to where joints will occur.

19. When applying asphalt shingles, the ridge cap is frequently made from regular shingles.

20. If asphalt shingles are adequately nailed it is unnecessary to glue them.

12 EXTERIOR DOORS AND WINDOWS

EXTERIOR DOORS

Exterior doors come in a variety of sizes, designs and materials. They can be very basic or combined with windows at the side called sidelights to form a feature of design. Main entrance doors in residential construction must be at least 44 mm thick; secondary entrance doors may be 35 mm thick if they are of solid wood, solid core, or stile and rail construction. *Weatherstripping* of metal, wood or fabric (or a combination of these materials), should be installed at the perimeter of all exterior door openings. Main entrance doors should be 810 mm wide and 2030 mm high, which is the standard height. Exterior doors are usually the flush type. If they are made of wood the exterior facing is usually plywood and the core either solid wood or rigid insulation. Exterior doors made of metal and filled with rigid insulation are becoming very popular.

EXTERIOR DOOR FRAMES

In residential dwellings, door frames are usually made of wood with either metal or wood doors. The basic exterior door frame consists of a *sill, jambs* and exterior *moulding*. The sill is the horizontal member at the bottom of the door frame; since it is subject to wear from foot traffic, edge grain fir or oak is used. The jambs consist of side and head pieces that provide a finished frame for the door assembly. The solid lumber used for the door jamb is rabbeted to receive the door. The exterior trim, commonly called *brickmould,* is attached to the outside of the jamb and is used as a means of attaching the door frame to the exterior wall. Figure 12-1 shows a typical wood exterior door frame.

Today, most exterior door frames are factory built. The door is frequently hinged and often even the lockset is installed. It is still necessary for a carpenter to lay out, cut and assemble the exterior door frame using basic lumber stock. The materials required will be jamb stock, sill stock and exterior moulding. In choosing jamb stock, the total thickness of the exterior wall must be known as that will be the width of the jamb stock (Fig. 12-2). The rabbet in jamb stock equals the thickness of the door times 12 mm deep.

The sill stock should have a bevel of 5° on either side. The width of the sill stock should equal the width of the jamb stock, plus the thickness of the exterior moulding, plus approximately 12 mm (Fig. 12-3). A *drip groove* is cut on the underside of the sill near the outside edge to prevent water from seeping back against the wall.

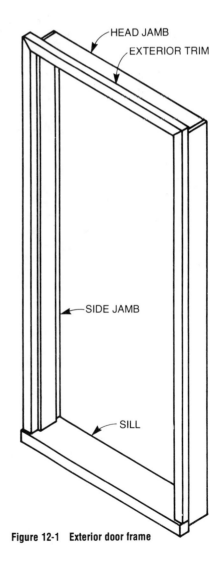

Figure 12-1 Exterior door frame

211

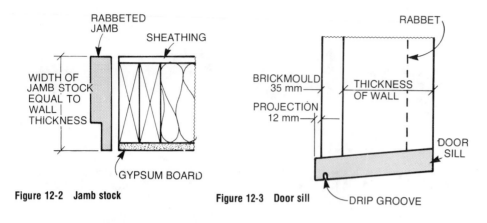

Figure 12-2   Jamb stock

Figure 12-3   Door sill

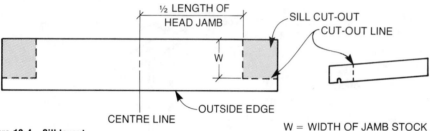

Figure 12-4   Sill layout

W = WIDTH OF JAMB STOCK

in the head jamb). From this point along the shoulder of the door rabbet, lay out the top of the sill dado at an angle of 5°. The slope must be drawn to the outside. Jambs should be laid out together to ensure that they are marked in pairs. The depth of the rabbet will be the same as that of the door rabbet, the width will be equal to the thickness of the sill being used (Fig. 12-6).

After the dado cuts have been made, the frame is ready to be assembled. Dry fit the members and check the inside measurements of the frame to see if the door fits properly. If all the measurements are correct the frame can be assembled. Nail the side jamb to the head jamb, being sure to keep the shoulders of the door rabbets flush. Nail the side jambs to the sill, keeping the inside edges of the jamb and sill flush. To square the door frame, measure the diagonals. Install temporary bracing to maintain squareness until the door frame has been installed. Install the braces diagonally from the side jambs to the centre of the head jamb (Fig. 12-7).

The exterior casing is now ready to be installed. Leave a 12 mm margin or *reveal* (Fig. 12-8). To ensure an even reveal, mark the jamb at convenient intervals using a sharp pencil and a combination square.

After the door jamb and sill stock have been selected and machined to the correct widths, the next step is to cut the head jamb to length. The length of the head jamb equals the width of the door to be hinged plus 4 mm for clearance. The next step is to lay out and cut the sill. The rough length of the sill should equal the width of the door plus 150 mm. From a centre line on the sill, measure one half the length of the head jamb in each direction, and square a line across the sill at these points. The depth will be equal to the width of the jamb stock. Draw a line parallel with the edge of the stock to the ends of the sill (Fig. 12-4). This will be cut out to receive the jamb stock.

After the sill is cut to receive the side jambs, the top of the sill must be flattened where the door comes to rest. The width of the flattened portion will be the thickness of the door (Fig. 12-5).

The next step is to lay out and cut the side jambs. Cut a rabbet at the top of each side jamb to receive the head jamb. The depth of the rabbet will be the same as that cut for the door, the width being the thickness of the head jamb stock. Lay out for the sill dado by measuring from the intersection of the head jamb rabbet and the shoulder of the door rabbet. Measure along the shoulder of the door rabbet a distance equal to the length of the door, *less 8 mm.* (Length of the door, plus 2 mm clearance at the top, plus 2 mm clearance at the bottom, less the 12 mm door rabbet

THICKNESS OF DOOR

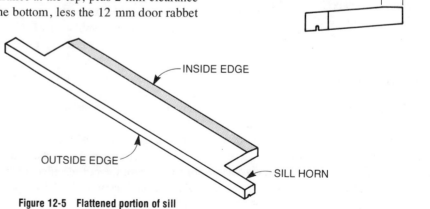

Figure 12-5   Flattened portion of sill

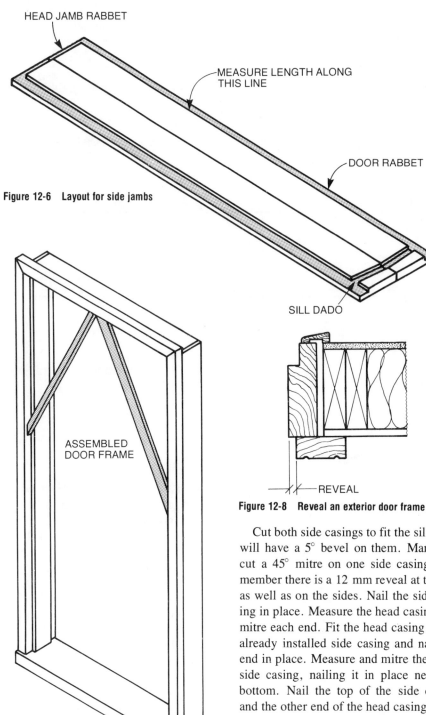

HEAD JAMB RABBET

MEASURE LENGTH ALONG THIS LINE

DOOR RABBET

Figure 12-6   Layout for side jambs

SILL DADO

ASSEMBLED DOOR FRAME

Figure 12-7   Temporary bracing on door frame

REVEAL

Figure 12-8   Reveal an exterior door frame

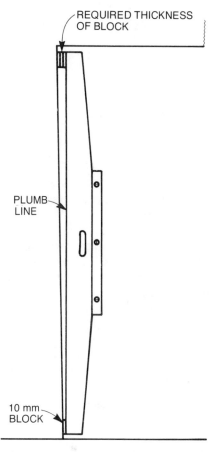

REQUIRED THICKNESS OF BLOCK

PLUMB LINE

10 mm BLOCK

Figure 12-9   Plumbing hinge side of rough opening

Cut both side casings to fit the sill; they will have a 5° bevel on them. Mark and cut a 45° mitre on one side casing. Remember there is a 12 mm reveal at the top as well as on the sides. Nail the side casing in place. Measure the head casing and mitre each end. Fit the head casing to the already installed side casing and nail the end in place. Measure and mitre the other side casing, nailing it in place near the bottom. Nail the top of the side casing and the other end of the head casing, after any minor adjustments have been made to ensure a tight fit. Trim the *sill horns* flush with the casing to facilitate the application of exterior siding.

Installing the Door Frame

The door frame is now ready to be installed in the rough opening. The rough opening should be approximately 20 mm wider than the outside measurement of the frame and approximately 10 mm higher. Apply a 300 mm strip of breather-type building paper around the rough opening. Plumb the hinge side of the rough opening by nailing a 10 mm block to the trimmer stud at the bottom of the opening. Plumb up from this block and nail the required thickness of block to the trimmer stud at the top of the opening (Fig. 12-9).

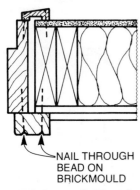

Figure 12-10   Door frame nailed through exterior moulding

NAIL THROUGH BEAD ON BRICKMOULD

Figure 12-11   Two hinge-door

275 mm   175 mm   TOP OF DOOR →

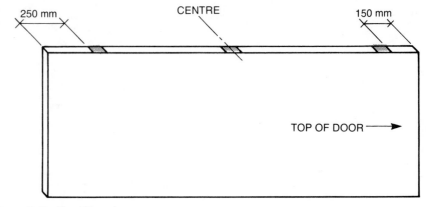

Figure 12-12   Three hinge-door

250 mm   CENTRE   150 mm   TOP OF DOOR →

Install the door frame, wedging the jamb tight against the plumbed blocks. The sill should be tight to the subfloor. Be sure that the frame braces have remained square by checking the head jamb for level. Nail the door frame in place through the exterior moulding (Fig. 12-10) at both the top and bottom corners. Straighten the side jambs with a straight edge, wedging where necessary. Finish nailing the frame in place through the exterior moulding. Space nails at 300 to 400 mm intervals. Cut and install wedges under the sill to prevent it from moving. A protection board is often nailed to the sill to prevent marring it during further construction.

Hanging an Exterior Door

Check the door for a bow. If there is a slight bow, place the concave side toward the door rabbet. Fit the door to the opening, allowing a 2 mm clearance all around. If the door frame was made to receive the door you are now hanging, no planing should be required. Place the door in the frame with the proper margin at the head jamb and the sill. Wedge it tight against the jamb on the side that will receive the hinges. Standard layout for the hinges are: on a two hinge door, the top hinge will be 175 mm down from the top of the door and the bottom hinge will

be 275 mm up from the bottom of the door to the bottom of the hinge (Fig. 12-11).

On a three hinge door, the top hinge will be 150 mm down from the top of the door and the bottom hinge will be 250 mm up from the bottom of the door to the bottom of the hinge with the third hinge centred between the other two (Fig. 12-12).

Mark the door and the frame for the position of the hinges at the same time using a 6 mm chisel. Place a small "x" on both door and frame where the hinges are to be located. Remove the door from the frame and mark the hinges on the door using a butt gauge or butt marker. When marking for the hinges on the jamb, the distance

between the shoulder of the door rabbet and hinge gain must be 2 mm wider than the distance between the hinge gain and the edge of the door (Fig. 12-13). If the hinge gains have been improperly located and the door hits the shoulder of the rabbet, the door is said to be *stop bound*.

Gain out for the hinges using a sharp wood chisel. Apply the hinges, driving all screws into the hinge leaf on the door, but drive *only* the centre screw into the hinge leaf on the frame. This allows the hinge leaf to be tapped up or down slightly for alignment if necessary. Hang the door, inserting the top pin first, then open the door and drive the remaining screws into the frame, making minor adjustments as required. If the hinge gain is too deep, the door could be *hinge bound*. A door is said

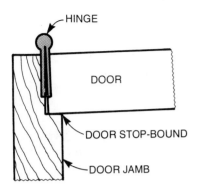

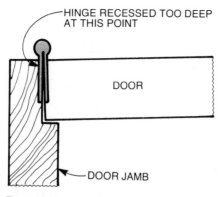

HINGE RECESSED TOO DEEP AT THIS POINT

Figure 12-14   Hinge bound

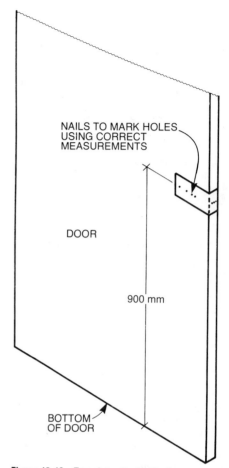

NAILS TO MARK HOLES USING CORRECT MEASUREMENTS

DOOR

900 mm

BOTTOM OF DOOR

Figure 12-16   Template attached to door

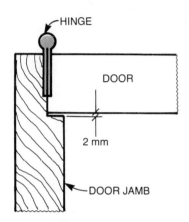

2 mm

Figure 12-13   Allowance to prevent door from being stop bound

to be hinge bound when one of the leaves has been gained too deep and the face of the hinges come together before the door is properly closed (Fig. 12-14). This situation is easily remedied by shimming the hinge leaf out.

If the door strikes the jamb on the lockside, it is probably because the hinge leaves have not been gained deep enough. The face of the hinge leaf should be flush with surface it is gained into. It may also be that the screws are not fully countersunk into the hinge. This will happen if the screws are driven in at an angle.

INSTALLATION OF LOCKSETS

There are many different types of locks available on the market today. If you

know how to install a lockset, and you follow the manufacturer's instructions precisely, any type of lock can be easily installed. The most popular type used for residential buildings is the *tubular lockset*. With every lockset a template is enclosed or incorporated as part of the packaging (Fig. 12-15). This template will give information about the correct positioning of holes for the *barrel* on the edge of the door, and for the *cylinder* on the face of the door. It will also give the sizes these holes should be bored. The standard height for locksets in residential use is 900 mm from the floor. Determine the correct backset to be used and tack the *template* to the face of the door at the correct height (Fig. 12-16). Tack the tem-

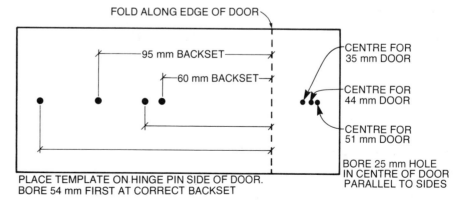

FOLD ALONG EDGE OF DOOR

95 mm BACKSET

60 mm BACKSET

CENTRE FOR 35 mm DOOR

CENTRE FOR 44 mm DOOR

CENTRE FOR 51 mm DOOR

BORE 25 mm HOLE IN CENTRE OF DOOR PARALLEL TO SIDES

PLACE TEMPLATE ON HINGE PIN SIDE OF DOOR. BORE 54 mm FIRST AT CORRECT BACKSET

Figure 12-15   Template

plate to the edge of the door through the correct centre, depending on the thickness of the door.

Bore the holes to the required size as stated on the template. The hole through the face of the door must be bored first. Care must be taken when boring the hole through the edge of the door for the barrel, making sure that it is parallel to the face of the door and perfectly horizontal. If this is not done, the lockset may not work freely. Mortise for the face plate and fit the barrel in position, then install the cylinder of the lock assembly and fix in position, with the key notches up. Install the *striker plate* to the jamb at the same height as the face plate on the edge of the door. The striker plate should be the same distance away from the shoulder of the door rabbet as the face plate is from the inside edge of the door. Finally, try the door and make any necessary adjustments.

TYPES OF DOORS

Due to construction costs, contractors frequently order door units completely assembled, with the doors hung and the lockset installed, and ready to be placed in the rough opening. Sometimes, the units ordered are to be assembled on the site. The front door unit is often used as a feature of design and sidelights added on one or both sides. These sidelights can be either fixed or opening units. The secondary entrance is usually a single door.

The flush-type door is probably the most common. It may have glass installed to provide light, viewing and to generally improve its appearance.

Doors are available in both wood and metal. Wood doors for exterior use must be solid wood, solid wood with stile and rail construction, or solid core. This solid core can be either laminations of wood or particleboard (Fig. 12-17).

A new product developed in recent

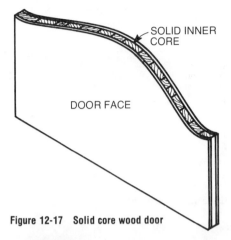

Figure 12-17   Solid core wood door

years that is gaining popularity is the insulated steel door unit. The steel surfaces of the door are hot dip galvanized for better rust and corrosion resistance. They are permanently bonded to a rigid urethane core, and crimped and glued around wood stiles to stop splitting, checking and delamination. Exposed wood edges provide a thermal barrier and allow the installation of hinges and a variety of latches without the use of special tools. The door thickness is 44 mm. The greatest advantage is the insulating quality afforded. This type of insulated steel door has an

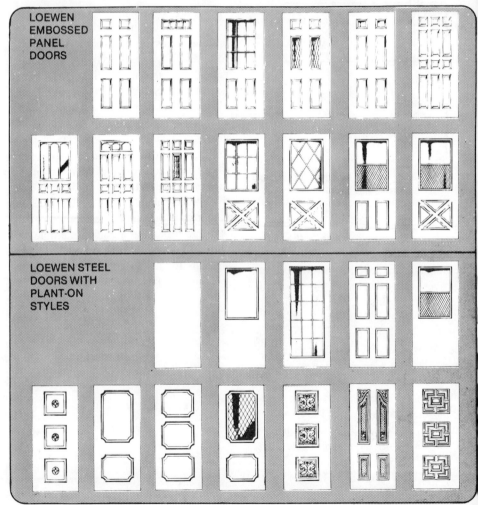

Figure 12-19   Exterior doors

Courtesy Loewen Millwork

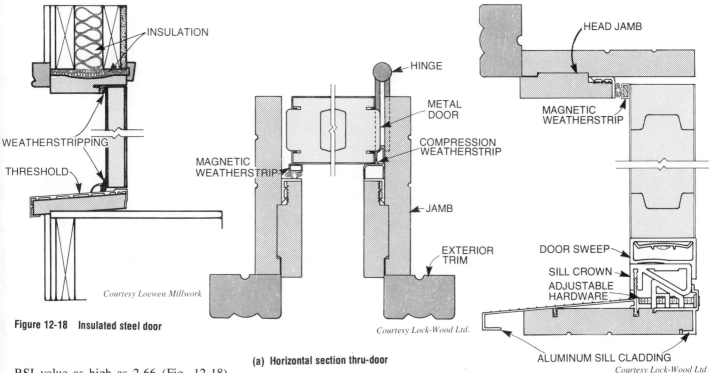

INSULATION

WEATHERSTRIPPING

THRESHOLD

Courtesy Loewen Millwork

Figure 12-18   Insulated steel door

HINGE

METAL DOOR

COMPRESSION WEATHERSTRIP

MAGNETIC WEATHERSTRIP

JAMB

EXTERIOR TRIM

Courtesy Lock-Wood Ltd.

(a) Horizontal section thru-door

Figure 12-20

HEAD JAMB

MAGNETIC WEATHERSTRIP

DOOR SWEEP

SILL CROWN

ADJUSTABLE HARDWARE

ALUMINUM SILL CLADDING

Courtesy Lock-Wood Ltd.

(b) Vertical section thru-door showing magnetic weatherstripping

RSI value as high as 2.66 (Fig. 12-18). Figure 12-19 shows some of the styles of exterior doors available.

Metal, plastic, rubber, wood, or fabric weatherstripping, or a combination of these materials should be installed at the perimeter of all exterior door openings.

A new type of weatherstripping is now on the market for use with metal doors. This type is not unlike that used on refrigerator doors, where a magnetic strip is used to make a tight seal. Some manufacturers use this type of magnetic seal on the side jambs, head jamb and sill. Figure 12-20 shows the type used by Lock-Wood Ltd. of Scoudouc, N.B., where the head and lock jamb have magnetic weatherstrip while the hinge jamb uses a compression weatherstrip. The door bottom is sealed by a tubular weatherstrip and a leg that catches on the sill crown as the door closes. The sill crown is adjustable both up and down to make possible a perfect fit (adjustable sill patented by Lock-Wood Ltd.).

Another type of door used on residential buildings is the sliding patio door. Patio door units are available in both wood and metal, and are glazed with either double or triple insulating glass. Figure 12-21 shows a cross-section of a wood patio door unit.

When ordering pre-built door units that are ready to be installed in the wall opening, it is important that the door swings in the proper direction. This is termed *hand of the door*. The hand of the door is determined by the side of the door that the hinges are on, as seen from the outside of the door, regardless of whether the door swings in or out. The outside of a door is the street side of an entrance door, or the corridor side of a room door. The outside of a closet door is the room side. Figure 12-22 shows the hand of doors.

WINDOWS

Window units are available in three basic types, and in a variety of styles and sizes. The three basic types of window units are; *fixed, sliding* and *hinged*. Fixed windows range in size from large picture windows to the small panels used in *sidelights* of door units. Fixed sashes are often combined with opening sashes to form a window unit. Window units with sliding sashes can be either horizontal sliding or vertical sliding. There are several different kinds of hinged sashes: casements are hinged on the side; awnings are hinged at the top and swing inward; hoppers are hinged at the bottom and swing inward.

When selecting the type, style and size of window desired, many factors must be

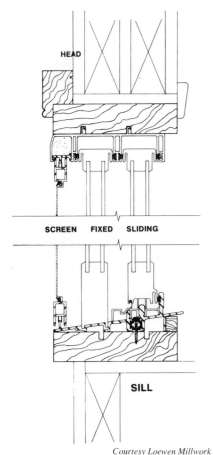

Courtesy Loewen Millwork

Figure 12-21   Cross-section of wood patio door

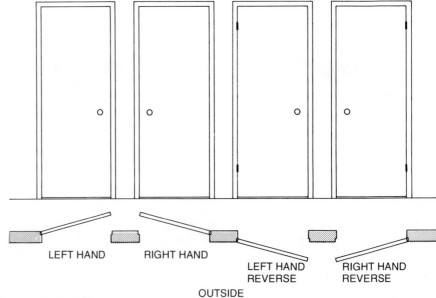

Figure 12-22   Hand of doors

considered: will the general appearance suit the architectural design of the building, to what degree is view or privacy a factor, will the window be needed for ventilation, how much natural light is desired, and perhaps the most important factor is whether the window unit will be energy-efficient.

With the advent of building products produced in metric measurements, the manufacturers of wood window units have endeavoured to build window units in metric modules. The module for the outside dimension of the window frame is 200 mm. There are to be basically ten different window styles, resulting in 262 different combinations. A series of preferred wood window sizes based on the 200 mm module range in size from 600 to 1800 mm for both height and width. A complete range of sizes is not offered in all of the ten different styles. Bow windows may be available in a 2000 mm height, and basement units in a 400 mm height. These preferred window sizes, however, are continually under review and may change to suit the needs of industry.

WINDOW TYPES

Fixed window units come with insulating double glazed glass with one 12 mm airspace, or triple glazed with two 12 mm air spaces. Figure 12-23 shows a fixed unit with triple glazed insulating glass.

Frequently, fixed windows are used along with opening windows to provide a more practical window unit. Figure 12-24 shows some possible combinations.

Of the **hinged window units,** the *awning* and outswinging *casement* are perhaps the most popular due to the fact that

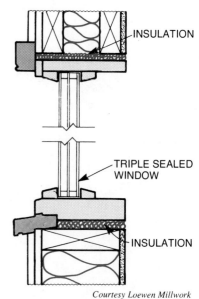

Courtesy Loewen Millwork

Figure 12-23   Triple glazed window unit

they may be opened without interfering with the operation of drapes and curtains. Figure 12-25 shows double and triple insulating glass available in an awning unit.

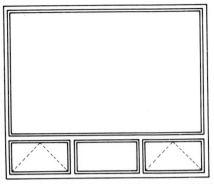

Figure 12-24   (a) Fixed window (sealed unit) with opening awning windows

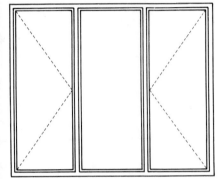

Figure 12-24   (b) Fixed window (sealed unit) with side casements

Figure 12-26 shows a casement unit, with a cross-section of a fixed unit and casement combination.

Bay and **bow windows** have been used as a design feature over the years. These windows can either be supported on the foundation wall or on a cantilevered floor area. A bay consists of three units while a

TRIPLE INSULATING GLASS

DOUBLE INSULATING GLASS

LEVER LOCK WITH TORSION BAR

Courtesy Loewen Millwork

Figure 12-25   Awning unit

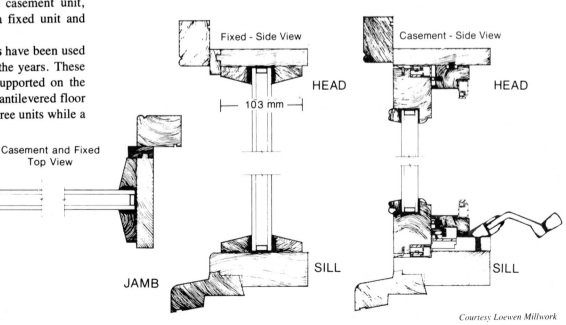

Casement and Fixed Top View

Fixed - Side View

HEAD

|← 103 mm →|

SILL

Casement - Side View

HEAD

SILL

MULLION   JAMB

Figure 12-26   Casement unit

Courtesy Loewen Millwork

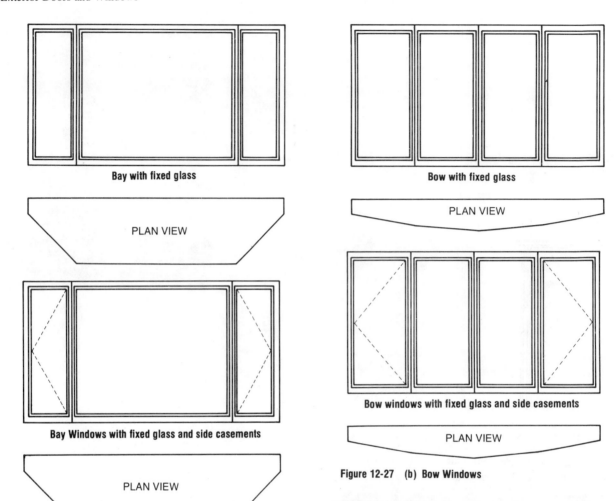

Bay with fixed glass

PLAN VIEW

Bow with fixed glass

PLAN VIEW

Bay Windows with fixed glass and side casements

PLAN VIEW

Bow windows with fixed glass and side casements

PLAN VIEW

Figure 12-27   (b) Bow Windows

Figure 12-27   (a) Bay Windows

Figure 12-28   Bay window   *Courtesy Loewen Millwork*

Figure 12-29   Bow window   *Courtesy Loewen Millwork*

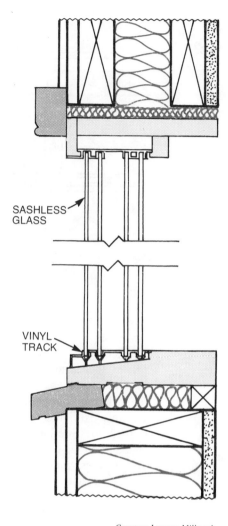

Courtesy Loewen Millwork
Figure 12-30   Horizontal sliding window

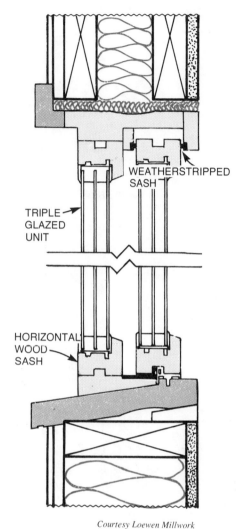

Courtesy Loewen Millwork
Figure 12-31   Wood sash window

bow consists of four or more units (Fig. 12-27). These window units may be all fixed or used in combination with casement units.

Figures 12-28 and 12-29 show bay and bow windows used as a feature of design on house construction.

There are two kinds of **sliding window units**, *horizontal* sliding and *vertical* sliding. Sashes for sliding windows may be wood, metal, or plastic. In some cases no sash is used. Most sashes slide in vinyl tracks. Figure 12-30 shows a horizontal sliding window where the glass is sashless, sliding in a vinyl track at the top and bottom.

With sashless glass sliding windows the glass is of plate thickness. Wooden sash windows are usually double or triple glazed with insulating glass. Figure 12-31 shows a horizontal sliding window with wood sashes.

Vertical sliding units have been used for years. They are commonly known as *double hung* windows. In double hung windows the upper sash is fixed and the lower sash is movable. Double hung window construction is similar to that of horizontal sliding windows where wooden sash is used.

Skylights are now beginning to be used extensively in residential housing. They provide light to interior rooms where wall

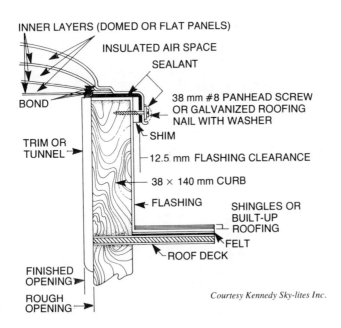

INNER LAYERS (DOMED OR FLAT PANELS)

INSULATED AIR SPACE

SEALANT

BOND

38 mm #8 PANHEAD SCREW OR GALVANIZED ROOFING NAIL WITH WASHER

TRIM OR TUNNEL

SHIM

12.5 mm FLASHING CLEARANCE

38 × 140 mm CURB

FLASHING

SHINGLES OR BUILT-UP ROOFING

FELT

ROOF DECK

FINISHED OPENING

ROUGH OPENING

Courtesy Kennedy Sky-lites Inc.

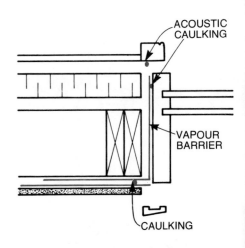

ACOUSTIC CAULKING

VAPOUR BARRIER

CAULKING

Figure 12-33   Vapour barrier applied to door or window frame (exploded view)

Courtesy Kennedy Sky-lites Inc.

Figure 12-32   Skylights

units are not possible. Skylights are now available that are double glazed, triple glazed and, for exceptional energy-efficient installation, quadruple glazed (Fig. 12-32).

INSTALLATION OF WINDOWS

In the past, a great deal of heat loss occurred in houses around the outside of the window frames. An improved method of installing windows has been devised in which a bead of acoustic caulking is applied around the entire window frame slightly away from the exterior trim. A polyethylene vapour barrier 300 to 400 mm in width is then stapled to the window frame, completing the seal between the caulking and the vapour barrier. The vapour barrier must be installed loosely at the corners so that it can be folded back to form a seal with the vapour barrier applied to the interior wall. This method of sealing is also recommended for exterior doors. The installation of windows is similar to that already described for exterior doors. A bead of caulking on the back of the exterior trim aids in forming an airtight seal. After the window frame is securely fastened in place, insulation is placed between the vapour barrier and the rough opening. The vapour barrier is left loose until the exterior wall is insulated and the main vapour barrier applied. It is then folded back, caulked and stapled to form a complete seal between vapour barriers (Fig. 12-33).

REVIEW QUESTIONS

12—EXTERIOR DOORS AND WINDOWS

Answer all questions on a separate sheet.

A. Select the correct answer

1. Which window types use hinges?
   - (a) Casement
   - (b) Check-rail
   - (c) Double hung
   - (d) Sashment

2. On what type of window do you find a check rail?
   - (a) Casement
   - (b) Double hung
   - (c) Hopper
   - (d) Sealed unit

3. A window that is hinged on top and swings out is known as a(n):
   - (a) Slab window
   - (b) Awning window
   - (c) Double mullion
   - (d) Fixed window

4. A single sash fastened solidly and permanently into a frame is called a:
   - (a) Transom sash
   - (b) Single casement
   - (c) Fixed sash
   - (d) Single light
   - (e) Checkrail window

5. When the two leaves of a butt hinge come together before the door is properly closed, it is said to be:
   - (a) Stop bound
   - (b) Butt stopped
   - (c) Hinge bound
   - (d) Jamb stopped

6. The reveal for the exterior casing of an exterior door frame is:
   - (a) The same as the reveal for the interior casing
   - (b) The same as the depth of the rabbet
   - (c) Half the thickness of the door jamb stock
   - (d) Half the thickness of the door
   - (e) The same as the reveal on the window

7. What moulding is usually used for outside casing of windows?
   - (a) Crown mould
   - (b) Brick mould
   - (c) Cove mould
   - (d) Gothic casing

8. Why is the sill on an exterior door sloped toward the outside?
   - (a) For easier walking
   - (b) For the water to run off
   - (c) For better construction
   - (d) So that the interior door will fit

9. Standing on the outside of a door that opens towards you, the hinges are on the right. It is a:
   - (a) Right hand door
   - (b) Left hand door
   - (c) Left hand reverse door
   - (d) Right hand reverse door

10. A door is hinged on the left hand side and swings away from you as you enter a room. This is a:
    - (a) Left hand door
    - (b) Left hand reverse door
    - (c) Right hand door
    - (d) Right hand reverse door
    - (e) None of these

B. Write full answers

11. Where are the hinges located on a casement window?

12. Why is it necessary to cut a groove in the underside of the door sill?

C. Replace the Xs with the correct word

13. The distance from the edge of the door to the centre of the door knob is termed the XXXXXXX.

D. Write full answers

14. What is meant by the term *stop-bound*?

15. Is weatherstripping required on all exterior door openings?

16. What are the three categories which all types of windows are classified under?

E. Mark as either TRUE or FALSE on a separate sheet.

17. The part of the lockset that is attached to the door jamb is called the striker plate.

18. On a lockset, the distance from the edge of the door to the center of the knob is called the setback.

19. The vertical side pieces of a door or window frame are called stiles.

20. When there is a slight crown or bow in an exterior door the concave should be towards the door rabbet.

13 EXTERIOR WALL FINISH

There are many types of exterior wall finish in use today, each with its own advantages. When selecting a wall finish, consideration should be given to climate, initial cost, maintenance and appearance. Exterior sidings are usually applied over a layer of *building paper*. This building paper must be of the breather type, that is, it is moisture-proof but not vapour-proof. If the exterior siding is not allowed to breathe properly, this will result in paint peeling on the surfaces soon after installation.

TYPES OF EXTERIOR WALL FINISHES

Lumber

Because exterior siding is continually subjected to both wet and dry conditions, a species of wood must be used that will withstand the elements. The species most commonly used are **cedar** and **redwood**, and **pine** to some extent. Cedars and redwoods are the most practical. Lumber siding is produced in four basic patterns: **bevel, tongue and groove, channel**, and **board** and **batten**. Sidings are available with a smooth or saw-textured surface, and in flush-joint or V-joint. They come in various widths and thicknesses, as well as in a simulated log cabin pattern. Bevel siding is available with plain or rabbetted bevel and with straight or wavy edges. Siding is manufactured in thicknesses of 12, 15, and 19 mm and in widths from 100 to 300 mm.

Wood Shingles and Shakes

Cedar shingles and shakes can be applied to exterior walls as well as roofs. A large

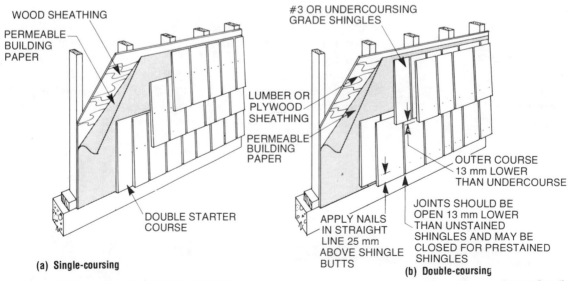

(a) Single-coursing

(b) Double-coursing

Figure 13-1   Single and double coursing cedar shingles and shakes

Courtesy Council of Forest Industries

224

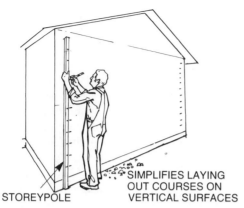

SIMPLIFIES LAYING OUT COURSES ON VERTICAL SURFACES

STOREYPOLE

Figure 13-2   Storeypole   *Courtesy Council of Forest Industries*

selection of specialty shingles are available, some of which are factory painted or stained. Specialty shingles are given a distinctive appearance during manufacture. This is done by shaping the butt of standard dimension shingles into various patterns. There are two basic methods of wall application; *single-coursing* and *double-coursing* (Fig. 13-1).

The use of a *storeypole* aids in keeping rows straight and ensures that all rows will be of equal exposure (Fig. 13-2).

Shingles and shakes may be **butted** to wood strips or **laced** at both interior and exterior corners (Fig. 13-3).

Plywood

Exterior-type plywood is used quite extensively as exterior wall finish. Patterns are many and varied, ranging from plain sheets to grooved panels, in endless combinations. Plywood is also available in a stucco finish. Most panels have ship-lapped edges, providing ease of installation (Fig. 13-4).

Plywood is also available with a resin-impregnated kraft paper, which is laminated to the face and provides a smooth surface that is moisture-resistant and especially suitable for painting.

Hard-Press Fibreboard

Hard-pressed fibreboard is available in panels or strips which can be applied in the form of lap siding. Usually a paint finish is applied at the factory. This finish is available in a variety of colours.

Stucco

Stucco is made from **Portland cement, sand** and **lime**, and is applied in three coats over a wire mesh that has been fastened to the exterior sheathing. The first two coats of stucco are base coats with the third coat providing a suitable finish. The finish coat may be left natural or can be tinted a colour. If a stone-dash finish is desired, mineral chips picked up in a hand scoop are thrown into the fresh mortar of the second base coat. When a stone-dash finish is used, the third coat is not required.

Metal

The most popular metals used for siding are **aluminum** and **steel**. They are available with or without an insulation base in a variety of shapes and sizes suitable for either vertical or horizontal installation.

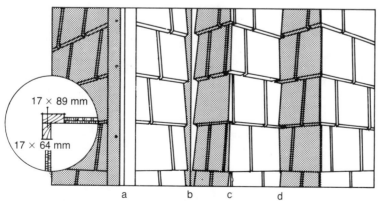

17 × 89 mm

17 × 64 mm

a) SHINGLES BUTTED AGAINST CORNER BOARDS
b) SHINGLES BUTTED AGAINST SQUARE WOOD STRIP ON INSIDE CORNER, FLASHING BEHIND
c) LACED OUTSIDE CORNER
d) LACED INSIDE CORNER WITH FLASHING BEHIND

a   b   c   d

Figure 13-3   Treatment of interior-exterior corners   *Courtesy Council of Forest Industries*

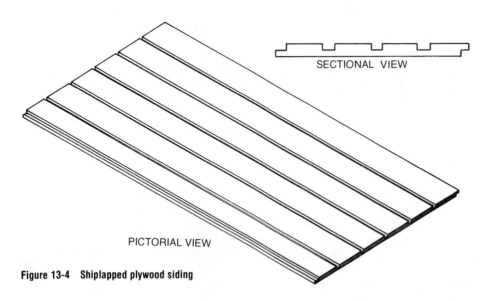

SECTIONAL VIEW

PICTORIAL VIEW

Figure 13-4   Shiplapped plywood siding

residential homes. With most of these products, detailed manufacturer's instructions are included and if conscientiously followed, they can be easily installed, and will give years of service with little maintenance.

One of the more common types of siding in use today is bevel wood siding. It is one of the few materials used in which little or no instructions are given with regard to installation, and which still requires the skill of a carpenter to apply.

Bevel siding is approximately 5 mm thick at the top and 12 mm thick at the butt for widths 184 mm or less and 14 mm thick for sidings wider than 184 mm. Bevel siding should not be more

The metal is factory painted with a long-lasting finish and if applied according to the manufacturer's instructions (which are enclosed with the siding), will give a pleasing appearance for many years with little or no maintenance.

Plastic

Siding is now available in vinyl, and in shapes and forms similar to metal sidings. Like metal sidings, manufacturer's instructions must be followed, especially with regard to contraction and expansion due to temperature change.

Masonry Veneer

Brick or stone veneer may be used for the entire exterior finish or as a feature wall to accent a front window and main entrance. When brick veneer is used, special attention must be given the foundation wall, as shown in Figure 13-5.

These are only a few of the more common materials used for exterior finish on

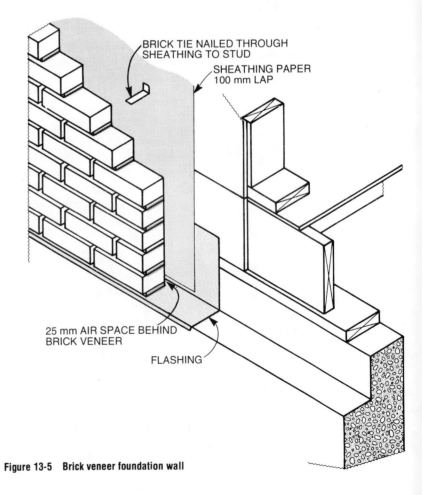

BRICK TIE NAILED THROUGH
SHEATHING TO STUD

SHEATHING PAPER
100 mm LAP

25 mm AIR SPACE BEHIND
BRICK VENEER

FLASHING

Figure 13-5   Brick veneer foundation wall

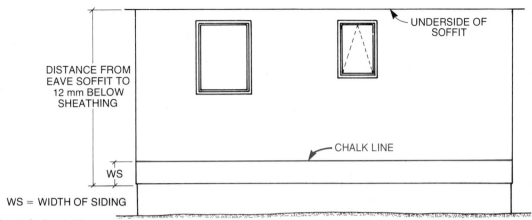

DISTANCE FROM
EAVE SOFFIT TO
12 mm BELOW
SHEATHING

UNDERSIDE OF
SOFFIT

CHALK LINE

WS

WS = WIDTH OF SIDING

Figure 13-6   Layout for bevel siding

than 286 mm wide and should lap at least 25 mm over the previous row. Bevel siding should extend approximately 12 mm below the exterior sheathing (Fig. 13-6).

For reasons of appearance, all rows of bevel siding should have the same exposure. To ensure this, some calculation is necessary. The bevel siding will extend from 12 mm below the sheathing to the underside of the soffit. Determine the maximum exposure possible for the bevel siding that is to be used. To do this, subtract 25 mm from the actual width of the siding. For example, if 235 mm wide bevel siding is to be used, the maximum exposure would be 210 mm. The next step is to find the number of rows required. This is done by dividing the total height of wall by the maximum exposure. If the total wall height to be covered with bevel siding is 2388 mm, and the maximum exposure for the siding is 210 mm, the number of rows would be 11.37.

$$210\overline{)2388}^{\,11.37}$$

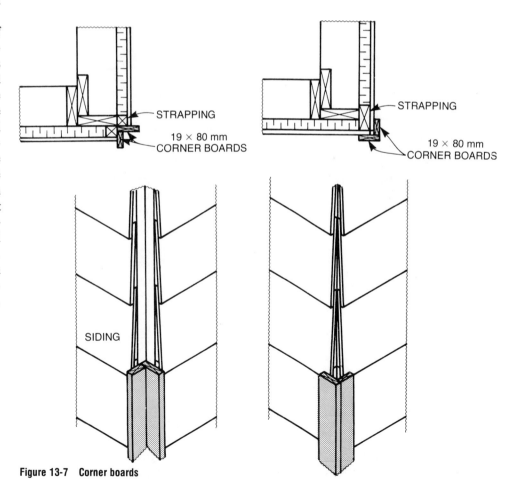

STRAPPING

19 × 80 mm
CORNER BOARDS

STRAPPING

19 × 80 mm
CORNER BOARDS

SIDING

Figure 13-7   Corner boards

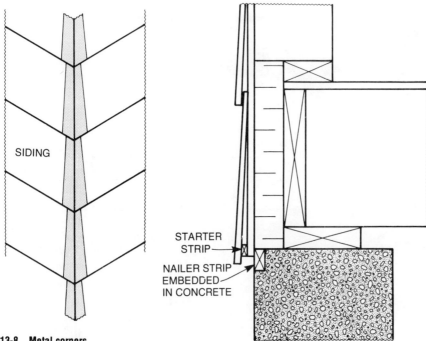

Figure 13-8   Metal corners

SIDING

STARTER
STRIP

NAILER STRIP
EMBEDDED
IN CONCRETE

Figure 13-9   Starter strip

Since all the rows are to be the same exposure, it is necessary to count the portion of the row (.37) as a full row, bringing the total to **12 rows**. Once the number of rows is determined, the actual exposure can be calculated by dividing the total wall height by the number of rows.

$$\begin{array}{r} 199 \\ 12\overline{)2388} \\ \underline{12} \\ 118 \\ \underline{108} \\ 108 \\ \underline{108} \end{array}$$

The actual exposure for this example will be 199 mm.

To facilitate application of bevel siding, the storeypole referred to in Figure

13-2 should be used. Made from a straight grained strip of wood, it must be as long as the total height of the wall that is to be covered with the siding. Mark out the actual exposure and use it as a guide at the ends of the wall to aid in keeping the rows straight and uniform. You will find

that a storeypole at each end of the wall will speed up the application.

There are several methods of dealing with the exterior corners when applying bevel siding; siding may be **butted** to corner boards (Fig. 13-7), have **metal corners** made of light-gauge metal such as aluminum applied over the exposed corner (Fig. 13-8), or the corners can be **mitred**. When the corners are mitred, the ends of the siding meet each other at a 45° angle. Accurate layout and skilled cutting is required to obtain a good joint.

APPLICATION OF BEVEL SIDING USING MITRED CORNERS

Each row of bevel siding is lapped over the top of the row below, giving the surface of each row the same slope in relation to the vertical. However, this is not true for the first row. Therefore, a *starter strip* made of the same thickness as the top portion of the siding that is covered by the lap, is required to bring the row to the same slope as the rest of the rows (Fig. 13-9). This starter strip may be cut from a sheet of plywood of the appropriate thickness.

With the starter strip installed, determine where the top of the first row of siding will be and snap a line (Fig. 13-10).

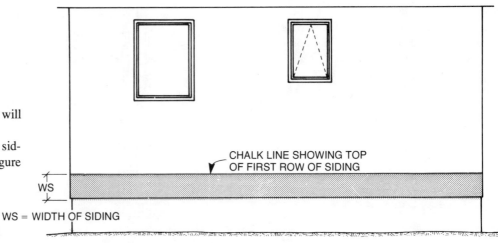

CHALK LINE SHOWING TOP
OF FIRST ROW OF SIDING

WS

WS = WIDTH OF SIDING

Figure 13-10   Chalkline showing top of siding

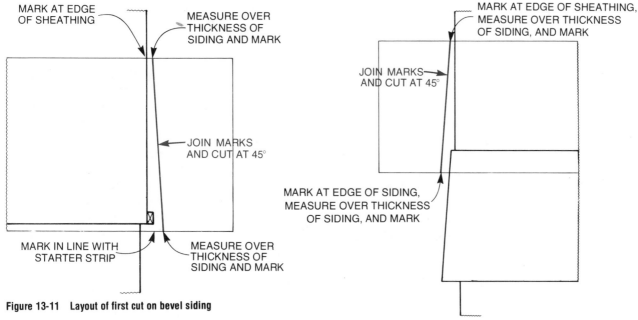

Figure 13-11   Layout of first cut on bevel siding

Figure 13-13   Layout for subsequent rows of siding

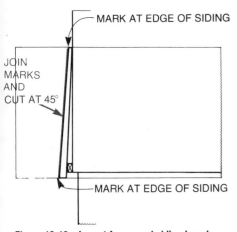

Figure 13-12   Layout for second siding board

Tack a siding board on the wall, with the top of the board even with the chalkline, as shown in Figure 13-10. Mark the top of the siding flush with the edge of the sheathing, and measure over the thickness of the top of the siding, making a second mark. Mark the bottom of the siding in line with the outer edge of the starter strip. Measure over the thickness of the siding at the bottom and make a second mark. Join the outside marks and cut at a 45° angle (Fig. 13-11).

Snap a chalkline on the adjacent wall to correspond with the chalkline on the first wall. Tack a siding board on the wall. The top of the board should be even with the chalkline. Be sure the top of the first siding board is tight against the sheathing, then mark the second siding board at the top and bottom, in line with the outer edge of the siding (Fig. 13-12). Join these two marks and cut at a 45° angle. Nail in place.

Using the storeypole, mark the ends of the wall where the bottom of the second row will be positioned. Snap a chalkline to ensure that the siding is kept straight. The first board in the second row is marked as shown in Figure 13-13, and the matching board on the adjacent wall is marked the same as the second board in the first row.

Nail the siding in place with galvanized or electro-coated siding nails that are 55 to 65 mm long. When nailing siding, **only one** nail is required at **each stud**.

The nail is placed at the bottom of the siding high enough so that it will miss the top of the siding below. This allows the siding to expand or contract freely without cracking or warping.

Continue to lay out, cut and nail the siding for the remainder of the wall in the same manner, using the storeypole to ensure the rows remain straight. The top row of siding will have to be cut to fit. The width will be the actual exposure determined when laying out the wall. Siding is butted against the exterior moulding of doors and windows, and slight planing with a block plane may be necessary to ensure a tight fit. Where the siding is butted against the door and window mouldings, it should be caulked to seal the joint. It is also good practice to put caulking in the mitre joint at the corner, wiping off the excess after the siding is nailed in place.

When it is necessary to join lengths of siding to complete a row, the ends should be carefully butted and the joints occur over solid backing or studs.

REVIEW QUESTIONS

13—EXTERIOR WALL FINISH

Answer all questions on a separate sheet.

A. Write full answers

1. Name the four basic patterns that lumber sidings are produced in.
2. Name the two basic methods of applying shingles as a side wall covering.
3. What aid can be used to keep rows of siding straight?

B. Replace the Xs with the correct word(s)

4. To provide ease of installation, the edges of the plywood panels used for siding are often XXXXX.
5. When brick veneer is used as exterior finish the foundation wall must be XXXXXXXXXXXX.

C. Write full answers

6. If the total wall height to be covered with 235 mm bevel siding is 2400 mm high, how many rows of siding will there be and what will be the actual exposure?
7. Name three methods of providing a satisfactory finish for exterior corners using bevel siding.
8. Why is it important to use galvanized or electro-coated nails when applying siding?
9. Why is it necessary to use a starter strip under the first row of bevel siding?
10. List four things that should be taken into consideration when selecting an exterior finish.

D. Mark as either TRUE or FALSE on a separate sheet.

11. Aluminum siding is available for application either horizontally or vertically.
12. Stucco can be applied only to a rough masonry wall.
13. One disadvantage of metal siding is that it must be painted annually.
14. Vinyl siding is not subject to expansion and contraction due to temperature change.
15. When a stone-dash finish is used with stucco finish a third coat is not required.

14 INSULATION AND VAPOUR BARRIER

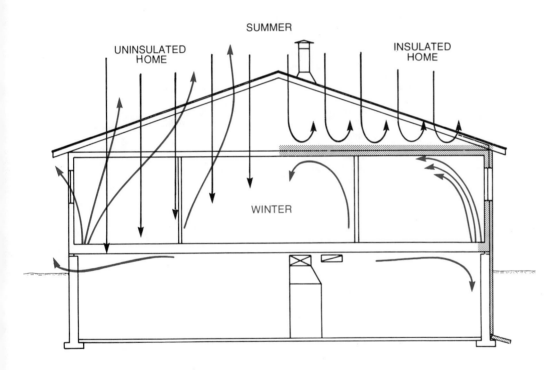

Figure 14-1 Heat loss in the winter versus heat gain in the summer

Changes in building materials and techniques have been stressed throughout the book, with respect to energy conservation. However, few areas have seen such a change in thinking and practice as we have seen in the areas of insulation, vapour barriers and their application.

HEAT LOSS

A basic law of physics states that heat (energy) flows from warm to cold. Heat that is added to the home will in cold weather eventually flow out of the house, and heat buildup outside the home in the summer will eventually flow *into* the home. Nothing can stop it! (Fig. 14-1) However, we slow down that heat transfer through insulating.

Heat flow travels by one of three processes:

231

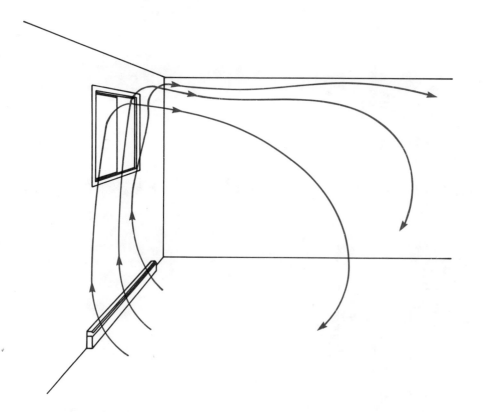

Figure 14-2 Distribution of heat by radiation

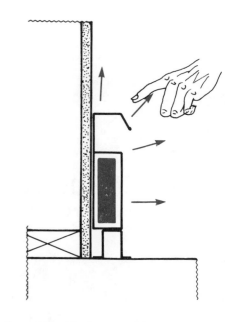

Figure 14-3 Transfer of heat by conduction

Radiation is the transfer of heat across space by electro-magnetic waves. Any warm object will give off heat in exactly the same way as the sun gives off (or radiates), heat (Fig. 14-2).

Conduction is the transfer of heat through any medium. All building materials will transmit heat in this way. Of course, some materials will conduct heat better than others (Fig. 14-3).

Convection is the transfer of heat by a medium such as a gas (air) or a liquid. Heat is transported from one surface to another by the motion of this medium. Assuming there is a difference in temperature between the inside and outside surface of the wall material, the air next to the material experiences a change in density, setting it in motion. The circular mo-

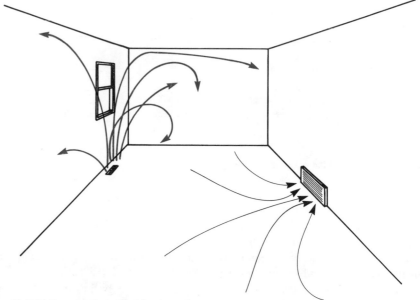

Figure 14-4(a) Transmission of heat by convection

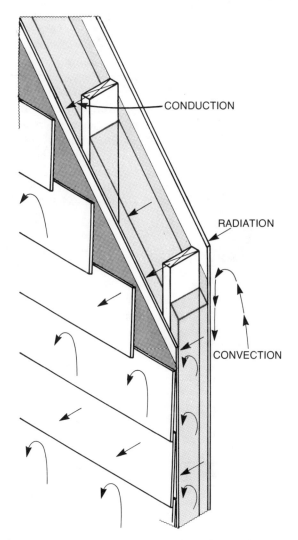

Figure 14-4(b) Heat flow through a wall by radiation, convection and conduction

convection is effectively minimized.

Insulation is manufactured and sold by its **thermal resistance** called ''R'' value with SI indicating a metric value. The RSI of any given material is a measurement of insulation's ability to resist the flow of heat. The greater the value, the better the resistance to heat transmission. This RSI value is measured using the metric term m²·°C/W, meaning the amount of heat loss per metre squared per degree Celsius per watt. Always remember that it is the RSI value of any given insulation package, not the thickness of the material, which ultimately determines the quality of the insulation in the home.

TYPES OF INSULATION

Insulation can be obtained in one of the following forms: **batt or roll, loose fill**, or **rigid board**. Each type of insulation has the same purpose; to fill the defined air space in the home, independent of its structural make-up; to resist the outward flow of heat during the winter months and to keep the house cool during the summer months.

Batt or Rolled Insulation

Batt or **rolled** insulation is manufactured from *glass, slag,* or *rock* spun into micron-sized fibres. These fibres are coated to allow bonding to take place in the form of batts or rolls. Batt insulation comes in two types. The **''friction fit''** type is an unfaced batt designed to be held in place by friction alone when installed between the framing members of floors, walls or ceilings.

The other batt insulation available is **kraft-faced batt**. This has an *asphalt-kraft paper* facing glued to the insulation. It has two flanges on the paper edge which enable the installer to staple the batt in place. This asphalt-paper membrane is also designed to help retard the flow of vapour under average humidity

tion or convection current carries the heat with it to warm the surfaces of the material and the volume of air which is contained in it (Fig. 14-4).

INSULATION

Insulation is placed in the wall cavities surrounding the living area of a house, wrapping it in a layer of material to partially block the loss or exfiltration of heat

by the three processes described. Therefore, the loss of heat can be attributed to one or more of these processes at any given time. Insulation materials *fill* or *sub-divide* an area into thousands of tiny air spaces. The material used for insulating is not as important as its ability to create the necessary air spaces. Air in itself is a poor conductor of heat. Therefore, air reduces heat loss due to conduction. Since the air cannot circulate because of the insulation, heat loss by

Figure 14-5(a)   Friction fit batts

Courtesy Fiberglas Canada Ltd.

Figure 14-5(b)   Rolled insulation (kraft-faced vapour barrier attached)

conditions. Remember however, that the facing can be highly flammable and must be covered for protection.

Roll insulation is available in the asphalt-kraft paper described above. Used as an alternative to the kraft-faced batts, it provides a minimum of breaks in the vapour barrier besides giving insulation continuity (Fig. 14-5).

When choosing batt or rolled insulation, consideration must be given to the insulating job. Batt insulation tends to be easier to handle than rolled because of the length of the individual pieces. Comparing mineral fibre to glass fibre insulation, the insulation made from glass fibre tends to be easier to work with and usually fills a given area more effectively than mineral fibre. On the other hand, mineral fibre tends to have a slightly higher RSI value in a confined space (see Table 14-1).

Loose Fill Insulation

Loose fill insulation is sold through retail outlets in bag form and is manufactured from **glass, rock, slag, pulverized paper, expanded mica, wood** and **oil by-products**. Loose fill is well suited for non-standard or irregular joist and stud spacing. Properly installed, it fills the cavities with few air pockets or voids and provides effective insulation. Generally, all loose fill insulations will settle over a number of years. They are used mainly in ceiling areas, and are not normally recommended for wall application.

Glass and mineral fibre come in both *pouring* and *blown* types of loose fill. Like their counterparts in batt, glass and mineral fibre are fire and moisture resistant. They can be used in both new and existing houses with relative ease. Keep in mind that in vertical spaces some settling will occur in the future. In horizontal areas application must be made level and even to ensure a uniform insulation job (Fig. 14-6).

Pulverized paper (cellulose fibre) is made from *recycled paper products*. A fire retardant is added during manufacture to make the material fire resistant and non-corrosive. However, it will absorb water so must not be used in areas where water or water vapour can come in direct contact with the insulation. Like glass fibre products, it is available in poured or blown form. It must be fluffed up when applied to create the tiny air spaces required. Its advantages and limitations are similar to those of glass-mineral fibre. Their settlement and RSI value are comparable (Fig. 14-7).

Expanded mica or **vermiculite** insulation comes in a poured-in-place granular form. It has an extremely high fire resistance, but like cellulose paper, will absorb water, making it ineffective if used in areas where water may be a problem.

TABLE 14-1   TECHNICAL DATA ON BATT AND ROLL INSULATION

FRICTION FIT BATTS

RSI Value	Nominal Thickness	Standard Widths	Standard Lengths
1.4	70 mm	380,584 mm	1.22 m
1.7	89 mm	380,584 mm	1.22 m
2.1	89 mm	380,584 mm	1.22 m
2.4	89 mm	380,584 mm	1.20 m
3.5	152 mm	380,584 mm	1.22 m
4.9	216 mm	406,610 mm	1.22 m
5.4	286 mm	406,610 mm	1.22 m
6.1	305 mm	406,610 mm	1.22 m

BATTS WITH AN ATTACHED VAPOUR BARRIER

RSI Value	Nominal Thickness	Standard Widths	Standard Lengths
1.4	70 mm	380,584 mm	1.22 m
1.7	89 mm	380,584 mm	1.22 m
2.1	89 mm	380,584 mm	1.22 m
3.5	152 mm	380,584 mm	1.22 m

ROLL INSULATION WITH AN ATTACHED VAPOUR BARRIER

RSI Value	Nominal Thickness	Standard Widths	Standard Lengths
1.4	70 mm	380,584 mm	24.4 m
1.7	89 m	380,584 mm	17.08 m

Courtesy Fiberglas Canada Ltd.

Figure 14-6   Blown glass mineral wool insulation

Settling also occurs when it is used for vertical applications (Fig. 14-8).

Wood shavings were once a widely used form of insulation in the home. Today they are used in local areas where they can be a viable alternative (i.e.-close to a sawmill). Wood shavings should be treated with a fire retardant before installation. Application of shavings in a vertical wall is not suggested due to the large amount of settling. However, in ceilings they do an adequate job, providing enough depth can be achieved.

Polystyrene scrap shredded into granular form can be an excellent loose fill insulation. The RSI value is quite high in comparison to other poured-in-place or blown-in insulations. Water resistance is also quite good, but due to its oil base it has a high fire rating and must be protected by a fire-resistant finish on the interior of the home. Settling will also take place when it is used in a vertical application.

Choosing and using the *best* loose fill insulation will require looking at all the variables involved. Each must be appraised on its own merit and whatever the choice remember that strict adherence to the manufacturer's instructions is important. If you plan to install the insulation yourself, a poured-in-place type must be purchased. The blown-in type of loose fill must be installed by machine, unless otherwise stated (Table 14-2).

Rigid Insulation

Rigid insulation is sold through retail outlets in **sheet** or *panel* form. Standard sizes vary from 300 × 2400 mm to 1200 × 2400 mm while thicknesses range from 25 to 140 mm depending on the material and manufacturer. Rigid insulation is presently being made from **wood by-products, glass,** and **oil-based materials.**

Figure 14-7   Cellulose fibre insulation

Courtesy Celufibre Industries Ltd.

Figure 14-7(b)   Cellulose fibre insulation blown in the attic area

TABLE 14-2 TECHNICAL DATA FOR LOOSE FILL INSULATIONS		
Type of Fill	RSI Value	Given Thickness
Mineral wool	1.2	57 mm
Mineral wool	1.4	64 mm
Mineral wool	1.7	76 mm
Mineral wool	2.1	95 mm
Mineral wool	3.5	152 mm
Cellulose fibre	.6	25 mm
Mica or vermiculite	.4	25 mm
Wood shavings	.4	25 mm

Figure 14-8   Expanded mica or vermiculite insulation

Courtesy Grace Co. of Canada

Cellular glass or **fibreglass** rigid board insulation is manufactured in a similar way to batt insulation. It is then compressed into board form. A vapour membrane may or may not be applied to one side of the board. Edges are left square, forming a butt joint, or may be shiplapped for ease and continuity (Fig. 14-9).

Fibreboards used for insulation purposes should not be confused with the many higher density hardboards on the market. Fibreboard insulations generally use a petroleum-base product as a binding medium and do not have the rigidity of hardboards. Like their counterparts (cellular glass), they can be purchased with butt or shiplap edges and in various sizes and thicknesses.

Rigid foam plastics (polyurethane and polystyrene) have many advantages. Along with their extremely high insulation value are their light weight, rigidity, sound absorption ability and vapour barrier qualities. Rigid foam insulations not only contribute to flame spread but also produce heavy, dense smoke, and give off poisonous gases under fire conditions. Because of this, it is important that they are properly installed.

Rigid foam insulations have been approved for use in buildings under the following conditions: used inside the structure, they must be covered with 12.5 mm gypsum board. They may be used outside the structure on walls (con-

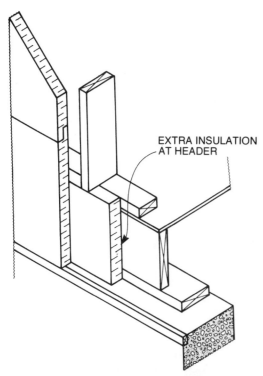

Figure 14-9(a)   Rigid insulation applied to the exterior of a framed wall

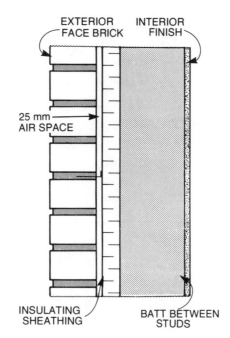

Figure 14-9(b)   Face brick and siding applied over rigid insulation

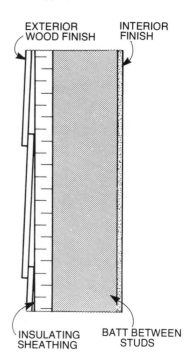

Courtesy Fiberglas Canada Ltd.

Figure 14-9(c)   Rigid insulation

crete and wood), on roof areas or under concrete floor slabs (Fig. 14-9).

Due to the composition of plastic foam insulation, it has a tendency to act like a vapour barrier under certain conditions. When applied to the exterior face of a framed wall, it may provide a natural trap for moisture, especially if the vapour bar- rier is in poor condition and allows humidity from the heated area into the frame wall. Over time this trapped mois- ture may result in decay of the framing members. To avoid this problem, it is rec- ommended that horizontal joints be pro- vided at intervals of 600 mm. This will allow any moisture that may penetrate the wall to escape out through the sheathing.

When used in place of the exterior wall sheathing, rigid insulation **does not** pro- vide the structure with the required rigid- ity. To provide this resistance, *let-in wood* or *steel bracing* is required beneath the insulation where structural sheathing has been omitted and substituted with in- sulation board. Table 14-3 indicates the RSI values of the various rigid insulations available.

TABLE 14-3 TECHNICAL DATA FOR RIGID INSULATION		
Type of Insulation	**RSI Value**	**Given Thickness**
Cellular glass	.5	25 mm
Glass fibre	.7	25 mm
Expanded polystyrene (smooth skin)	.9	25 mm
Expanded polystyrene (moulded bead)	.6	25 mm
Mineral fibre	.6	25 mm

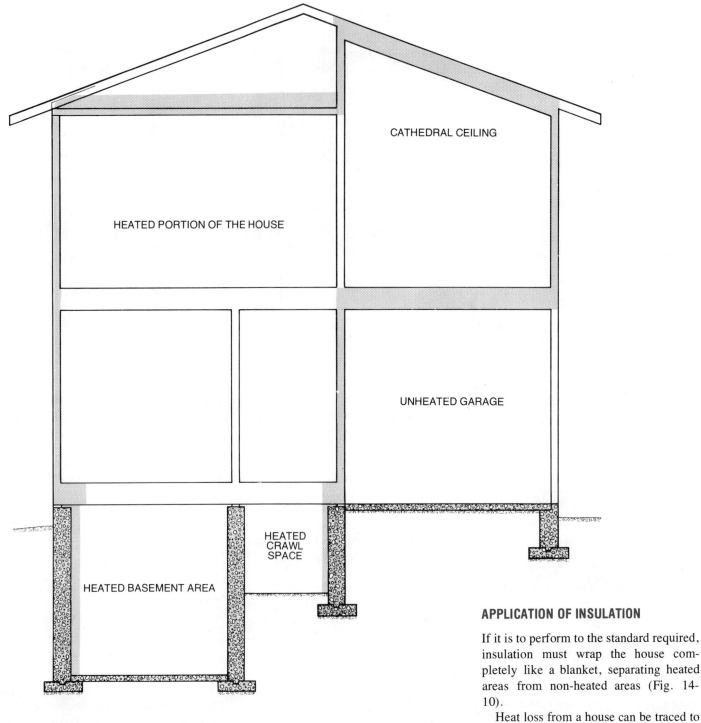

CATHEDRAL CEILING

HEATED PORTION OF THE HOUSE

UNHEATED GARAGE

HEATED
CRAWL
SPACE

HEATED BASEMENT AREA

Figure 14-10   Areas in a home requiring insulation

APPLICATION OF INSULATION

If it is to perform to the standard required, insulation must wrap the house completely like a blanket, separating heated areas from non-heated areas (Fig. 14-10).

Heat loss from a house can be traced to five specific areas. The **basement** area, exposed **floor area, walls, windows** and

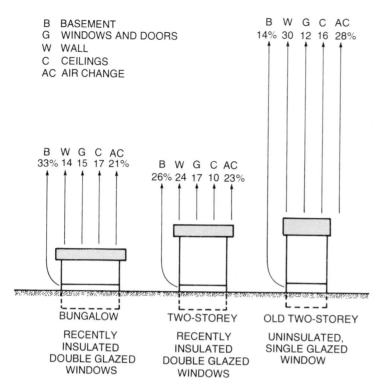

B   BASEMENT
G   WINDOWS AND DOORS
W   WALL
C   CEILINGS
AC AIR CHANGE

B   W   G   C   AC
14% 30 12 16 28%

B   W   G   C   AC
33% 14 15 17 21%

B   W   G   C   AC
26% 24 17 10 23%

BUNGALOW

RECENTLY
INSULATED
DOUBLE GLAZED
WINDOWS

TWO-STOREY

RECENTLY
INSULATED
DOUBLE GLAZED
WINDOWS

OLD TWO-STOREY

UNINSULATED,
SINGLE GLAZED
WINDOW

Figure 14-11   Approximate proportions of heat loss for homes and their components

doors, and the **ceiling** and **roof area** all contribute to heat loss. Add to this the normal air leakage which occurs because of wind pressure, which causes positive and negative pressures around the home. The result is air leakage through cracks in the building envelope. In the past, the practice was to treat many of these components with little or no regard to the amount of heat loss that was actually occurring. However, due to the energy crisis, both homeowners and the government have become more conscious of the need to conserve. This concern has led us to look at the parts of the home and at what we can do to prevent or reduce the amount of heat which is lost over a given time period (Fig. 14-11).

Basement Areas

Until recently this area of a home was completely ignored unless the basement was designed as an integral part of the living area of the house. The idea that a basement is unheated is not true. All basements receive large quantities of heat throughout the heating season. This comes from the floor above, and is circulated through the house by the furnace via the duct work, especially through heat outlets in the basement area. Even without this the basement qualifies as a heated area.

A basement with 300 to 450 mm of concrete foundation exposed above the ground can be one of the largest single contributors to heat loss. This proportion of heat loss increases as the upper portion of the home is insulated to a better standard. Figure 14-11 indicates that heat loss through the basement wall can be as high as one third of the total heat bill for a bungalow-style home. The area for the first 300 to 600 mm below grade can be

half as large as that above grade. As you go deeper, the insulation effect of the surrounding ground does improve, but only slightly.

The **choice** of insulating on either the inside or outside the basement wall will depend on the style of the home, the use of the basement area and where the best dollar value can be applied. Exterior insulation can be easier to install because there are fewer obstructions on the outer face of the wall. It can also help shed surface water away from the basement wall, while protecting the footing under the basement wall from frost.

Expanded polystyrene (waterproof type) should be used for exterior applications. The insulated area above the grade must be protected from damage and the risk of fire. This can be accomplished by using pressure-treated plywood fastened through the insulation at the top to the bottom wall plate. Expanded wire or stucco wire is then secured to the pressure-treated material and parged over to finish the area above ground. A metal flashing bent to accommodate the siding and covering the top of the insulation completes the junction of the materials. The bottom edge of the insulation and covering is held in place by the soil surrounding the basement. Being a very vulnerable area, the junction of the floor and basement should be insulated by extending the insulation up past the floor header and onto the wall to ensure that heat loss is restricted when insulating the exterior wall surface (Fig. 14-12).

Insulation applied to the inside of the foundation wall may be chosen if completion of the basement as a livable area is desired. Although in some areas only partial application of insulation is required, complete insulation of the interior basement wall is strongly suggested. If the inner face of the wall is only partially insulated, it is desirable to achieve the same

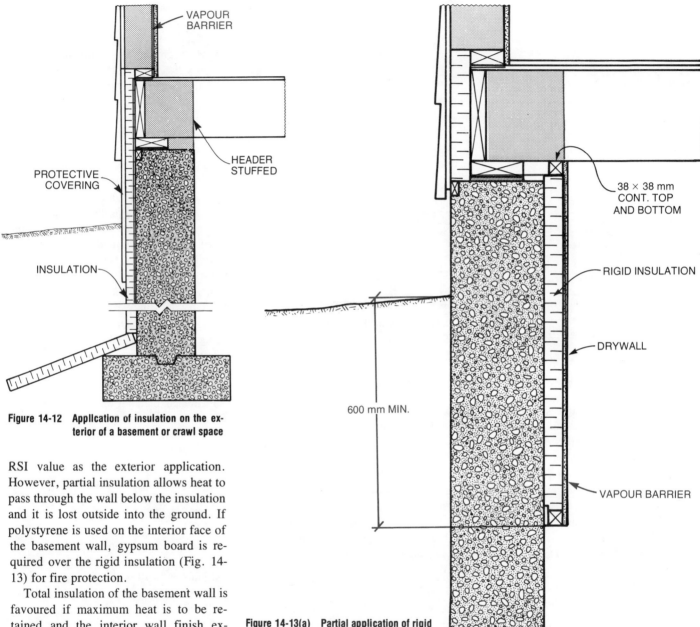

Figure 14-12 Application of insulation on the exterior of a basement or crawl space

Figure 14-13(a) Partial application of rigid insulation on the inner face of the basement wall

RSI value as the exterior application. However, partial insulation allows heat to pass through the wall below the insulation and it is lost outside into the ground. If polystyrene is used on the interior face of the basement wall, gypsum board is required over the rigid insulation (Fig. 14-13) for fire protection.

Total insulation of the basement wall is favoured if maximum heat is to be retained and the interior wall finish extended the full height. When this is desired, a *moisture barrier* should be applied to the foundation wall first. This moisture barrier can be polyethylene or regular foundation coating, but it must not go above ground level. Any moisture trapped behind the insulation must be able to *escape* at the top of the foundation. On

any interior insulation construction, the moisture barrier must also be sealed along the bottom edge of the framing to prevent the movement of air along the back of the wall. If air is allowed to enter, it will carry with it moisture which will con-

dense and ice will form during the winter months (Fig. 14-13(b), 14).

Exterior walls of **crawl spaces** under homes can be treated in the same manner as full basement walls, unless moisture or water problems require that the area be

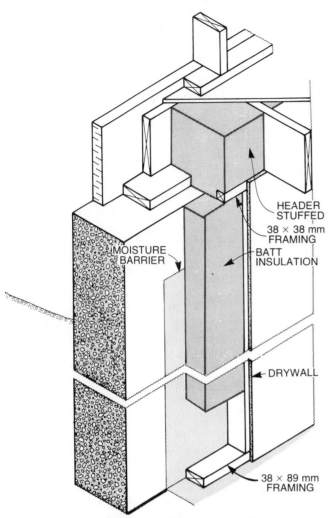

HEADER
STUFFED

38 × 38 mm
FRAMING

BATT
INSULATION

MOISTURE
BARRIER

DRYWALL

38 × 89 mm
FRAMING

**Figure 14-13(b)   Partial application of batt insulation on the
inner face of the basement wall**

ventilated both winter and summer. It is
generally accepted that rigid polystyrene
is best in this area because of its waterproof qualities. Wall insulation is considered better than floor insulating because it
eliminates the need to cover the ductwork
and plumbing in the crawl space. This
area effectively becomes heated space,
leading to a drier area and warmer floors
above.

Insulation used outside should extend
down past the footing to ensure that frost
cannot cause heaving of the foundation.
This will not be necessary if the footing is
below frost line. Inside, the insulation
should be applied to the wall and continue

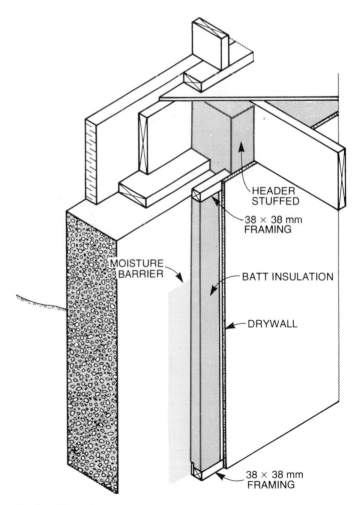

HEADER
STUFFED

38 × 38 mm
FRAMING

BATT INSULATION

MOISTURE
BARRIER

DRYWALL

38 × 38 mm
FRAMING

**Figure 14-14   Complete application of insulation
on the inner face of a basement wall**

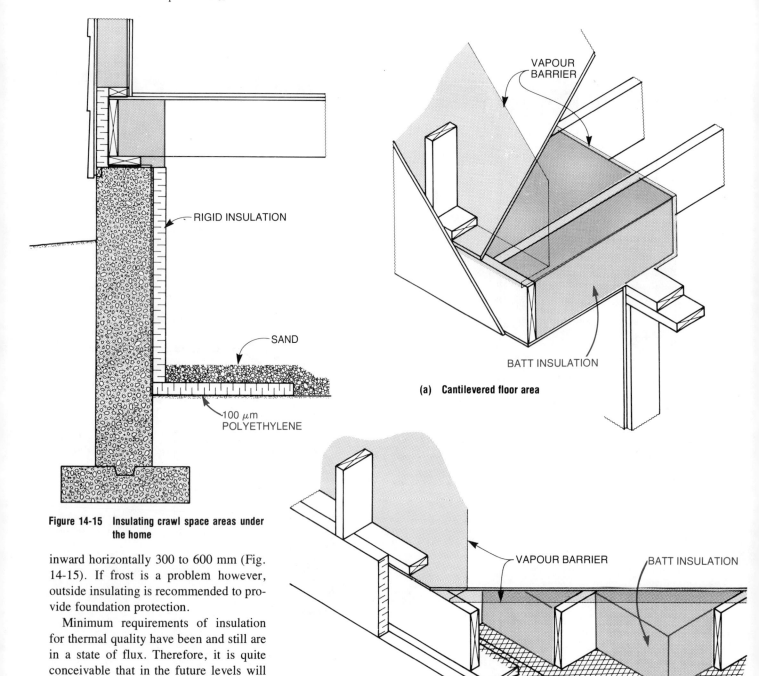

**Figure 14-15   Insulating crawl space areas under
the home**

RIGID INSULATION

SAND

100 μm
POLYETHYLENE

VAPOUR
BARRIER

BATT INSULATION

(a)   Cantilevered floor area

VAPOUR BARRIER

BATT INSULATION

(b)   Unheated crawlspace

CHICKEN
WIRE

Figure 14-16   Insulation of exposed floor areas

inward horizontally 300 to 600 mm (Fig.
14-15). If frost is a problem however,
outside insulating is recommended to provide foundation protection.

Minimum requirements of insulation
for thermal quality have been and still are
in a state of flux. Therefore, it is quite
conceivable that in the future levels will
equal or exceed maximums used in our
homes today.

Exposed Floors

Floors within the perimeter of the basement area require no insulation. How-

ever, partially or fully exposed floors must be fully insulated. Floors of this type are found under rooms above **unheated garages** or **carports** over **crawl spaces** or **cellars** that require year-round ventilation, and under any **upper storey projections**.

The exposed underside of the floor area is generally accessible and insulating can be done quite easily using mineral wool batts or blown-in insulation. The RSI value of the insulation should be at least that required for other exposed areas but should preferably be as much as the area will contain. If a vapour barrier is attached to the insulation, it must be installed up against the floor deck. An applied vapour barrier must extend over the insulation on the *warm* side only. If the underside is not being covered with ceiling material, wire strung 450 mm apart or chicken wire may be used to ensure that the insulation stays in place (Fig. 14-16).

Exterior Walls

As drastic changes in energy supplies have occurred, causing prices to increase repeatedly, one of the beneficial spin-offs that has occurred has been the change in home building techniques. The low energy home of today, compared to those built to the minimum standards in existence, requires approximately one-third or less energy to heat.

This **reduction** is achieved by **first** increasing the **wall construction thickness** by using a wider stud assembly or by going to a **double wall system** (see Chapter 2—House Styles, and Chapter 8—Wall Framing). By increasing the wall thickness, insulation thicknesses increase as well, resulting in RSI values at least double the existing government standards. In addition to these two changes and just as important is the application of the vapour barrier to seal the home (discussed later in this chapter).

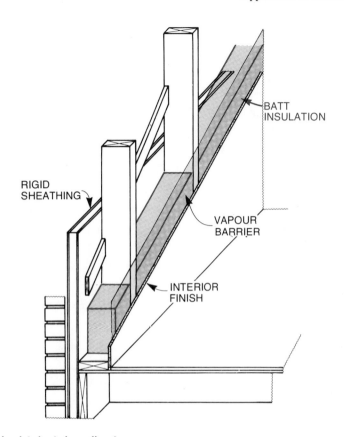

Figure 14-17   Insulated exterior wall systems

Framed walls with or without masonry exteriors are treated in the same way. The wall space between the studding is filled with batt, roll, or loose fill insulation. Where non-standard stud spacings exist, cut the insulation (if the batt type is used) 25 mm wider than the space to be filled. Make sure that all electrical outlets and pipes in the walls are insulated behind the outlets rather than in front. If this is not done, air leakage can occur around the electrical outlet; while pipes can freeze during extremely cold weather. All cracks around window and door frame units must be stuffed with insulation or *oakum* to ensure little or no air leakage can occur. Exterior rigid insulation can be applied as an exterior sheathing or over the sheathing. The air space between the masonry facing and the wall framing must be maintained (Fig. 14-9(b), 14-17).

Because of the way they are built, plank-framed walls have no cavity into which insulation may be added. Therefore, either interior or exterior insulation applications are required, depending on the decor of the home. If plank construction exposed on the exterior is desired, interior insulation would be required and vice versa. Where it is desirable to have both interior and exterior surfaces of the plank or log exposed, the method used and described in Chapter 2 (pg. 38) can be employed. Application and types of insulation used on plank walls are similar to those used on regular framed walls (Fig. 14-18).

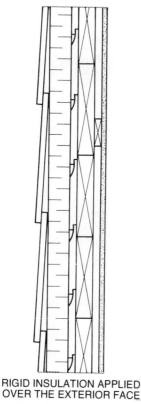

RIGID INSULATION APPLIED
OVER THE EXTERIOR FACE

BATT INSULATION APPLIED
OVER THE INNER FACE

Figure 14-18   Insulated plank wall construction

Ceilings and Roofs

Types of roofs used in residential construction fall in three different categories.
Those having an accessible **roof space** or
an attic are generally framed with a truss
or conventional type rafter. **Flat** or **low
slope** roofs, although having no accessible roof space, can be insulated between
the roof joists. **Cathedral ceilings** with
exposed wood or plank deck have no usu-
able space for insulating.

In a home using a conventional rafter,
the insulation is applied just above the
ceiling finish. Batt, roll and loose-fill insulations can all be used for this area quite
effectively. Batt or roll type insulation is
inserted between the ceiling joists. If a
vapour barrier is attached to the insula-

tion, this must be placed next to the interior (warm side) of the home. The insulation should extend out as far as possible
over the exterior wall plate, but *should
not* block off the eave vents (Fig. 14-19).
Ends of individual batts should be pressed
tightly together forming a continuous
layer of insulation. When using batt insulation, if the first layer fills the joist space
entirely and more is required, the second
layer should be laid at right angles to the
first.

Where **recessed lights** are installed in
the ceiling area, *do not* insulate closer
than 760 mm around the fixture as heat
build-up could lead to a fire hazard.
Rather than leaving this area around the
fixture uninsulated, consideration should
be given to building a box over the fixture, lining it with gypsum board and then
insulating over the top to the required
depth (Fig. 14-20). Insulation placed

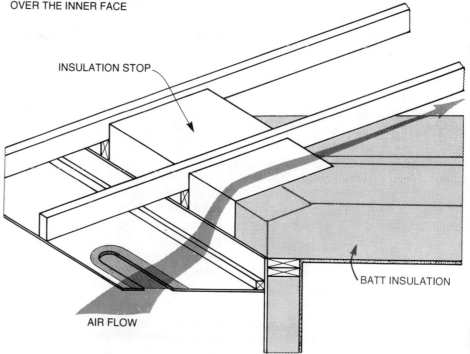

INSULATION STOP

BATT INSULATION

AIR FLOW

Figure 14-19(a)   Installation of insulation in the ceiling area

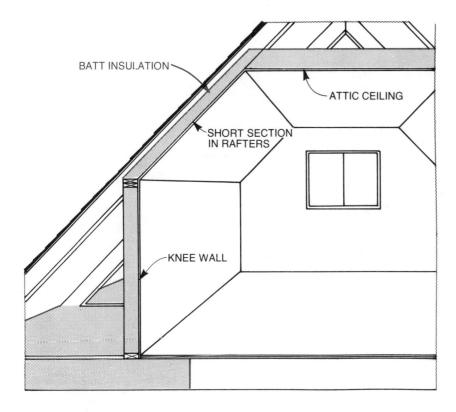

Figure 14-19(b)   Insulating the ceiling area of a one and one-half storey home

around the chimney should be a non-combustible type.

Another area in the ceiling requiring special attention is the **attic hatch or access**. It should be treated in the same way as recessed fixtures, except that the box should be set above the existing ceiling joist, allowing a build-up of insulation around it. Don't forget to insulate inside it and weatherstrip the lower hatch to prevent air leakage (Fig. 14-21).

When loose fill insulation is being used in a ceiling area, it is either blown or poured to the required depth. Like batt installation, similar precautions must be taken around recessed lights, the attic hatch, and the chimney opening. At the outer wall, baffles made of wood or cardboard must be placed between the rafters to insure that ventilation into the attic area is not restricted or blocked (Fig. 14-22).

Flat roof design is commonly used on duplexes, row housing, and to some extent, detached houses. The design of this roof lends itself to two types of insulating. One method is to either insulate between the joists with either batt or roll insulation or use blown-in loose fill insulation. The other method is to apply the insulation over the roof sheathing. A rigid type of insulation is used for this purpose and the roofing membrane is placed over it (Fig. 14-23).

The **cathedral style** ceiling with exposed wood deck is generally insulated above the deck with the roofing membrane (shingles) applied over it. In some cases it may be necessary to install the insulation between the exposed beams. If this occurs rigid insulation should be covered with gypsum board. An air space must be left between the top of the insulation and the deck to allow any condensation to escape through the roof vents (Fig. 14-24).

The true cost of keeping our homes warm and comfortable is based on the fol-

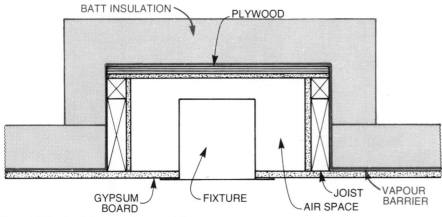

Figure 14-20   Insulating around recessed fixtures

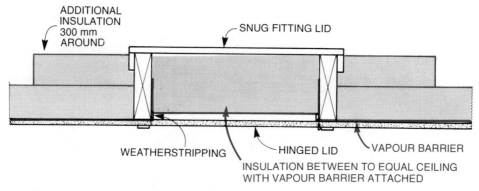

Figure 14-21   Construction and insulation of the attic hatch or access opening

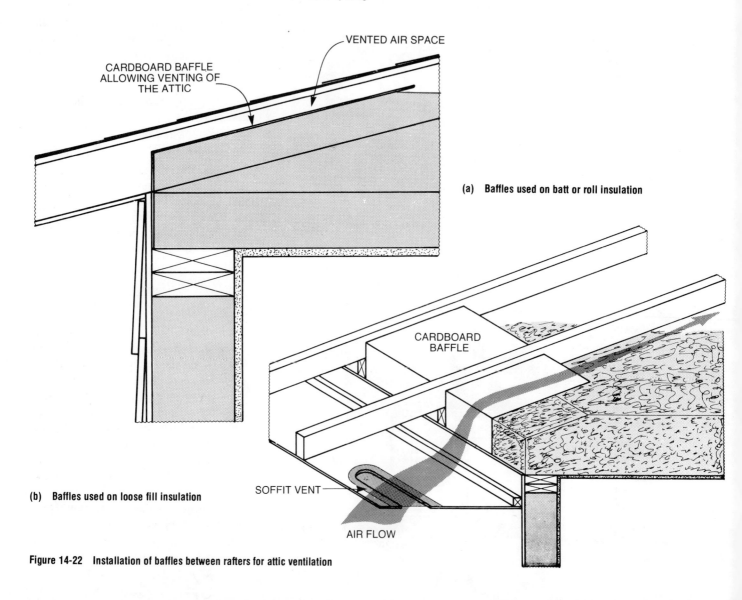

(a)   **Baffles used on batt or roll insulation**

(b)   **Baffles used on loose fill insulation**

Figure 14-22   Installation of baffles between rafters for attic ventilation

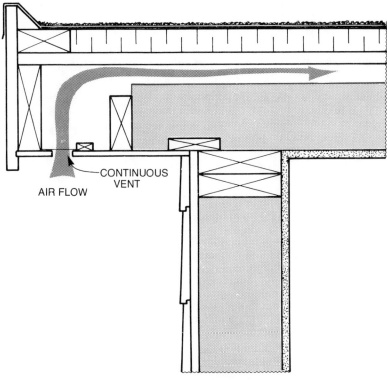

(a)   Insulation installed between the joist on a flat roof

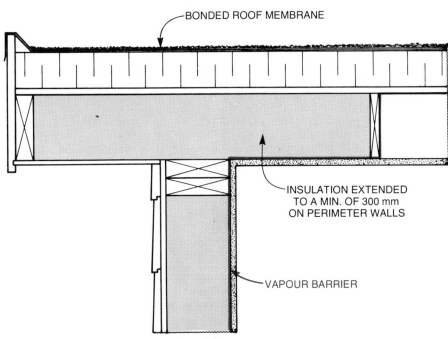

(b)   Insulation applied over the roof sheathing and covered by a roofing membrane

Figure 14-23   Insulation applied to flat roofs

lowing criteria: the cost of the energy, be it oil, gas, coal, electricity or solar, to replace the heat lost from our homes during the heating season; the cost of the energy-saving materials we build into our homes in the form of a *thermal barrier* produce an energy-efficient home. The ultimate aim should be to arrive at the most favourable point in fuel savings, where the greatest return for the least cost in insulation materials is achieved. However, as fuel costs continue to rise, what may be adequate insulation values today could be inadequate tomorrow.

VAPOUR BARRIER

We have established the value of providing a high level of insulation around the heated area of a house. With the addition of environmentally blending the house into the surrounding area, heating costs can be drastically reduced while summer-time cooling requirements become almost non-existent. What is often not understood and quite often poorly applied is the air-vapour barrier. This results in the insulation functioning inefficiently. The part that the air-vapour barrier plays in the management of the air within the house, and the protection it must provide proves that the insulation package is of prime importance.

Air movement or **leakage** out of the house occurs because of differentials in pressure between the inside and outside of the house. As the warm air rises within the house envelope, some of it is lost through natural convection while some is lost through numerous outlets to the atmosphere outside. This in turn creates negative pressure in the lower level of the house, which creates a vacuum effect. This in turn draws cold air into the home.

Areas in the home which contribute to

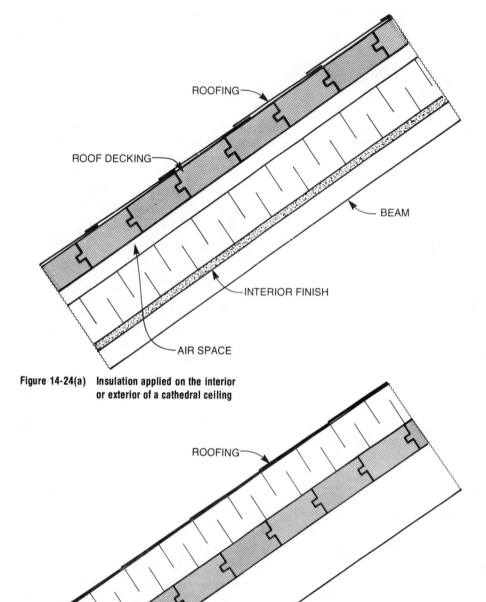

Figure 14-24(a)   **Insulation applied on the interior or exterior of a cathedral ceiling**

Figure 14-24(b)   **Insulation applied to the interior or exterior of a roof deck**

this leakage include: poorly fitted **doors,** loose **windows, attic hatches,** electrical **outlets** and **fixtures,** plumbing **stacks,** and **chimney openings,** along with numerous joints at wall and ceiling junctions. Combined, this represents a complete air change every one to two hours during the heating season in a conventional home. This also represents a good portion of the heating bill as the incoming fresh air must be heated to room temperature.

Meanwhile, **water vapour** is present in the air at all times. The amount of water vapour that air can hold is dependent on the temperature of that air. Therefore, warm air is capable of containing a much larger amount of water than cool air. **Condensation** occurs when the temperature of air falls to a point where it can no longer hold the water in the form of vapour. It begins to condense into water or frost and may appear as surface or concealed moisture.

It is very important to control the air leakage and the moisture from inside the home. Since the movement of this warm moist air is the *crucial factor* in preventing condensation, the installation of a true *air-vapour barrier* is of top priority. Ideally, this barrier should be a layer of material which is impervious to both the flow of air and moisture. It must envelop the building on the **heated** side of the insulation, thereby keeping the air and moisture inside the home (Fig. 14-25).

The air flow *restricting* effect of the vapour barrier not only reduces the amount of water vapour escaping; it drastically *reduces* the rate of air change within the building. This restriction of air becomes an important means of *controlling* heat loss as well. Once the air vapour loss has been reduced (in low energy homes to as low as two air changes per day), the problem becomes one of **first** controlling moisture within the home and **secondly,**

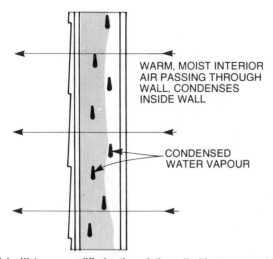

Figure 14-25(a)   Water-vapour diffusion through the wall without a vapour barrier

WARM, MOIST INTERIOR AIR PASSING THROUGH WALL, CONDENSES INSIDE WALL

CONDENSED WATER VAPOUR

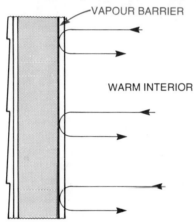

VAPOUR BARRIER

WARM INTERIOR

Figure 14-25(b)   Water-vapour diffusion restricted with the use of a vapour barrier

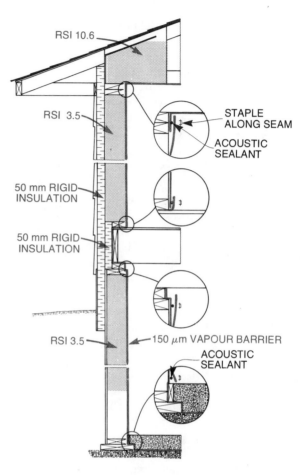

RSI 10.6

RSI 3.5

STAPLE ALONG SEAM

ACOUSTIC SEALANT

50 mm RIGID INSULATION

50 mm RIGID INSULATION

RSI 3.5   150 μm VAPOUR BARRIER

ACOUSTIC SEALANT

Courtesy National Research Council of Canada

Figure 14-26   Typical wall application of the vapour barrier

managing the air in the home to prevent the accumulation of foul air.

APPLICATION OF THE VAPOUR BARRIER

The installation of an air-tight membrane on the warm side of the exterior walls and ceiling is the first step to controlling the air. The most common, and by far the most effective barrier available is **polyethylene film**, which can be applied with relative ease over large areas.

Although fifty micrometres (50 μm) is the minimum thickness required for a va-pour barrier, the use of a thicker, 150 μm film will tend to ward off premature damage to the barrier until such time as it has been covered and protected by the interior finish. The application of the polyethylene requires a *solid backing* to which to attach the edges, a *continuous bead of sealant* that will not harden or peel away from the film when joining all edges, and a good *covering material*. With these three essential items, the procedure for applying the film is quite simple. What is needed most of all is *care* in application.

With the **single stud assembly** shown in Figure 14-26 there are a number of differences in the placement of the vapour barrier. All joints are caulked and stapled to a firm backing, either at the plates or studding. To ensure the continuity of the vapour barrier, strips of film are placed on the top of all partitions where they meet the ceiling and all exterior wall and partition junctions while the framing is taking place. The strips of film should provide at least 100 mm of overlap with the large sheets placed later on.

Prior to the floor and upper wall placement the outer area of the header and floor joist are also wrapped to ensure both continuity and a good seal (Fig. 14-26). **Note** the use of smaller plates (size drops one standard width). This allows for more insulation outside the vapour barrier without weakening the structure in any way. In this drawing the film continues down the warm side of the pressure-treated wood basement wall.

The major *disadvantage* with the single wall system is the necessity to cut through the vapour barrier to install **electrical outlets**. One way of treating this problem while assuring a good seal is shown in Figure 14-27. Using either a *commercial polypan* or one formed from a piece of polyethylene, a box is formed around the outlet. 38 × 89 mm blocking is installed, ensuring that all edges of the pan are secured. The electrical wire is brought in through the blocking or wall stud and acoustical caulking applied around the wire to ensure sealing is complete. The electrical box is attached to the wall stud and wiring is completed. The area is then insulated, the outer edge caulked and the inner film placed over the entire area with a hole cut in the vapour barrier to expose the outlet.

Surface **ceiling fixtures** can be treated in a similar manner. However, recessed fixtures must be assembled as mentioned earlier (Fig. 14-20). **Plumbing vent stacks** penetrating the ceiling barrier should be carefully cut to the outside diameter of the pipe. Plastic pipe is used extensively today. An expansion joint should be placed just below the upper plates so that expansion and contraction can take place without disturbing the acoustical sealant between the pipe and the wall where it penetrates the vapour barrier.

Window and **door** areas create another problem. To seal this area, it is recom-

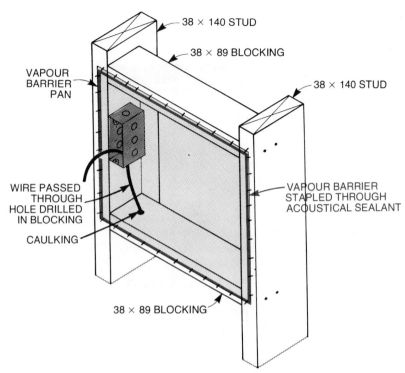

Courtesy National Research Council of Canada

Figure 14-27　Installation of the vapour barrier around electrical outlets

mended that polyethylene film be secured to the window or door frame, then brought back into the inner wall face and secured to the inside film. Figure 14-28 shows a window unit installed in a double wall system. The technique is similar in the single wall. For more information on this installation, refer to Chapter 12—Exterior Doors and Windows (Fig. 12-30).

In the **double wall system** (Fig. 14-29), the *major* difference is the placement of the vapour barrier on the wall section. By placing it on the outside of the inside wall studding, it is away from the inner face, this *minimizes* the amount of electrical wiring and plumbing which must penetrate it. The vapour barrier may be placed in this position without fear of condensation forming as long as a minimum of two-thirds of the insulation package is placed outside the vapour barrier.

Like doors and windows, other open-

ings in both systems are treated similarly. **Attic hatches** can be sealed as shown in Figure 14-21, but as suggested in Chapter 2, it would be much simpler to remove it from the inside of the house and place it in the gable end. Where the **chimney** penetrates the vapour barrier, fire-resistant insulation can be used while the vapour barrier should be securely attached to the framing around the chimney. With the edge of the metal chimney stop caulked to the framing, a reasonably good seal can be attained. A flexible cement (muffler cement) applied to the chimney wall at the metal fire stop will help to seal this area.

AIR-TO-AIR HEAT EXCHANGER

Once the escape of air from the house has been effectively controlled, some provision must be made to remove the **stale air** along with any **excess moisture** that ac-

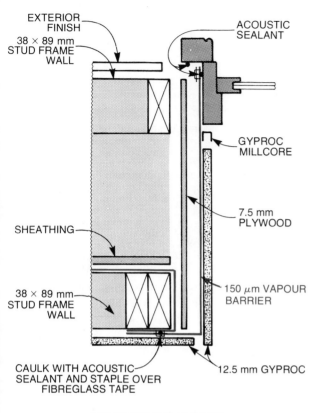

EXTERIOR FINISH

38 × 89 mm STUD FRAME WALL

ACOUSTIC SEALANT

GYPROC MILLCORE

7.5 mm PLYWOOD

SHEATHING

38 × 89 mm STUD FRAME WALL

150 μm VAPOUR BARRIER

CAULK WITH ACOUSTIC SEALANT AND STAPLE OVER FIBREGLASS TAPE

12.5 mm GYPROC

(ENLARGED FOR CLARITY)

Courtesy National Research Council of Canada

Figure 14-28 Application of the vapour barrier to the window frame on a double wall system

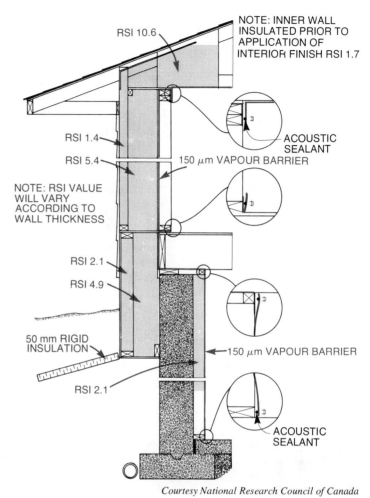

RSI 10.6

NOTE: INNER WALL INSULATED PRIOR TO APPLICATION OF INTERIOR FINISH RSI 1.7

RSI 1.4

RSI 5.4

ACOUSTIC SEALANT

150 μm VAPOUR BARRIER

NOTE: RSI VALUE WILL VARY ACCORDING TO WALL THICKNESS

RSI 2.1

RSI 4.9

50 mm RIGID INSULATION

150 μm VAPOUR BARRIER

RSI 2.1

ACOUSTIC SEALANT

Courtesy National Research Council of Canada

Figure 14-29 Placement of the vapour barrier on the double wall system

cumulates. The amount of moisture (humidity) in our homes results from *washing and drying clothes, cooking, washing dishes, taking baths* or *showers,* and other assorted daily activities. Since the restriction of air changes can lead to excessive humidity levels during the colder months of the year, condensation will collect on the interior of window surfaces or other cool areas of the walls and floors. Left unchecked, it will cause mould and eventual rotting over a period of time.

The need then for **mechanical ventilation** has come about because of increased improvements in insulation, a *true* air-

tight vapour barrier, and the desire to *retain* as much heat as possible within the home. To control this moisture and manage the air entering the house, an air-to-air heat exchanger is required. Figure 14-30 shows a schematic drawing of the operational features of such a device.

The heat exchanger replaces *all* exhaust fans found in the home. Using this system, a fan is used to exhaust the stale air from a number of sources including *bathrooms, kitchen* and *utility* areas

through the heat exchanger to the outside of the house. As the warm *exhaust* air travels through the heat exchanger, the heat energy is **transferred** to the incoming *fresh* air. Through conduction within the heat exchanger the incoming fresh air is warmed up to within 20% to 25% of the inside house temperature before it is distributed within the house, replacing the stale air being removed. Fig. 14-31 shows a commercially built air-to-air heat exchanger, available for home use.

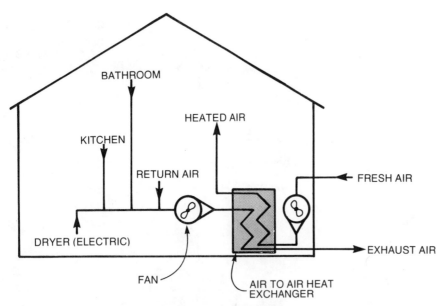

Figure 14-30   Operational schematic of an air-to- air heat exchanger

It is important to **remember** that for an air-to-air heat exchanger to function properly, it **must** be installed in an energy-efficient house with a low air change ratio. To install such a device in a conventional home where the air change is at twelve or more times per day would not conserve energy and would not even be required as the air is changing often enough to prevent any build-up of stale air.

Pointed out at the beginning, heat moves in all directions via three processes. It *flows* from warm areas to cold areas. With the use of high levels of insulation and a *true* air-vapour barrier this flow can be greatly reduced. However, in reducing the air flow, moisture and air management must be considered and extra ventilation in the form of an air-to-air heat exchanger for the comfort of the occupants is required.

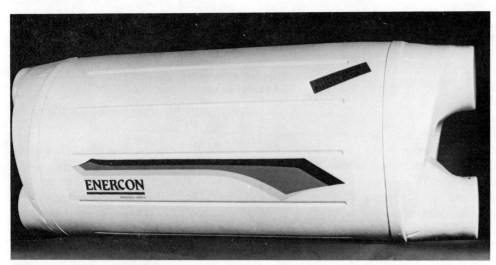

Figure 14-31   A fresh air heat exchanger

Courtesy Enercon Consultants Ltd.

REVIEW QUESTIONS

14—INSULATION AND VAPOUR BARRIER

Answer all questions on a separate sheet

A. Replace the Xs with the correct word

1. Energy (heat) flows from hot to cold. This equalization of heat is called XXXXXXX.
2. Three modes which allow heat to equalize are XXXXXX, XXXXXXX and XXXXXXX.
3. Correctly placed, insulation XXXXXX large areas into tiny XXXXXX.
4. RSI value is rated by the insulation's ability to resist XXXXXX, not by its thickness.

B. Write full answers

5. List the types of insulation used and give an example of where each type can be applied.
6. Insulation is made from numerous common materials. Make a list of those mentioned in the chapter.
7. Choosing the *best* insulation involves a number of variables. Is there a *best* insulation? Defend your answer with several statements which back up its validity.
8. Heat loss can be traced to a number of areas in the house. List these areas.
9. Plastic foam insulations must be fireproofed when applied to the interior face of the house. Why is this necessary?
10. When compressed, insulations lose a good portion of their insulation value. What causes this to happen?

C. Select the correct answer

11. "Friction fit" insulation
    - (a) Is held in place by the vapour barrier
    - (b) Is slightly compressed to fill an airspace
    - (c) Is available in rolls
    - (d) Is available in sheets
12. Insulating the interior face of the foundation wall can cause
    - (a) Frost to penetrate deeper along the exterior of the wall
    - (b) Moisture to accumulate on the wall
    - (c) Accelerated heat loss in the upper floors
    - (d) Basement floors to become warmer
13. Insulating in areas where electrical boxes or plumbing pipes occur requires
    - (a) Insulation to be placed over this equipment
    - (b) Insulation cut to fit around the equipment
    - (c) Insulation to be foamed into place around the equipment
    - (d) Insulation be inserted behind all such equipment
14. Insulating ceiling areas over the exterior wall requires special consideration to provide
    - (a) Unrestricted air flow into the attic area
    - (b) A restricted area to prevent air flow into the attic area
    - (c) Insulation over the exterior wall plates
    - (d) Warmer ceiling perimeters
15. The air-vapour barrier is applied to the "warm" side of the wall to
    - (a) Prevent the insulation from moving before the interior finish is applied
    - (b) Prevent the movement of moisture and air through the wall
    - (c) Prevent heat loss from within the warm area
    - (d) Reduce the use of caulking around the exterior at door and window openings

D. Mark as either TRUE or FALSE

16. The edges of the air-vapour barrier must be sealed and fastened to the backing to achieve a "true" barrier.
17. The air-vapour barrier is applied over electrical boxes, windows and doors. At a later date during construction it is cut out to expose these areas.
18. In the double wall system, the air-vapour barrier may be applied to the outside face of the inner wall, providing that one-half of the insulation package is outside the barrier.

E. Replace the Xs with the correct word

19. In a low energy house, stale air is removed and fresh air introduced into the house through an XXXXXX.
20. By using a mechanical control to remove stale air from a conventional house, the homeowner will not save energy, because XXXXXXX.

15 INTERIOR WALL AND CEILING FINISH ▬▬▬

TYPES OF INTERIOR FINISH

Interior finish is classified as any material used to cover interior walls and ceiling framing.

For years, the most common type of wall and ceiling finish for residential use was **plaster**. Plaster was applied in two or three coats over *wooden lath*, which was nailed to the framing members. The laths were spaced about 10 mm apart, and this provided a bond when the plaster was forced through the spaces. This was known as a key. Wood lath was then replaced by *gypsum board lath,* which had a gypsum core faced with paper on each side. Gypsum lath came in sheets 400 mm wide and 1200 mm long and was applied horizontally over the framing members. When gypsum board lath was used, usually only one coat of plaster was required to finish the surface.

Although still used to some extent, plaster has virtually been replaced by materials known as **drywall finish**. Drywall finish is any finish that requires little or no water for application. This would include plywood, fibreboard and other finishes that are applied in dry form. Drywall however, has come to mean almost exclusively gypsum board that has been filled at the joints to give the appearance of a plaster wall. Gypsum board is a sheet material made up of a

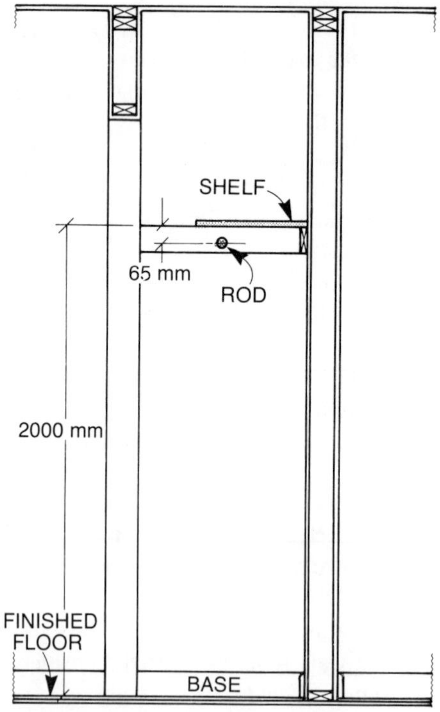

Figure 15-1   Height of clothes closet rod

gypsum filler between two layers of paper. These sheets are 1200 mm wide and come in a standard length of 2400 mm but are available in lengths up to 4800 mm. Gypsum board comes in thicknesses of 9.5, 12.7 and 16 mm, but 12.7 mm is most commonly used for residential work. It is normal practice to use sheets as long as the wall is wide and

apply the sheets horizontally, thus eliminating extra joints.

Gypsum board drywalling has become a specialized operation and is rarely applied by the average carpenter today. Therefore, this method will not be dealt with to any great extent.

Because there is no holding power in gypsum board, it is very important, that where necessary, *backing* be installed to provide support for clothes closet rods, shelves, etc. Figure 15-1 shows the standard height for clothes closet rods; backings should be installed accordingly.

Figure 15-2 shows shelving arrangement for linen closets; provisions should be made for supporting the shelves.

Backing is also necessary where shower curtain rods will be fastened to gypsum board (Fig. 15-3).

Hollow wall fasteners There will be occasions when it is necessary to fasten objects or fixtures to gypsum board where there is no backing. In situations like this, the use of **hollow wall fasteners** is recommended. A common hollow wall fastener, known as a toggle is available in two types–*gravity* and *spring*. **Gravity toggle bolts** (Fig. 15-4) are long slender bolts (1) with a toggle head pivoted off-centre so it falls by gravity into a cross position when pushed through the hole made in the gypsum board (2). The full length of the toggle

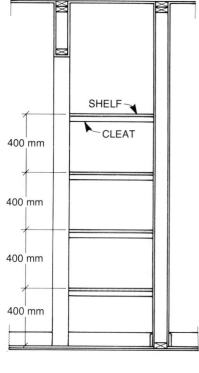

SHELF

CLEAT

400 mm

400 mm

400 mm

400 mm

Figure 15-2   Linen closets

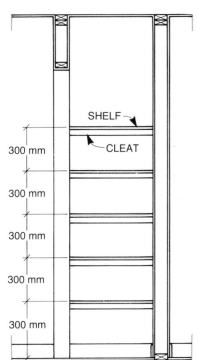

SHELF

CLEAT

300 mm

300 mm

300 mm

300 mm

300 mm

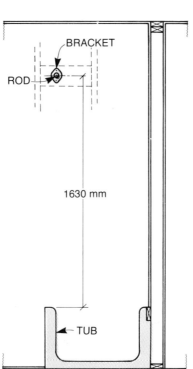

BRACKET

ROD

1630 mm

TUB

Figure 15-3   Backing for shower curtain rods

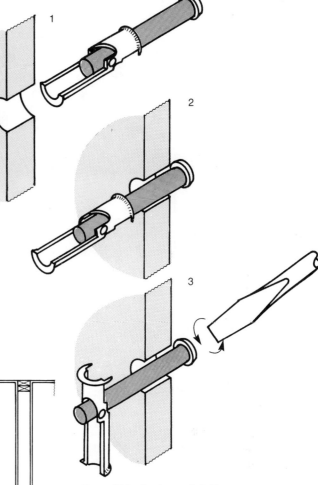

1

2

3

Figure 15-4   Gravity toggle bolt

head then bears against the inner wall surface (3). Gravity toggle bolts can be used only on vertical locations where the toggle will fall into position.

Spring toggles act similarly to gravity toggles; the toggle is equipped with a spring which expands it after it is inserted through the hole (Fig. 15-5). This type is used on floors, walls and ceilings. One disadvantage of this type of hollow wall fastener is that when the screw is removed the toggle part of the fastener is lost, making the fastener useless.

There are **screw anchors** available which are cylinders, threaded at one end

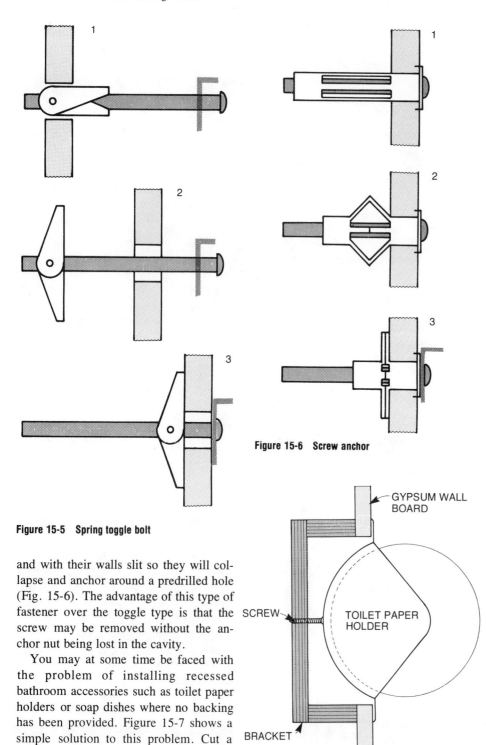

Figure 15-5　Spring toggle bolt

Figure 15-6　Screw anchor

GYPSUM WALL BOARD

SCREW

TOILET PAPER HOLDER

BRACKET

Figure 15-7　Bathroom accessories

and with their walls slit so they will collapse and anchor around a predrilled hole (Fig. 15-6). The advantage of this type of fastener over the toggle type is that the screw may be removed without the anchor nut being lost in the cavity.

You may at some time be faced with the problem of installing recessed bathroom accessories such as toilet paper holders or soap dishes where no backing has been provided. Figure 15-7 shows a simple solution to this problem. Cut a hole in the gypsum board the size of the back of the holder, make a bracket out of

plywood, insert it in the hole and hold it in place with a wire or string until the holder can be installed and fixed in place by screws. It is a good idea to pre-drill the bracket to receive the screws.

PANELLING

Wall finish in the form of sheets made from **plywood** or **hardboard** are frequently used as interior finish for an entire room or as a feature wall. Panels are available in a great variety of colours and patterns in both plywood and hardboard. Plywood panels are available in a variety of wood species either pre-finished or requiring finishing after installation.

APPLICATION OF PANELLING

Panelling is best applied so the first panel on the wall section will be the same as the last; this is called *balancing* (Fig. 15-8). One method of achieving this is to strap furring strips to the wall either horizontally or vertically. Panelling may be applied to walls that are rough-sheathed either with plywood or boards; no strapping is required in such cases.

Panels should be selected so the colour and grain will match one another when applied side by side. It is important that the first panel be applied perfectly plumb. If it is not plumb, it will affect all subsequent panels on that wall. Panels should be *scribed* to the ceiling and adjacent walls; this can be done using a scriber as shown in Figure 15-9, or with a small block of wood as in Figure 15-10.

After the first panel is fitted to the ceiling and adjacent wall, secure it in place either with nails or panel glue. If panel glue is used, be sure to follow manufacturer's instructions supplied with the glue. If nails are used, a definite nailing pattern should be decided on and followed. Attach subsequent panels, paying special attention to the edge joint. Edges

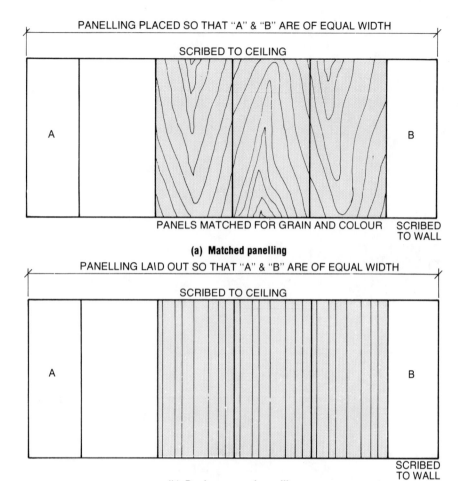

PANELLING PLACED SO THAT "A" & "B" ARE OF EQUAL WIDTH

SCRIBED TO CEILING

A

B

PANELS MATCHED FOR GRAIN AND COLOUR   SCRIBED TO WALL

(a) Matched panelling

PANELLING LAID OUT SO THAT "A" & "B" ARE OF EQUAL WIDTH

SCRIBED TO CEILING

A

B

SCRIBED TO WALL

(b) Random grooved panelling

Figure 15-8   Balanced panelling

may have to be slightly *back planed* to ensure a tight fit. It will be necessary to cut and fit the last panel of the wall section. Measure the width of the last panel, then rough cut the sheet to be used approximately 12 mm wider than this measurement. Measure back from the edge of the last panel 19 mm, and make a mark at both top and bottom of the wall (Fig. 15-11). To avoid marking the surface of the finished panelling, a strip of masking tape placed on the surface of the panel can be used to mark on. Tack the last panel in position along the marks on the previous panel and scribe the corner edge with a 19 mm block of wood. Be sure the wall panelling on the adjacent

wall is securely fastened to the framing before attempting to scribe the corner.

If it is necessary to apply panelling to external corners and a mitre joint is required, a simple jig can be made to facilitate the marking of panels at the corner (Fig. 15-12).

Tack the corner panels in place one at a time (Fig. 15-13). Then scribe along the edge of the panel using the scribing jig (Fig. 15-14). Cut along the scribe line at a 45° angle.

Attach the first panel in place, apply glue along the mitre joint and apply the second panel. Sand the mitre joint immediately as the glue will dry very quickly. Always sand with the grain, keeping the

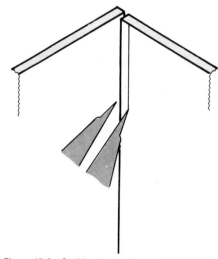

Figure 15-9   Scribing, using a scriber

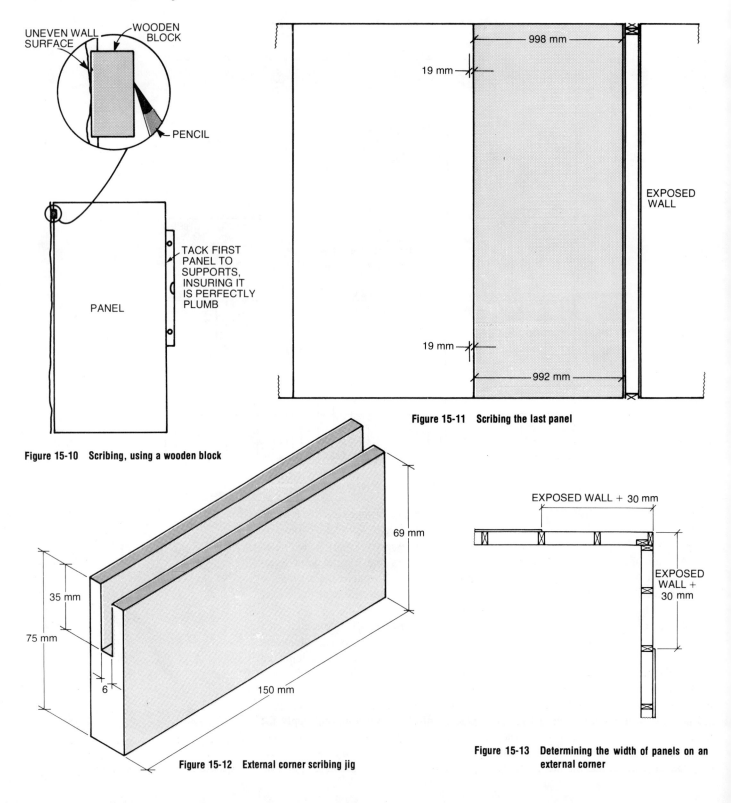

UNEVEN WALL SURFACE

WOODEN BLOCK

PENCIL

TACK FIRST PANEL TO SUPPORTS, INSURING IT IS PERFECTLY PLUMB

PANEL

Figure 15-10   Scribing, using a wooden block

998 mm

19 mm

EXPOSED WALL

19 mm

992 mm

Figure 15-11   Scribing the last panel

69 mm

35 mm

75 mm

6

150 mm

Figure 15-12   External corner scribing jig

EXPOSED WALL + 30 mm

EXPOSED WALL + 30 mm

Figure 15-13   Determining the width of panels on an external corner

sanding block flat on the surface while sanding. Remove the *arris* (sharp edge) by using very light strokes with the sanding block. This sanding method can only be used on panels that are not pre-finished.

CEILING FINISH

The most common material used for ceilings in residential construction is gypsum board. This is often given a *textured finish*. The joints are filled and then an acoustical plaster mix is sprayed on the ceiling. As well as having a pleasing appearance this textured ceiling will provide some sound-proofing.

Another type of ceiling finish used quite frequently is **fibreboard tile**. These tiles come in a range of sizes from 300 mm square to 400 × 800 mm. They come pre-finished in a wide variety of patterns. Tiles may be *stapled* to strapping which has been nailed to the underside of the ceiling framing members. It is possible to *glue* them in place if they are applied over gypsum board or similar material. For years, tiles have been used in commercial buildings using a **suspended ceiling system**. This consists of a light metal framework suspended from the ceiling joists with the tiles dropped in place between the grid work. Used in residential work, this system is particularly adaptable to basements and lower living areas where it is necessary to have access to plumbing, heating ducts and electrical wiring.

APPLICATION OF CEILING TILE

When ceiling tiles are used as an interior finish, they, like wall panelling, look best when *balanced*. That is, there should be an equal portion or border on each side of the room (Fig. 15-15).

Find the dimensions of the ceiling to be tiled and divide these dimensions by the

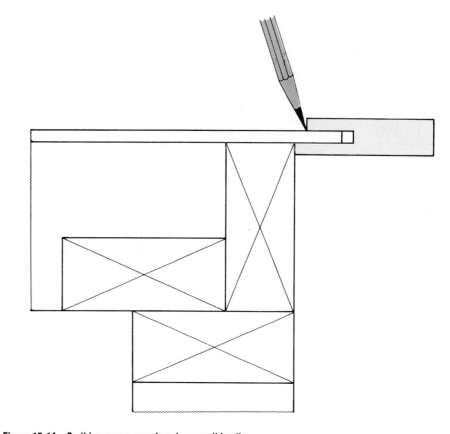

Figure 15-14   Scribing corner panels using a scribing jig

size of the tile to be used. For example, if the room size is 3404 × 4344 mm and the tile is 300 mm square, divide these dimensions by 300.

$$\begin{array}{r}
11 \\
\text{Width} \quad 300{\overline{)3404}} \\
\underline{300x} \\
404 \\
\underline{300} \\
104
\end{array}$$

There would be 11 full tiles, with 104 mm divided in two for the starting tile and the last tile. This would mean that there would be 52 mm on each side of the room. With only a 52 mm border, if the room is either off-square or the first tile is positioned wrong, the border would be very noticeable even were it to vary only

10 to 20 mm from one end to the other. In this case it is advisable to use only 10 full tiles and divide the extra tile in two, adding it to the existing border. This would mean the border would be 150 + 52 or 202 mm. A slight variation in width would not be as noticeable with this wide a border tile. The same procedure would be followed for the length of the room.

$$\begin{array}{r}
14 \\
\text{Length} \quad 300{\overline{)4344}} \\
\underline{300x} \\
1344 \\
\underline{1200} \\
144
\end{array}$$

There would be 14 full tiles with 144 mm divided in two for the borders. Each border in this case would be 72 mm;

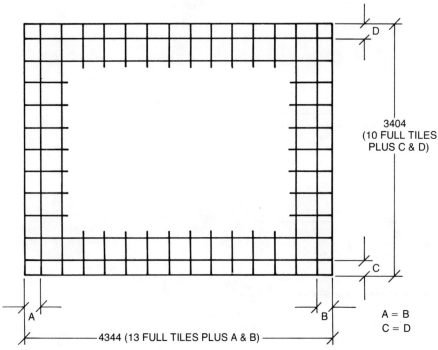

3404
(10 FULL TILES
PLUS C & D)

A = B
C = D

4344 (13 FULL TILES PLUS A & B)

Figure 15-15   Ceiling tile installed keeping edges balanced

once again it is a good idea to take a full tile and divide it in two and add to the existing border, 150 +72 or 222 mm.

Once the borders are determined, the 19 ×70 mm strapping is nailed on at right angles to the underside of the framing members. The first piece of strapping is centred on the border tile measurement of 202 mm. It is fastened into place with succeeding rows of strapping determined by the size of ceiling tile to be used. Where the underside of the framing members are uneven, *shimming* the strapping down to a point level with the rest of the strapping will eliminate any imperfections and produce a level ceiling when completed.

When the ceiling tile is stapled directly to the strapping, a chalkline is snapped on the first piece of strapping at 202 mm plus the width of the stapling flange. This will ensure straight lines as the tiling proceeds. Beginning in the right hand corner of the room, the first tile will be cut to the size determined by the room measurements of 202 × 222 mm. It is then positioned along the chalkline and stapled into place through the stapling flange. Succeeding border tile are then cut to the chalked line and fastened securely. With the first row of tile in position, the remainder of the room can be completed. The last row of tile is again cut to fit the remaining space (202 mm) along the opposite wall from where you began.

REVIEW QUESTIONS

15—INTERIOR WALL AND CEILING FINISH

Answer all questions on a separate sheet

A. Replace the Xs with the correct word

1. For years the most common type of wall and ceiling finish for residential use was XXXX.
2. The standard height for clothes closet rods is XXX mm.
3. The standard height for a shower curtain rod above the top edge of the bathtub is XX mm.
4. Describe two methods of scribing paneling on interior corners.
5. For best results the exterior corners of panelling should be XXXX.
6. A room 3560 × 7060 mm is to be finished with 300 mm square ceiling tile. What will be the size of the borders and the number of full rows in each direction?

B. Mark as either TRUE or FALSE

7. When plaster is applied over wooden laths, the plaster is secured to the wall by means of a key.
8. One disadvantage of toggle bolts is that the screw cannot be removed without causing it to be useless.
9. One advantage of a screw anchor is that the screw can be removed without the anchor nut being lost in the wall cavity.
10. If wall panelling is scribed into the corners, panels need not be applied plumb.

16 INTERIOR DOORS AND TRIM

There are many different types of doors used in residential construction today. The most common is the *conventional door,* made of wood and hinged at the side. Wooden doors now come in a standard height—2030 mm. They vary in width and thickness. The standard thicknesses for interior doors are 35 and 45 mm. Standard widths are 455, 610, 710, 760, 810 and 910 mm.

There are two general types of wooden doors, **panel** and **flush**. A *panel* door

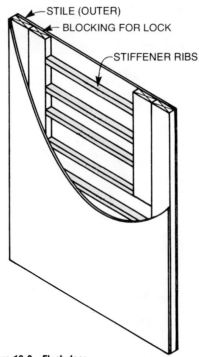

Figure 16-2   Flush door

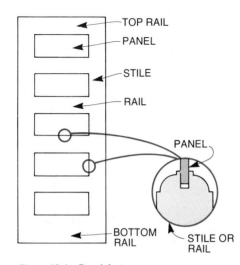

Figure 16-1   Panel door

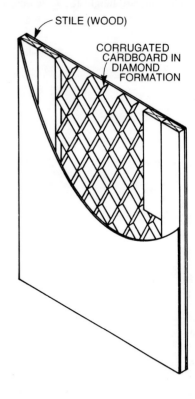

consists of exposed *stiles* and *rails* in a wide variety of panel combinations (Fig. 16-1). Panels consist of solid stock moulded to give the door a distinctive pattern, or they may be constructed of plywood.

Flush doors use a basic framework consisting of *stiles* and *rails* that is cov-

ered with a sheeting material of either *wood* or *fibreboard*. Most interior doors have a hollow core; that is, the door is basically hollow between the outer sheets except for ribs of wood or heavy cardboard placed between the stiles to give strength and rigidity to the door (Fig. 16-2).

262

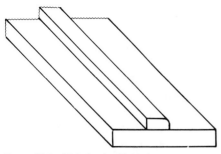

Figure 16-3   Plain jamb with stop

Doors that are conventionally hinged at the side may be installed with two different types of jambs, **plain** or **rabbeted**. Plain jamb stock is usually 19 mm thick and as wide as the thickness of the wall in which it is placed. A **door stop** is attached to the plain jamb after the door is hung (Fig. 16-3). This door stop is sometimes referred to as a *planted-on* stop.

Rabbeted jamb stock, although mainly used for exterior doors, may also be used for interior doors. Stock for interior doors is usually milled to a lesser thickness than exterior jamb stock (Fig. 16-4).

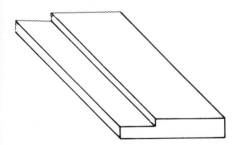

Figure 16-4   Rabbeted jamb

When space does not allow for a conventional hinged door, a door may be used that will slide into the wall; this is known as a **pocket door**. Years ago this would have meant building two separate walls with the door sliding on a cumbersome track between them. This results in an extremely thick wall. Manufacturers today have produced pocket door hardware which allows the door to slide on a narrow overhead track between specially constructed split studs that can be installed in a conventional wall (Fig. 16-5). Installing the metal frame in the opening and the hanging of the door can be easily done by following manufacturer's instructions enclosed with the hardware.

Folding doors are also used where space is limited. These doors consist of

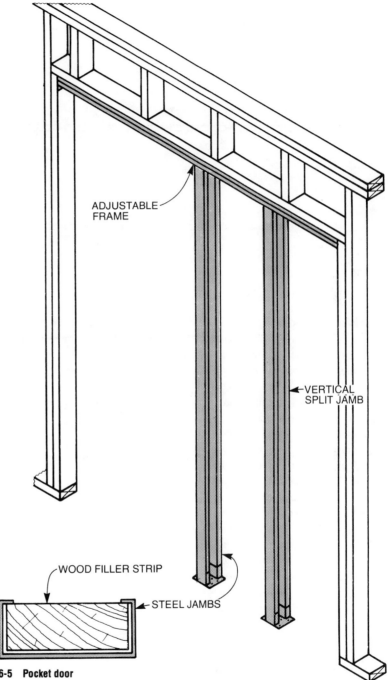

Figure 16-5   Pocket door

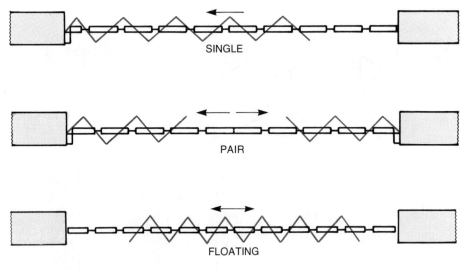

Figure 16-6   Folding doors

narrow panels which fold along an over-head track much like an accordion. When the door is open, the panels are stacked to one side of the door opening using approximately 50 mm of space for each 300 mm of door. This means that a 910 mm door would use 150 mm of space when opened (Fig. 16-6).

Folding doors are available with *wooden* or *fabric* covered panels. The wood used for folding doors is very carefully selected; it must be completely dry and free from all imperfections. The most common types of wood used for the manufacture of these doors are cedar and mahogany. Fabric doors are usually made of vinyl material over wooden panels on a metal framework. Fabric doors are available in a wide selection of colours and finishes. This type of door is easily installed if the manufacturer's instructions are carefully followed.

A wide variety of doors are available for use on closets. All of the doors previously mentioned can be used as well as a number of others. A type of door that has become increasingly popular for closets over the last few years is the **bi-fold**.

With this type of door the panels hinge in the centre and fold to one side on a special overhead track. They are available in either two or four panel units (Fig. 16-7).

Note: With all styles of doors which use some form of hardware supplied by a manufacturer, it is important that the installation instructions supplied be carefully followed.

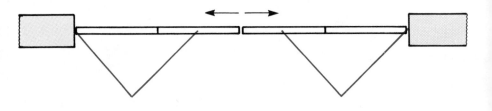

Figure 16-7   Bi-fold doors

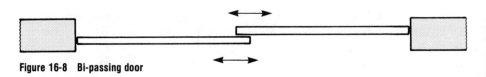

Figure 16-8   Bi-passing door

Another type of closet door used is the *bi-passing sliding* door. With this type of door, panels simply slide past one another on an overhead track (Fig. 16-8). One disadvantage with this type of door is that only one half of the closet opening is accessible at a time.

Panels for this type of door are often plywood or panel. They are, however, available as completely mirrored panels, which give a pleasing effect as well as being practical.

INSTALLATION OF DOOR JAMBS

Most interior doors are installed using plain jambs. The first step in preparing jamb stock is to be certain the edge of the jamb stock is perfectly straight; it may be necessary to true the edge using a jointer. The jamb stock must then be cut to the correct width. The width of the stock will be the same as the thickness of the wall (Fig. 16-9). When cutting the jamb stock on the saw, remember to leave sufficient width so that the edge of the stock may be jointed to produce a smooth finish. If the edges are slightly beveled, this will ensure a tight fit when the trim is installed.

The side jambs are then cut to length

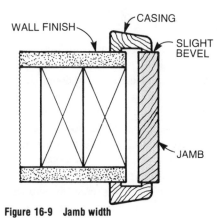

Figure 16-9 Jamb width

and rabbeted to receive the head jamb. Check the side jambs for bow. If there is one, place it towards the trimmer stud as it is much easier to straighten the jamb through the use of wooden wedges than to try to draw it straight using nails (Fig. 16-10).

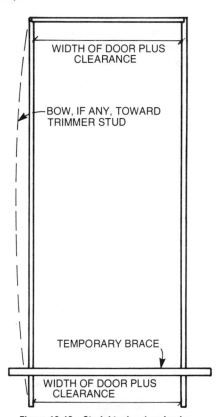

Figure 16-10 Straightening door jambs

Side jambs should be of sufficient length to allow for clearances both at the top and the bottom of the door. When allowing for clearance on the bottom of the door, consideration must be given to the finish floor covering. For example, is the floor finish to be 3 mm tile or 19 mm carpet? This is important as the door jambs are usually set before the floor covering is installed. The type of heating system must also be taken into consideration, since a forced air system with cold air returns located in a central hallway will necessitate a space of at least 15 mm between the finish floor and the bottom of the door for the system to work properly.

Check if the floor is level at the door opening; if it isn't, it will be necessary to have one side jamb longer than the other

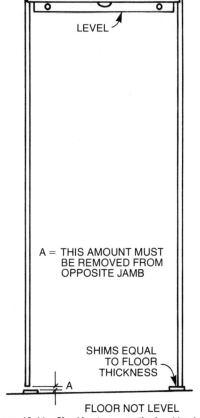

Figure 16-11 Checking to ensure the head jamb is level

to ensure that the head jamb remains perfectly level (Fig. 16-11).

Cut the head jamb to length, allowing for the depth of rabbets on the side jambs and the side clearance. Nail the side jambs to the head jamb. A tight joint will be obtained if the head jamb is nailed through the side and top as shown in Figure 16-12. The nail through the top should be driven first.

Rough openings are usually framed so they are 25 mm wider than the outside measurement of the door frame. Plumb wedges on the hinge side by tacking a 12 mm block at the bottom of the trimmer, and nail wedges at the top plumb with the lower block (Fig. 16-13).

Place the door jamb in the opening and nail the hinge side jamb to the plumbed wedges. Straighten the jamb between the upper and lower wedges with a straight edge. Nail the jamb in place using wedges where necessary. Wedge the lock side jamb at the top and nail in place. If the door is a pre-hung unit, set the door in place on the hinges and nail the lock jamb in place, wedging where necessary to provide correct clearance. If the door is not pre-hung, a spreader the same length as the door width plus clearances can be

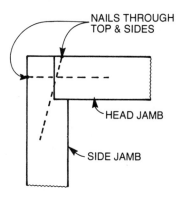

Figure 16-12 Nailing the head jamb

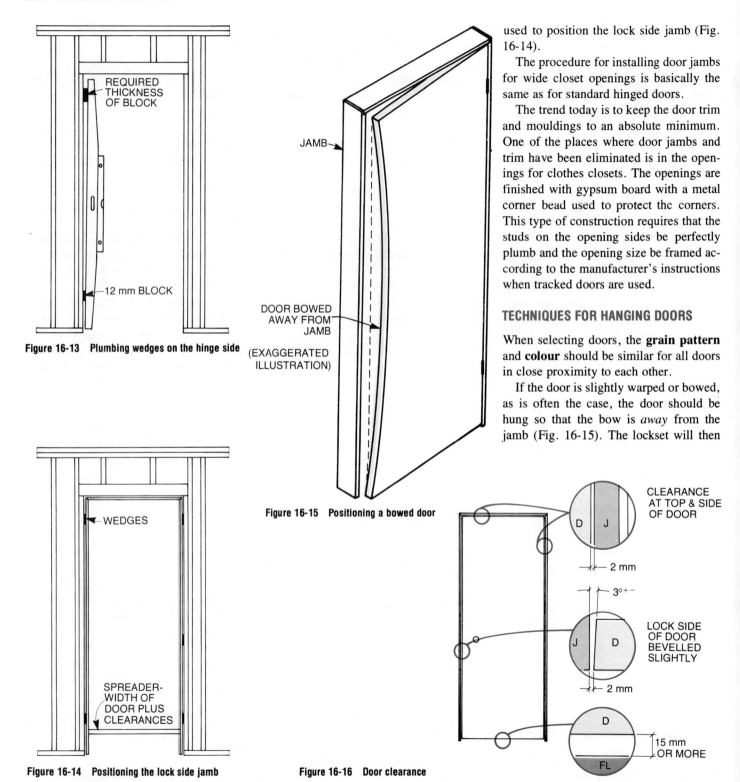

Figure 16-13   Plumbing wedges on the hinge side

REQUIRED
THICKNESS
OF BLOCK

12 mm BLOCK

JAMB

DOOR BOWED
AWAY FROM
JAMB

(EXAGGERATED
ILLUSTRATION)

Figure 16-15   Positioning a bowed door

WEDGES

SPREADER—
WIDTH OF
DOOR PLUS
CLEARANCES

Figure 16-14   Positioning the lock side jamb

CLEARANCE
AT TOP & SIDE
OF DOOR

D   J

2 mm

3°+−

LOCK SIDE
OF DOOR
BEVELLED
SLIGHTLY

J   D

2 mm

D

15 mm
OR MORE

FL

Figure 16-16   Door clearance

used to position the lock side jamb (Fig. 16-14).

The procedure for installing door jambs for wide closet openings is basically the same as for standard hinged doors.

The trend today is to keep the door trim and mouldings to an absolute minimum. One of the places where door jambs and trim have been eliminated is in the openings for clothes closets. The openings are finished with gypsum board with a metal corner bead used to protect the corners. This type of construction requires that the studs on the opening sides be perfectly plumb and the opening size be framed according to the manufacturer's instructions when tracked doors are used.

TECHNIQUES FOR HANGING DOORS

When selecting doors, the **grain pattern** and **colour** should be similar for all doors in close proximity to each other.

If the door is slightly warped or bowed, as is often the case, the door should be hung so that the bow is *away* from the jamb (Fig. 16-15). The lockset will then

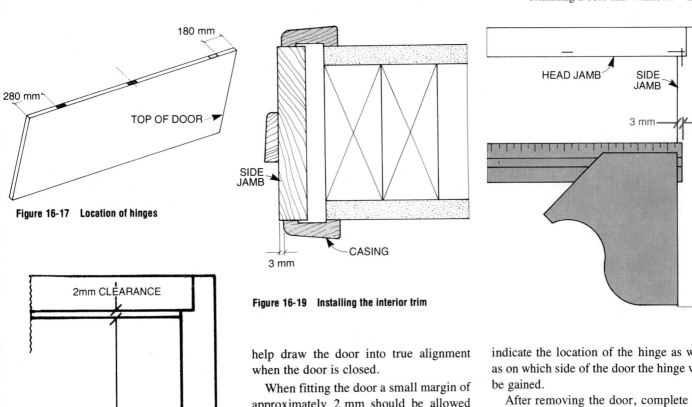

Figure 16-17   Location of hinges

Figure 16-19   Installing the interior trim

Figure 16-18   Marking the hinge location

help draw the door into true alignment when the door is closed.

When fitting the door a small margin of approximately 2 mm should be allowed between the door and the jambs on both sides and at the top. This will allow the door to close properly. This clearance will vary slightly depending on the type of finish applied to the door and the jamb. If the lock side of the door is slightly beveled, it will fit better. A bevel of approximately 3° is recommended (Fig. 16-16).

There is no hard and fast rule for the location of hinges; a generally accepted layout is 180 mm from the top of the door to the top hinge and 280 mm from the bottom of the door to the bottom of the hinge. If three hinges are required, the third hinge is centered between the other two (Fig. 16-17).

Set the door into the jamb and wedge the door tight to the hinge side, leaving the correct clearance on the top. With a sharp 6 mm chisel, mark the location of the hinges (Fig. 16-18). At this time, it is a good practice to place a small "x" where the hinge is to be gained. This will

indicate the location of the hinge as well as on which side of the door the hinge will be gained.

After removing the door, complete the hinge layout, chisel out for the hinges, and install them accordingly. After the hinges have been installed on both the door and jamb, set the door in place, inserting the top pin first. Make any necessary adjustments to ensure that the door swings freely. The procedure for installing the lockset is the same as that used on exterior doors.

TRIMMING DOORS AND WINDOWS

To provide a satisfactory finish between the door jamb and the wall finish, a form of *interior casing* must be applied. There are many types of casing available, narrow casings being the most common today.

Windows and doors are usually trimmed with the same casing throughout the home. In both doors and windows the trim is set back from the edge of the jamb a short distance (usually 3 mm). This is called the *reveal* (Fig. 16-19).

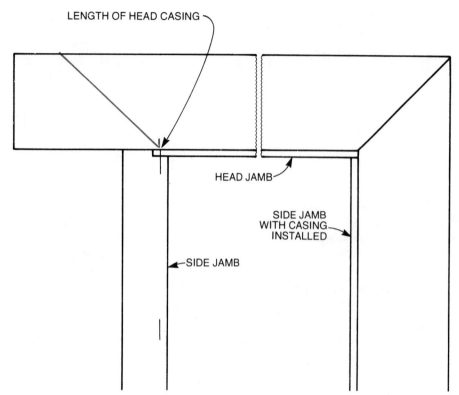

Figure 16-20   Marking the interior trim

Casing is **mitred** at the corners. Care should be taken when measuring the length of casing required. Be sure that the reveal is taken into consideration along with the required length of the opening (Fig. 16-20).

To complete the finish between the wall and the floor, a trim is required. This trim is referred to as the *baseboard*. Baseboards are usually installed before the finish flooring is installed, therefore another moulding is required between the baseboard and the finish floor, called a *carpet strip* (Fig. 16-21).

There is a type of baseboard available that combines the baseboard and carpet strip into one moulding. This is known as a *combination base*. It is also referred to as *hospital base* (Fig. 16-22). This baseboard is installed after the flooring is installed in the home. However, if this type of baseboard is installed before the carpet is laid, then the baseboard must be left 10 mm above the subfloor to allow the carpet layer to fasten the edges of the carpet correctly.

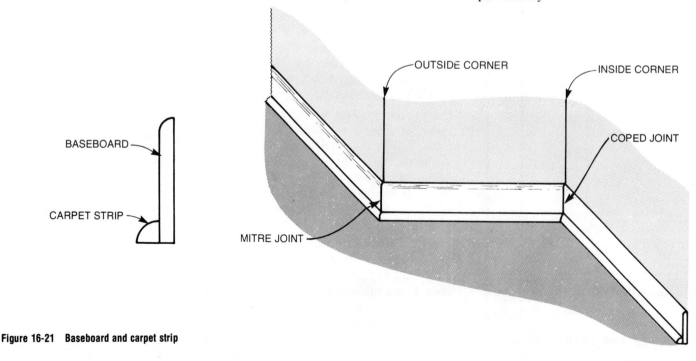

Figure 16-21   Baseboard and carpet strip

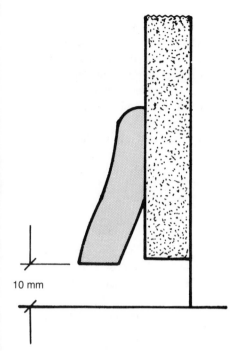

Figure 16-22   Combination base

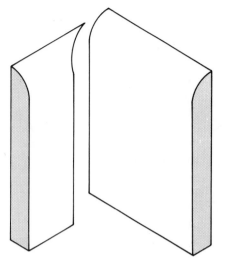

Figure 16-23   Coped joint

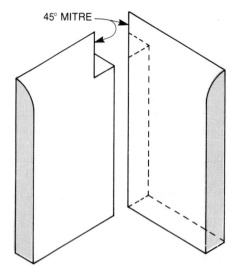

Figure 16-24   Combination butt and mitre joint

There are various types of joints that can be used to join the baseboard. At **internal corners**, the joint most commonly used is the *coped* joint (Fig. 16-23). This type is used particularly with combination base.

Another type of joint is a *combination butt and mitre* joint (Fig. 16-24). A straight mitre joint should **not** be used as nailing will cause the joint to open, leaving an unsightly crack.

The baseboard is joined at **exterior** corners using a *mitre* joint (Fig. 16-25). These joints should be glued and sanded.

Whenever possible, **intermediate** joints in baseboards should be avoided. However, in some cases this is impossible. When it is necessary to join baseboard along a wall, it should be as inconspicuous as possible. If the baseboard is to be stained and varnished, the grain should be matched. The joint itself should be *mitred* (Fig. 16-26).

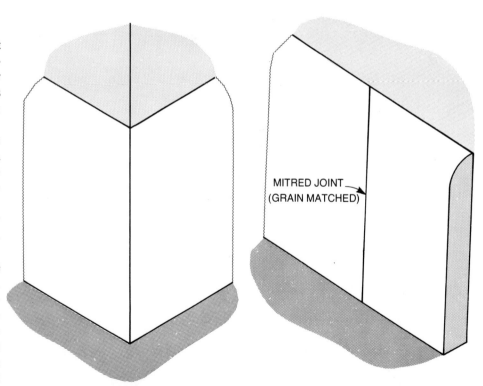

Figure 16-25   External baseboard corner

Figure 16-26   Intermediate joints

REVIEW QUESTIONS

16—INTERIOR DOORS AND TRIM

Answer all questions on a separate sheet.

A. Select the correct answer

1. Which type of door would be *most suitable* for a bathroom where floor area prevents the use of a standard hinged door?
   - (a) Bi-fold
   - (b) Pivoting
   - (c) Bi-pass
   - (d) Pocket
   - (e) None of the above

2. What dimension should be known when ordering bi-fold door hardware?
   - (a) The thickness of the wall structure
   - (b) The width of the finished opening
   - (c) The width of the door jamb
   - (d) The height of the finished opening
   - (e) The size of the head jamb stock

3. The vertical member of a panel door is called the:
   - (a) Upright
   - (b) Stile
   - (c) Mullion
   - (d) Rail

4. One type of interior sliding door is the pocket door; what is the name of another similar door?
   - (a) Casement
   - (b) Bi-fold
   - (c) Bi-passing
   - (d) Horizontal sliding

5. A door frame with planted-on stops is generally used for:
   - (a) Interior doors
   - (b) Exterior doors
   - (c) Overhead garage doors
   - (d) Double-acting doors

6. To install hinges you will need the aid of a:
   - (a) Butt marker
   - (b) Butt cutter
   - (c) Mallet only
   - (d) Mortise gauge

7. What is a "pre-hung" door?
   - (a) A knocked down assembly
   - (b) Everything is glued together
   - (c) A completely assembled unit
   - (d) A door jamb and casing assembly

8. On a panel door the horizontal members are known as:
   - (a) Muntins
   - (b) Panels
   - (c) Rails
   - (d) Stiles

B. Write a full answer

9. When selecting door jamb stock the width is governed by . . . .

10. Why is it unnecessary to use the hand level when setting the lock side of the jamb?

17 STAIR CONSTRUCTION

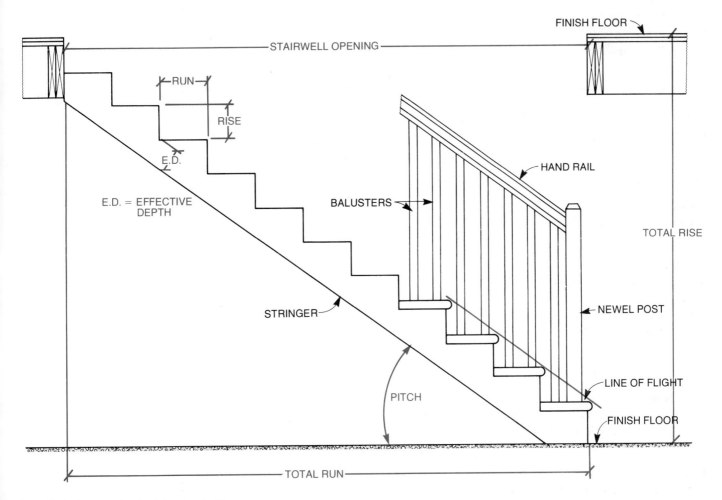

Figure 17-1 Component parts of stair construction

Construction of a flight of stairs is a job in which the builder must display great skill, as a set of stairs must have a finished appearance equal in quality to a set of kitchen cupboards. It is important that stairs be located in the plan of the house to facilitate the moving of large household furnishings to desired locations.

Stairs used in residential buildings can usually be divided into two types. Those that lead from a finished floor to another finished floor are referred to as *main* stairs and those leading to unfinished areas such as basements and attics are referred to as *utility* or *service* stairs. Because of their frequent use, main stairs are built to provide ease of travel, while utility stairs are often built to conform only to the minimum safety standards.

TERMS USED IN STAIR CONSTRUCTION

Carpenters must be familiar with the terms used in stair construction if they are to read a set of blueprints or follow a set of working drawings. Following are some of the terms used in stair construction:

Flight of Stairs
A series of steps leading from one landing to another.

Total Rise
The vertical distance from the finish floor of one storey to the finish floor of the next storey. It also means the height of one riser mutliplied by the number of risers.

Rise
The vertical distance between treads.

Total Run
The horizontal distance of a flight of stairs. Also the length of one run multiplied by the number of runs.

Run
The horizontal distance from the face of one riser to the face of the next riser.

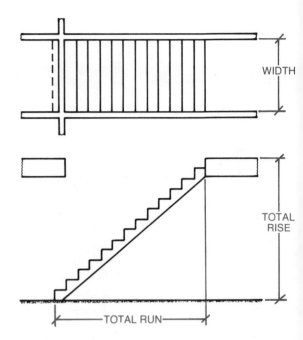

Figure 17-2   Straight flight of stairs

Pitch
The angle of inclination to the horizontal of a flight of stairs.

Step
A step consists of one tread and one riser.

Tread
The horizontal part of a step of a stair.

Stringer
The inclined member which supports the treads and risers of a stair, sometimes called a *stair carriage* or *stair horse*.

Stairwell
The space occupied by a flight of stairs. The opening left in the floor construction.

Line of Flight
A line drawn through the outer edge of the stair nosing.

Nosing
The outer edge of a tread that projects beyond the face of the riser.

Headroom
The vertical clearance from the open end of the stairwell to the line of flight.

Handrail
A rail running parallel to the inclination of the stairs, used as a support when ascending or descending a flight of stairs.

Baluster
A vertical member supporting the handrail of a staircase.

Newel Post
A post at the beginning and/or end of a handrail which supports the handrail.

Balustrade
The complete àssembly consisting of balusters, newel post and handrail.

Fliers
Treads that are uniform in width throughout their length.

Winders
Treads that are wider at one end than the other.

Effective Depth
The uncut portion of a cut or open stringer.

TYPES OF STAIRS

Stairways may have a straight continuous run without an intermediate landing or they may consist of two or more runs with changes in direction. It is common practice to have a landing when the stair changes direction. The length or width of any landing should not be less than the width of the stairs in which they occur. If space does not permit a landing the turn may be made with radiating treads called *winders*.

Straight Run

This is the least costly type of stair to build as it has no turns or landings. If it has a wall or partition on either side of the flight (boxed stair) it is even more economical since the partitions can be used for stair support, permitting the use of lighter weight stringers. A long, straight flight of stairs is not easy on the climber, nor does it have much to offer with regard to appearance (Fig. 17-2).

"L" Type

Because it has a landing, this type of stair is easier to climb as it provides a resting place. The landing allows the stair to turn 90°, thereby reducing the amount of space required in any one direction.

"L" type stairs may have a long "L" as in Figure 17-3, with the turn at either the top or the bottom. Where space is unusually cramped, winders may be used at the turn instead of a landing.

If the landing is placed near the centre of the flight about half way between the two floors it is known as a wide "L" stair (Fig. 17-4). This provides a compact installation suitable for a corner of the building.

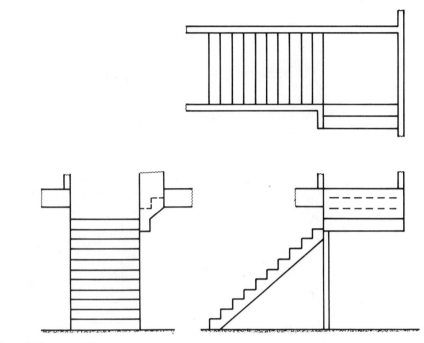

Figure 17-3   Long "L"

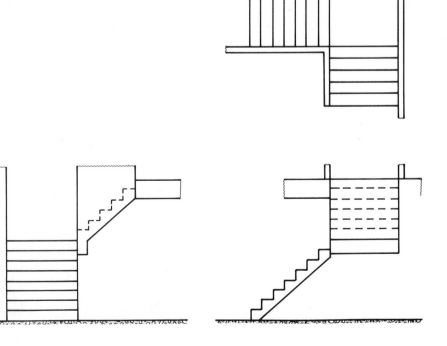

Figure 17-4   Wide "L"

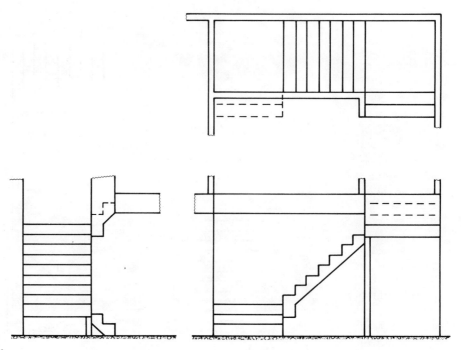

Figure 17-5   Double "L"

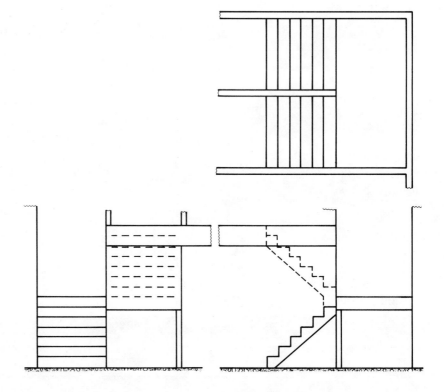

Figure 17-6   Narrow "U"

If there is a turn at the bottom as well as at the top, it is known as a double "L" (Fig. 17-5).

"U" Type

With this type of stair, a lower flight goes up to a landing twice the stair width and then proceeds up from the landing in the opposite direction. This is known as a narrow "U" stair (Fig. 17-6).

A different type of "U" stair is one which has two landings with a short flight between landings. This is called a wide "U" (Fig. 17-7), sometimes called an *open-newel*. The small boundary area enclosed on three sides is called the *well-hole*.

When grade entrances are used in construction, a landing is required and either an "L" type stair or a narrow "U" is often used.

TYPES OF STAIR STRINGERS

A stair stringer is the inclined member which supports the treads and risers of a stair.

Housed Stringer

With this type of construction, the treads and risers are dadoed or housed into the stringer to a depth of approximately 12 mm. The dadoes are cut in such a manner as to allow a wedge to be driven underneath the tread and behind the riser (Fig. 17-8). The wedges are made of hardwood and driven tightly into place, forcing the tread and the riser tight against the edge of the dado.

Open or Cut Stringer

This is the simplest type of stringer used. The rise and run are laid out on the stringer stock and then cut out (Fig. 17-9). Care must be taken in selecting the size of stringer stock to ensure that the

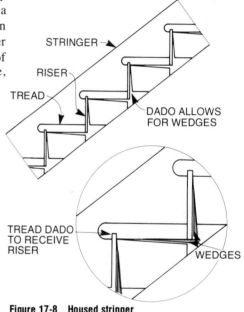

Figure 17-8   Housed stringer

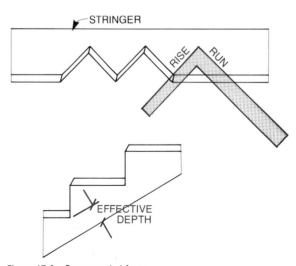

Figure 17-9   Open or cut stringer

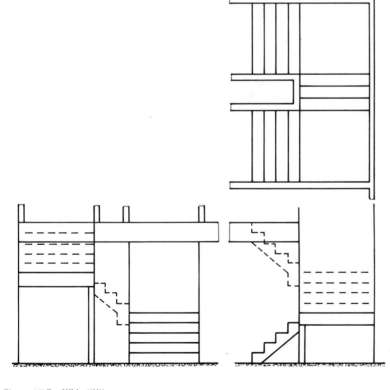

Figure 17-7   Wide "U"

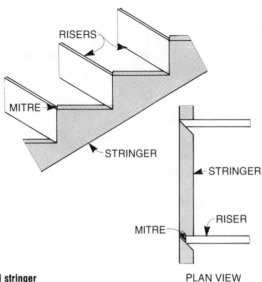

Figure 17-10   Mitred stringer

PLAN VIEW

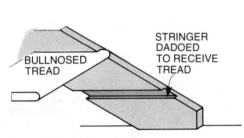

Figure 17-11   Dadoes stringer

portion beneath the rise and run cut-out (called *effective depth*), conforms to building standards.

Mitred Stringer

A mitred stringer is basically an open or cut stringer in which the risers are mitred into the stringer so that no end grain is exposed. (Fig. 17-10). This type of stringer, built in conjunction with a balustrade, can be a very attractive feature of design for a main flight of stairs. Frequently, one stringer will be mitred and the other will be housed.

Dadoed Stringer

Dadoed stringers are most often used for utility and basement stairs. The stringer

stock should be wide enough to support the full width of tread and dadoed approximately 12 mm deep (Fig. 17-11).

BASIC RULES IN STAIR CONSTRUCTION

A well designed stair is one that can be easily travelled both up and down. It must therefore be neither too steep nor inadequately *pitched*. The angle of a stairway with the horizontal should not be more than 50° or less than 20°. The preferred angle for greatest safety and comfort is between 30° and 35° (Fig. 17-12).

The relation between the risers and treads is especially important to ensure

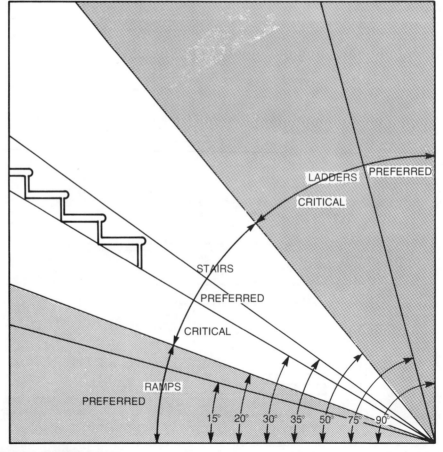

Figure 17-12   Pitch of a stair

safety and ease in travel. Several rules have been derived from experience and used in proportioning the rise and run of a stair. Three of these rules are:

Rule 1 — the rise plus run of a stair should be between 430 and 455 mm.
Rise + Run = 430 to 455 mm.

Rule 2 — the sum of twice the rise plus the run should be between 600 and 635 mm.
2(rise) + run = 600 to 635 mm.

Rule 3 — the product of the rise and the run should be between 45 000 and 48 000 mm.
Rise × run = 45 000 to 48 000 mm.

It is usually safe to use one of the forgoing rules, however, remember that they are only **rules of thumb** and that the actual rise and actual run must conform to the minimum and maximum standards set out by building code authorities. In public buildings the treads are generally wider and the risers lower than those for residential use. A good step for a public building is one having a 300 mm tread and a riser of 150 mm. For residential use where space is limited, a good tread width is approximately 255 mm with a riser of between 180 and 190 mm.

FLOOR FRAMING

The finished opening in the floor to receive a flight of stairs is known as a stairwell opening. The opening in the floor frame is known as a *rough opening*. The rough opening is larger than the dimension of the finished stairwell opening to allow for installation of finishing materials.

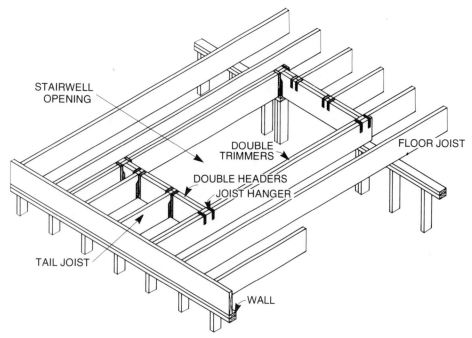

Figure 17-13 Stairwell framed parallel to the joists

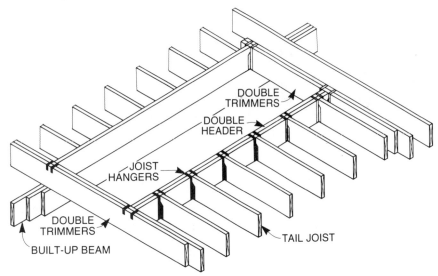

Figure 17-14 Stairwell framed perpendicular to the joists

METHODS OF GAINING HEADROOM IN A STAIRWELL

Headroom can be increased by making the stairwell opening larger. This is not too difficult if the floor joists are placed as in Figure 17-13. However, if they are as in Figure 17-14, it is more difficult. If the floor plan does not permit the lengthening of the stairwell opening, sometimes

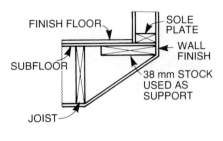

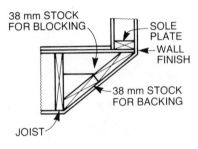

Figure 17-15   Angling stairwell opening

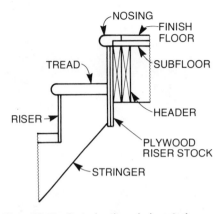

Figure 17-16   Fastening through riser stock

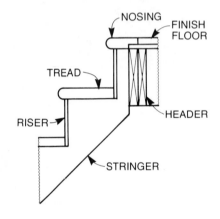

Figure 17-17   Support through stringers

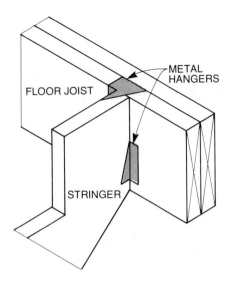

Figure 17-18   Stair fastened with metal hangers

enough headroom can be gained by simply angling the end of the stairwell (Fig. 17-15).

By reducing the unit run a slight increase in headroom will be gained. Care must be taken not to reduce the run less than the minimum set out by the building authorities. Another method of gaining headroom is to reduce the number of risers, thereby increasing the height of each remaining riser. Once again, care must be taken not to increase each riser to the extent that it no longer conforms to building standards.

METHODS OF FASTENING STAIRS IN PLACE

Stairs may be built in place or in a shop and then installed as a unit. Usually, stairs need only be fastened at the top and their own weight will keep them in place. They may be fastened through the riser stock (Fig. 17-16), particularly if the stock is plywood. With this method, the rough opening length need only be slightly larger than the desired finished opening.

Stringers may be constructed in such a manner as to have more support (Fig. 17-17). More allowance in rough opening length must be made.

Stringers may be constructed as in Figure 17-18 where more of the stringer is used as support and metal hangers can be added for additional strength.

CALCULATING THE ACTUAL RISE IN A FLIGHT OF STAIRS

Procedure

1. Determine the total rise either from the blueprint or by actual on-the-job measurement. As an example a total rise of 2900 mm will be used.

2. Divide the total rise by the *proposed rise*. The proposed rise is the height you would like the riser to be if it were possible. Proposed rise for this example—178 mm.

$$2900 \div 178 = 16.29$$

16.29 would be the number of risers required if the rise were 178 mm. Since there can't be a part of a riser, either 16 or 17 risers would be used. If the proposed rise is the maximum according to building standards, then the decimal fraction must be taken to the **next** whole number.

3. Divide the total rise by the number of risers decided upon in step 2.

$$2900 \div 16 = 181.2$$

181.2 mm will be the actual rise if there are 16 risers.

4. Check for accuracy by multiplying the actual rise (181.2) by the number of risers (16). The answer should be the same as the total rise (2900).

DETERMINING STAIRWELL OPENING LENGTH

During construction of the floor frame the length of the stairwell opening must be determined so that the rough opening in the floor can be framed. If the rough opening is framed into the floor frame correctly there is no need to adjust either the opening or the flight of stairs when it comes time to actually install them.

Procedure

1. Determine the total rise from the blueprint. For this example—2650 mm.
2. Determine the actual rise using the procedure previously described. For this example the proposed rise will be 200 mm which will be a maximum.

$$2650 \div 200 = 13.25$$
$$1650 \div 14 = 189.3$$

The actual rise will be 189.3 mm.

3. Determine the actual run by using one of the rules previously mentioned. For this example use rule no. 1 (rise plus run equals between 430 and 455 mm).

$$430 - 189.3 = 240.7$$

The actual run would be 240.7 mm.

Check at this time to see if the actual rise and actual run conform to the limitations set out by the building standards.

4. Determine the number of *uncovered risers* and treads by adding the required headroom to the upper floor construction (Fig. 17-19). For this example use 1950 mm as the headroom required and a floor construction of 300 mm.

$$1950 + 300 = 2250$$

After the headroom plus floor construction is calculated, divide this number (2250) by the actual rise found in step 2 (189.3)

$$2250 \div 189.3 = 11.88$$

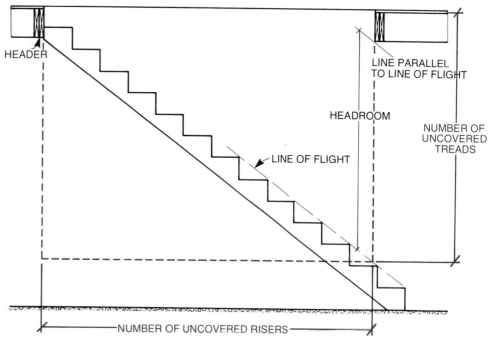

Figure 17-19 Stairwell opening

The answer (11.88) will be the number of uncovered risers and also the number of uncovered treads in the stairwell opening, (Fig. 17-19).

5. Determine the *workable* stairwell opening length. The workable opening means the length the stairwell must be before allowances for tread nosing and finishes are added. To find the workable opening, multiply the number of uncovered risers (11.88) by the actual run (240.7, step 3)

$$11.88 \times 240.7 = 2859.5$$

The workable opening for this example will be 2859.5 mm.

6. Determine the finished stairwell opening length by adding the width of one tread nosing to the workable opening. For this example, the width of nosing will be 25 mm.

$$2859.5 + 25 = 2884.5$$

The finished stairwell opening length for this example will be 2884.5 mm.

7. Determine the rough stairwell opening length by adding to the finished stairwell opening the allowances for **finishes**. The term finishes include such items as gyproc or wallboard at the end of the stairwell opening, the thickness of top riser stock and the method of fastening the staircase. If 75 mm is allowed, any type of finish can be used.

$$2884.5 + 75 = 2959.5$$

The rough opening for this example will be 2959.5 mm.

DETERMINING THE MAXIMUM RUN OF A STAIR WITH A FIXED OPENING

When planning the layout for a set of stairs the actual rise and actual run must be known. This can be determined when the opening is fixed by the following method.

Procedure

1. Determine the actual rise by using the total rise and the proposed rise: For this example the actual rise will be 190 mm.
2. Determine the number of uncovered risers or treads by adding the required headroom to the floor construction and then dividing it by the actual rise. For this example the required headroom will be 1950 mm and the floor construction 300 mm.

$$2250 \div 190 = 11.84$$

11.84 will be the number of uncovered risers, and also the number of uncovered treads for this example.

3. Determine the rough opening length from the blueprints or by actual measurement. Remember the rough opening is measured from the framing members. For this example, the rough opening length will be 2950 mm.
4. Determine the finished stairwell opening length by subtracting allowances for finishes from the rough opening length. For this example the allowance for finishes will be 75 mm.

$$2950 - 75 = 2875$$

The finished stairwell opening length for this example will be 2875 mm.

5. Determine the workable stairwell opening length by subtracting the width of one tread nosing from the finished stairwell opening length. For this example, the width of tread nosing will be 25 mm.

$$2875 - 25 = 2850$$

The workable opening will be 2850 mm.

6. Determine the maximum possible run by dividing the workable stairwell opening length (2850) by the number of uncovered risers (11.84).

$$2850 \div 11.84 = 240.7$$

The maximum possible run for this example is 240.7 mm.

SPLIT ENTRIES

A type of stair that is very common in house construction is one where there is a landing between floors used where there is a *grade entrance*. It is important that the carpenter understand stair construction because the plan for the stairs must be known as early as the forming for the foundation wall so as to put the grade level opening in at the correct height. In a split entry the landing should be situated so the actual rise for the upper flight will be the same as the actual rise of the lower flight. To accomplish this, the total rise must be known. Then, by using a proposed rise you can find the actual rise and the number of risers. The height of the landing will be a multiple of the actual rise. Following is the procedure, with an example for determining the rough opening lengths for a flight of stairs with a split entry where the landing is already fixed.

Procedure

1. Determine the rise of the upper flight of stairs by actual measurement or by subtracting the height of the landing from the total rise. For this example the total rise will be 2650 mm and the height of the landing will be 2050 mm (Fig. 17-20).

$$2650 - 2085 = 565$$

The total rise of the upper flight for this example is 565 mm.

2. Determine the actual rise of the upper flight by dividing the total rise (565) by the number of risers shown in the blueprint. If this is not possible, find the actual rise by using a proposed rise. For this example, the proposed rise will be 200 mm which will be the maximum.

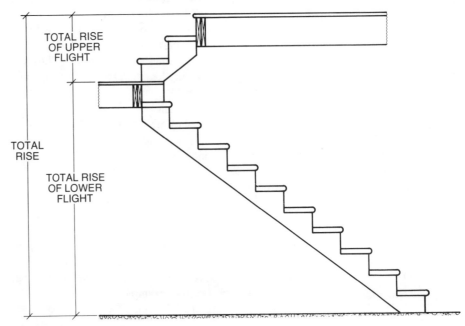

Figure 17-20   Split entry flight of stairs

$$565 \div 200 = 2.8$$
$$565 \div 3 = 188.3$$

There will be 3 risers with an actual rise of 188.3 mm.

3. Determine the actual run of the upper flight by using one of the rules previously described. For this example, use the rule that states one rise plus one run will equal 430 mm.

$$430 - 188.3 = 241.7$$

The actual run will be 241.7 mm.

4. Determine the total rough opening length for the upper flight by multiplying the actual run (241.7) by the number of treads. **Remember** there is always one less tread than there are risers.

$$241.7 \times 2 = 483.4$$

To this number, add the width of one tread nosing. For this example, 25 mm.

$$483.4 + 25 = 508.4$$

To this number add the allowances for finishes; for this example 75 mm.

$$508.4 + 75 = 583.4$$

To this number add the width of the landing, for this example 860 mm.

$$583.4 + 860 = 1443.4$$

The total rough opening length to be framed into the floor construction will be 1443.4 mm. When the landing is built allowance must be made to support the stringer (Fig. 17-21).

5. Determine the actual rise of the lower flight of stairs by dividing the height of the landing (2085) by the actual rise of the upper flight (188.3), because it is the proposed rise.

$$2085 \div 188.3 = 11.07$$
$$2085 \div 11 = 189.5$$

There will be 11 risers with an actual rise of 189.5 mm. If the landing had been the correct height the actual rise

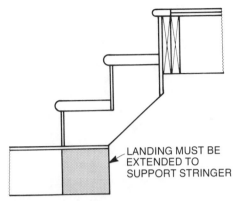

LANDING MUST BE EXTENDED TO SUPPORT STRINGER

Figure 17-21   Upper flight of stairs

of the lower flight would have been the same as the actual rise for the upper flight. For this example the difference in the height of risers for the two flights will be 1.2 mm, which would not be noticeable as you walk up them. However, with adequate planning they can be equal.

6. Determine the number of uncovered risers or treads in the lower flight by adding the required headroom, for this example 1950 mm, to the upper floor construction, for this example 300 mm.

$$1950 + 300 = 2250$$

From this number, subtract the total rise of the upper flight of stairs (565)

$$2250 - 565 = 1685$$

Divide this number by the actual rise of the lower flight (189.5)

$$1685 \div 189.5 = 8.89$$

The number of uncovered risers or treads in the lower flight of stairs will be 8.89.

7. Determine the workable stairwell opening length of the lower flight of stairs by multiplying the number of uncovered risers (8.89) by the actual run (241.7). The actual run for the lower flight of stairs should be the same as for the upper flight.

$$8.89 \times 241.7 = 2148.7$$

The workable opening will be 2148.7 mm.

8. Determine the finished stairwell opening length by adding the width of one tread nosing (25 mm) to the workable opening.

$$2148.7 + 25 = 2173.7$$

The finished stairwell opening will be 2173.7 mm from the edge of the landing. To get the total finished opening add the width of the landing (860).

$$2173.7 + 860 = 3033.7$$

The total finished opening length will be 3033.7 mm.

9. Determine the total rough opening length by adding the allowances for finishes (75 mm) to the finished opening length (3033.7)

$$3033.7 + 75 = 3108.7$$

The total rough opening length to be framed into the floor construction for the lower flight of stairs will be 3108.7 mm.

DETERMINING ACTUAL HEADROOM

There is a lot of planning that must be done before any construction can be started with regard to stairs. Given the finished stairwell opening and the total rise, which can be found on the blueprints, the carpenter should be able to determine the actual headroom there will be when the stairs are installed.

Procedure

1. Determine the total rise from the blueprint or by actual measurement if the opening is already framed. For this example, the total rise will be 2850 mm.

2. Determine the actual rise using a proposed rise. For this example, the proposed rise will be 190 mm.

$$2850 \div 190 = 15$$

In this case, the proposed rise will be the actual rise as the answer in the first step is a whole number. There will be 15 risers with an actual rise of 190 mm each.

3. Determine the actual run by using one of the rules previously mentioned. In this example, use the rule that states one rise plus one run equals 430 mm.

$$430 - 190 = 240$$

The actual run will be 240 mm.

4. Determine the number of uncovered risers or treads by subtracting the width of one tread nosing, for this example 25 mm; from the finished stairwell opening, for this example 2920 mm.

$$2920 - 25 = 2895$$

This number will be the workable opening. Next divide the workable opening by the actual run (240).

$$2895 \div 240 = 12.06$$

The number of uncovered risers or treads will be 12.06.

5. Determine the combined height of headroom and floor construction by multiplying the number of uncovered risers or treads (12.06) by the actual rise (190).

$$12.06 \times 190 = 2291.4$$

The headroom plus floor construction will be 2291.4 mm.

6. Determine the actual headroom by subtracting the thickness of the floor construction found on the blueprint or by actual measurement, from the combined height found in step 5. Floor construction for this example is 300 mm.

$$2291.4 - 300 = 1991.4$$

The actual headroom for this example will be 1991.4 mm.

There are two statements that can be used

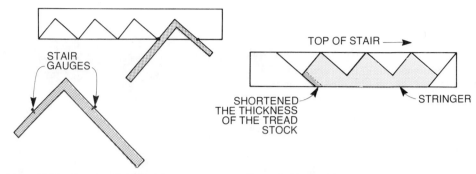

Figure 17-22   Stepping off using stair gages

Figure 17-23   Stair layout

together as a mathematical equation to find such measurements as headroom, actual run or workable stairwell opening. These two statements are:

(Headroom + Floor Construction) ÷ Actual Rise = No. of uncovered risers.

Number of uncovered risers × actual run = workable opening.

By substituting known measurements in the two statements a mathematical equation is set up, allowing you to determine the unknowns.

BUILDING A SET OF STAIRS

The two most common types of stringers used in stair construction are open or cut stringers and housed stringers. Following are methods employed in building stairs using these two types of stringers.

Open or Cut Stringer Stairs

1. Determine the total rise for the flight of stairs to be built.
2. Calculate the actual rise and actual run using a proposed rise and one of the rules to determine the actual run. In calculating the rise and run apply the information previously covered in this chapter with regard to headroom.
3. Select material to be used for stringers–be sure the stock is wide enough that when the rise and run are cut out there remains an effective

depth that will conform to the building standards.

4. Rough cut the stringer stock to length–the length will be the diagonal of the actual rise and actual run multiplied by the number of risers.
5. Clamp stair gauges to a framing square with the rise on the tongue and the run on the body. Step off with the square, marking along the body and the tongue of the square with a sharp pencil. Step off as many times as there are risers (Fig. 17-22).
6. Lay out the top and bottom cuts by extending the rise and run lines across the stock as shown in Figure 17-23.

To ensure that all the risers will be the same height when the treads are installed, the bottom of the stringer must be shortened the thickness of the tread stock (Fig. 17-23). If the bottom of the stringer rests on the subfloor, the stringer need only be shortened by an amount equal to the difference between the tread thickness and the finish floor thickness.

7. Lay out the second stringer to match the first stringer to make a pair.
8. Cut stringers and attach treads. Install in place (Fig. 17-24).

Housed Stringer Stairs

1. Determine the total rise for the flight of stairs to be built, then calculate the actual rise and the actual run as with open or cut stringers.

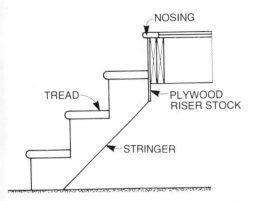

Figure 17-24   Open or cut stringer

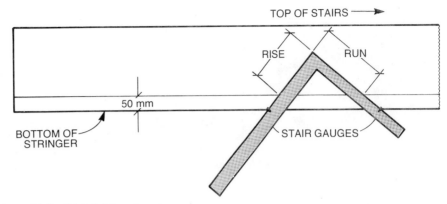

Figure 17-25   Housed stringer layout

2. Select stringer stock but be sure both stringers are exactly the same width.
3. Rough cut stringers to length.
4. With the open or cut stringer the layout was from the top edge of the stringer. With a housed stringer the layout is from the bottom edge of the stringer. Draw a line parallel to the edge of the stringer 50 mm from the bottom edge (Fig. 17-25) and place the framing square with the figures for the rise and run on this line. Clamp the stair gauges on the square as shown.
5. Step off as many times as there are risers. Start the layout from the bottom of the stairs. The rise and run lines must intersect on the parallel line. Mark both stringers making sure they are laid out in pairs.
6. Lay out for the tread dado by measuring down the rise line a distance equal to the thickness of the tread stock (Fig. 17-26) and draw a line parallel to the run line.
7. Lay out for the riser dado by measuring along the run line a distance equal to the thickness of the riser stock and draw a line parallel to the rise line (Fig. 17-26).
8. Lay out for wedges by measuring back from rise and run lines 10 mm

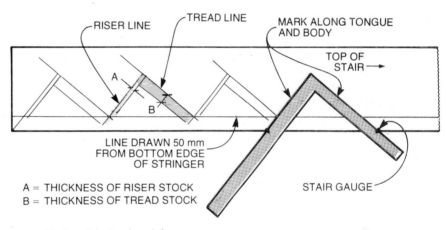

A = THICKNESS OF RISER STOCK
B = THICKNESS OF TREAD STOCK

Figure 17-26   Layout for treads and risers

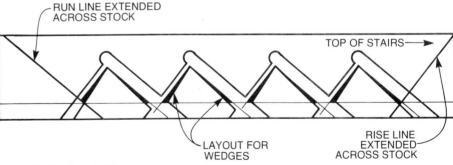

Figure 17-27   Housed stringer

along the edge of the stringers (Fig. 17-27).
9. Lay out the top and bottom cut lines by extending the rise and run lines across the stock (Fig. 17-27). The second cuts on the top and bottom will depend on the type of finish desired. An acceptable method is shown in Figure 17-28.

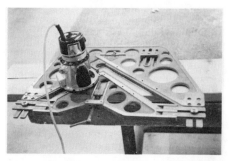
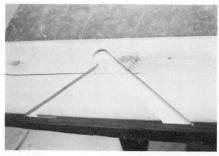
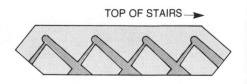

TOP OF STAIRS ⟶

Figure 17-28   Metal template for routing housed stringers

10. Lay out for nosing by extending the tread line past the riser a distance equal to the width of the tread nosing. If the nosing is round a hole the width of the nosing can be drilled.

11. Rout out for risers, treads and wedges using a router and template as shown in Figure 17-28, or use a portable electric handsaw and wood chisel or hand router (Fig. 17-29).

Depth of dadoes should be approximately 12 mm.

12. Cut treads and risers to length and install. Treads may be grooved at the front edge to receive risers. Nail treads in place through the stringer stock. Nail the bottom of the riser stock to the tread stock. Install wedges to ensure that treads and risers are tight.

Figure 17-29   Housed stringer using electric handsaw and hand router.

REVIEW QUESTIONS

17—STAIR CONSTRUCTION

Answer all questions on a separate sheet

A. Write full answers

1. What special consideration must be given when selecting dimensional stock for a cut or open type stringer?
2. A flight of stairs with fourteen risers would have how many treads?
3. Why must the total rise always be taken from finish floor to finish floor?
4. Why must all the risers in a flight of stairs be the same height?
5. Why is it necessary to deduct the thickness of the tread from the bottom of open or cut stringers?
6. With the bottom of the stringer resting on the subfloor, what deduction would be made on the bottom of an open or cut stringer if the underlay to be used on the floor was 10 mm thick and the finish hardwood floor to be used was 10 mm thick? Tread thickness is 38 mm.
7. If the actual rise is 180 mm, the headroom is 1950 mm and the upper floor construction is 275 mm, how many uncovered risers will there be?
8. If a flight of stairs has 11.38 uncovered risers or treads and the actual run is 250 mm, what will be the workable stairwell opening length?
9. What would be the rough opening length in Question 8 if a nosing of 25 mm is used and 50 mm allowed for finishes?
10. The distance from the finished basement floor to the finished main floor is 2650 mm, and the height of the landing is 2120 mm with 3 risers in the upper flight. What will be the actual rise of the upper flight? What will be the actual rise of the lower flight?
11. What are two methods that can be used to increase the actual headroom on a flight of stairs without changing the floor framing?

B. Replace the Xs with the correct word

12. Clear headroom is measured from the XXXXXXX to the XXXXXX.
13. The ends of the risers are sawn at a 45° angle when they are attached to a XXXX stringer.
14. The "preferred angle" from the horizontal for a flight of stairs is XXXX° to XXXXX°.
15. The angle of the stairway with the horizontal should not be more than XXX° nor less than XXX°.
16. If 10 mm were taken off each run of a 15 riser stairway, the total run would be reduced by XXX mm.

C. Solve the following

17. Given the following information for a straight flight of stairs: rough stairwell opening = 3000 mm; allowance for finishes = 75 mm; width of nosing = 25 mm; total rise = 2730 mm; proposed rise = 190 mm maximum; proposed run = 255 mm; headroom required = 1950 mm; upper floor construction = 390 mm; determine the following: actual rise; number of risers; number of uncovered risers or treads; and the actual run.
18. Given the following information for a straight flight of stairs, determine the actual headroom. Total rise = 2850 mm; actual rise = 190 mm; tread width = 260 mm; nosing = 25 mm; upper floor construction = 290 mm; finished stairwell opening length = 2810 mm.

D. Write full answers

19. Tell where or how to find the following:
    (a) total rise
    (b) number of risers
    (c) actual rise
    (d) actual run
    (e) total run
    (f) number of uncovered risers
    (g) workable stairwell opening
    (h) finished stairwell opening
    (i) rough stairwell opening
    (j) maximum possible run if the workable stairwell opening is known.
20. Sketch and name at least eight parts of a straight flight of stairs.

18 CABINETRY

The carpenter, as well as being able to perform the skills necessary to build quality cupboards, must also be able to plan and design kitchens. This is one area in a home in which carpenters are often asked to offer their personal suggestions and ideas. The carpenter should make it a point to keep abreast of new and improved appliances as well as new building products which will make the kitchen convenient and attractive. If a house has an attractive, work-saving kitchen, it will often be one of the deciding factors in selling the home.

KITCHEN PLANNING

Kitchens are usually designed around three main activities: **cooking** (stove), **clean-up** (sink), and **food storage** (refrigerator). The three work centres for these activities form the *kitchen work triangle*. Figure 18-1 shows five basic kitchen plans and their work triangles.

The location of each major component of the work triangle and its distance from the others will determine the efficiency of a kitchen. The total distance between these components should be between 5100 and 6600 mm. Since the work triangle will be the busiest area in the kitchen, try to plan it so that traffic will not pass through this area.

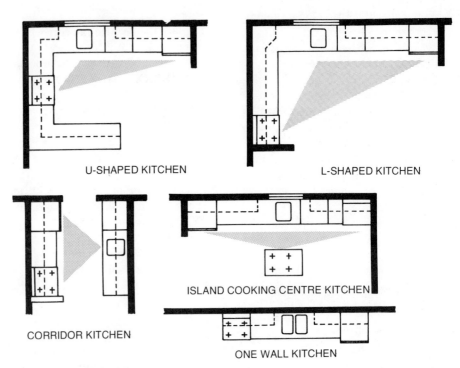

U-SHAPED KITCHEN

L-SHAPED KITCHEN

CORRIDOR KITCHEN

ISLAND COOKING CENTRE KITCHEN

ONE WALL KITCHEN

Figure 18-1   Five basic kitchen plans

Courtesy Merit Industries Ltd.

In planning a kitchen the number of activities that will be performed there have to be taken into consideration. The absolute minimum amount of activity in the kitchen will be the storage and preparation of food, as well as the cooking and clean-up of the area. If the kitchen is large enough, frequently an eating area will be included. Often a laundry area is added, and if space permits, a planning centre.

Food Storage and Preparation Centre

The major appliance in this centre is the *refrigerator*. It is wise to choose one with

286

Figure 18-2   Storage cupboard

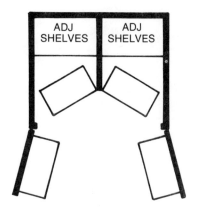

Courtesy Merit Industries Ltd.

is a cupboard designed for food materials not requiring refrigeration. A large amount of storage space that is readily accessible can be obtained by building shelves on the doors as shown in Figure 18-2.

Cooking Centre

The major appliances in this area will be the **cooking units**. This may include a **range** with the oven and surface elements together or a **built-in unit** with the oven and surface elements separate. As the oven unit is the least frequently used, it may be placed outside the work triangle. The surface elements should be within the work triangle–preferably near the sink.

A cooking appliance which has gained popularity during recent years is the *microwave oven*. Another appliance of similar size is the *convection oven*. Convenient methods of positioning these ovens are to build them into shelves below the upper cupboards (see Fig. 18-3). A space of approximately 225 mm left between this shelf and the lower counter top will make the oven readily accessible but still permit use of the counter top area. Care must be taken to follow manufacturer's instructions regarding necessary clearances at the top and back for ventilation, etc. when building-in these units.

a large freezer compartment, thus eliminating the need for a separate freezer in the kitchen. For ease in handling foods, the refrigerator door should open directly onto the counter. This is easily achieved as refrigerators are available with both left and right hand operated doors. For the refrigerator to function both efficiently and economically, it should be positioned away from sources of heat such as heating ducts, direct sunlight, the stove or the dishwasher.

When it is necessary to have the stove and refrigerator on the same wall, as often occurs in a **corridor kitchen**, a counter of at least 600 mm width is desirable between them. A counter space of between 1200 and 1650 mm long should be incorporated somewhere in the work triangle to be used as a preparation centre. Ideally this should be situated near the sink for vegetable cleaning, and near the refrigerator for storage access. Where space permits an ideal addition for kitchen storage

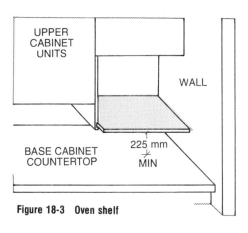

Figure 18-3   Oven shelf

Clean-Up Centre

The clean-up area may expand from a simple *sink* to the addition of other appliances such as a *dishwasher, food waste disposer,* and possibly a *trash compactor.* If the sink is the only component used in the clean-up area, it will often be a double sink unit. When a dishwasher is added to the clean-up area, the double sink is sometimes replaced by a large single bowl sink. There should be a minimum counter surface of 600 mm on either side of the sink. Wherever possible increase this measurement to 900 mm on one side. Where a permanent under-counter dishwasher is installed, it should be beside the sink. For a built-in dishwasher to be properly installed, it requires a countertop of at least 600 mm in width, with 625 mm being more desirable.

If a food waste disposer is to be added to the clean-up area, the sink into which it is connected must have a 100 mm drain. Plumbing and wiring must conform to local codes. Although normally installed near the sink, the trash compactor may be functional outside the work triangle.

Eating Area

If an eating area is to be included in the kitchen plan, a space 2400 × 2700 mm will be necessary for comfortable seating if a table and individual chairs are to be used. Space can be saved by having one side of the table against a wall or by using a counter arrangement. Allow 600 mm of table or counter space for each place setting.

Laundry Area

If space permits, the **washer** and **dryer** may be included in the kitchen. However, keep the laundry activity well separated from the food preparation area. Compact, stack-on washer and dryer units are available which conserve space and are often

Figure 18-4   Flush doors

hidden behind folding doors when not in use.

Planning Centre

Many kitchen designs include a **desk** built in with the kitchen cupboards. A desk should be between 700 and 750 mm high depending on the type of chair used. If a standard chair is used, the desk height should be 750 mm. The desk top should be 525 to 600 mm deep and should be about 900 mm long to be functional. It may be a simple counter or a more elaborate combination of drawers and shelves depending on the space available and the use it will be put to. In most cases, a telephone should be planned for this area.

TYPES OF CUPBOARDS

Cupboards are usually classified by the type of door used on them. There are cupboards with **flush** doors (Fig. 18-4), **rabbeted** doors (Fig. 18-5), and **overlay** doors (Fig. 18-6).

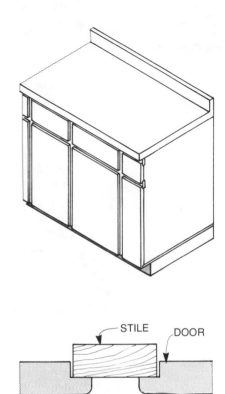

Figure 18-5   Rabbeted doors

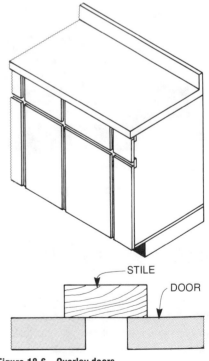

Figure 18-6   Overlay doors

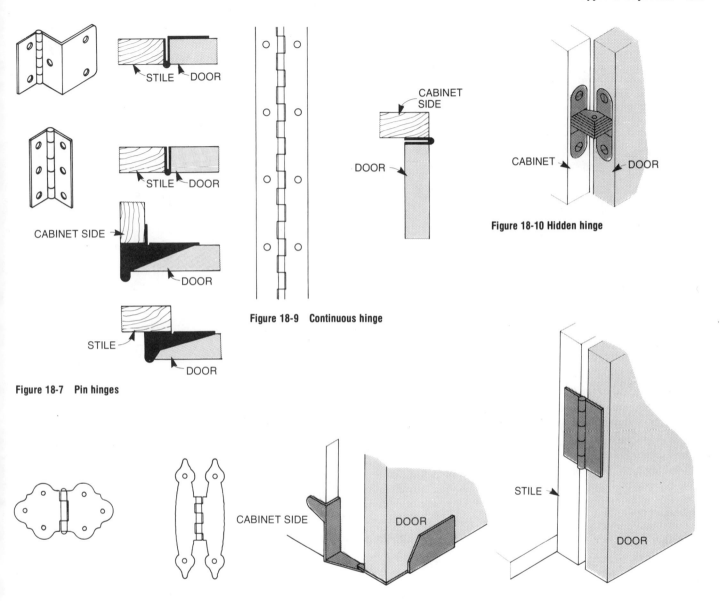

Figure 18-7   Pin hinges

Figure 18-8   Surface mounted hinges

Figure 18-9   Continuous hinge

Figure 18-10 Hidden hinge

Figure 18-11   Hinged flush door with or without a stile

There are many types of hinges available for use on the various types of doors, some of these being: *pin, offset, overlapping, surface, hidden* and *continuous.*

The door is usually attached to a vertical framing member called a *stile*.

Flush Doors

When doors are attached flush with the stiles, hinges may be pin, fixed or loose as shown in Figure 18-7, surface mounted as in Figure 18-8, continuous as in Figure 18-9, or hidden as in Figure 18-10.

Flush doors may be installed with or without the use of stiles using hinges as shown in Figure 18-11.

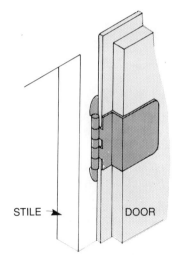

Figure 18-12   Hinged rabbeted door using the off-set hinge

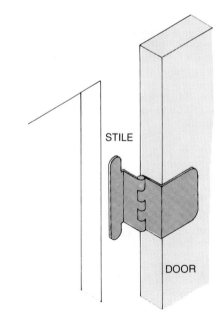

Figure 18-13 Hinged overlay door

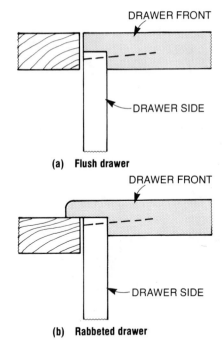

(a)   Flush drawer

(b)   Rabbeted drawer

(c)   Overlay drawer

Figure 18-14 Drawer fronts

Rabbeted Doors

When rabbeted doors are used, offset hinges are attached to the door, fitted into the rabbet and then fastened to the stile (Fig. 18-12). Hinges are available that have a spring incorporated into the pin that will keep the door in a closed position, eliminating the need for a door catch.

Overlay Doors

On overlay doors, a hinge similar to the offset hinge is used. This hinge is fastened to the inside face of the door and the stile (Fig. 18-13), and is available with a spring to keep the door closed.

Drawers

Drawer fronts are made to match the type of doors used. That is, they will fit **flush** with the stile, have a **rabbeted** front or **overlap** the stile. Regardless of the drawer front used, the basic construction of the drawer is the same. Drawer sides are either *dadoed* or *rabbeted* into the drawer front (Fig. 18-14).

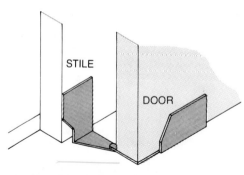

Figure 18-15(a)   Drawer construction

BACK RESTS ON BOTTOM

DRAWER WITH FLUSH FRONT

BOTTOM DADOED INTO SIDES AND FRONT

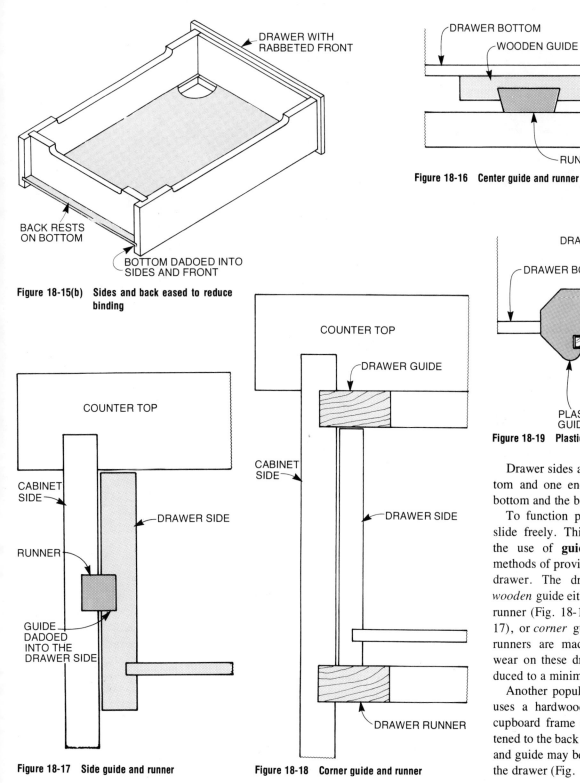

DRAWER WITH RABBETED FRONT

BACK RESTS ON BOTTOM

BOTTOM DADOED INTO SIDES AND FRONT

Figure 18-15(b)   Sides and back eased to reduce binding

DRAWER BOTTOM

WOODEN GUIDE

RUNNER

Figure 18-16   Center guide and runner

DRAWER BACK

DRAWER BOTTOM

PLASTIC GUIDE

HARDWOOD RUNNER

Figure 18-19   Plastic drawer guide and runner

COUNTER TOP

CABINET SIDE

RUNNER

DRAWER SIDE

GUIDE DADOED INTO THE DRAWER SIDE

Figure 18-17   Side guide and runner

COUNTER TOP

DRAWER GUIDE

CABINET SIDE

DRAWER SIDE

DRAWER RUNNER

Figure 18-18   Corner guide and runner

Drawer sides are dadoed along the bottom and one end to receive the drawer bottom and the back (Fig. 18-15).

To function properly, drawers should slide freely. This is accomplished with the use of **guides**. There are several methods of providing these guides for the drawer. The drawer may slide on a *wooden* guide either as a *centre* guide and runner (Fig. 18-16), *side* guide (Fig. 18-17), or *corner* guide (Fig. 18-18). If the runners are made from hardwood, the wear on these drawer guides will be reduced to a minimum.

Another popular type of drawer guide uses a hardwood track attached to the cupboard frame and a *plastic* guide fastened to the back of the drawer. This track and guide may be on the top or bottom of the drawer (Fig. 18-19).

Courtesy Merit Industries Ltd.

Figure 18-20 Metal drawer guides

There are also drawer guides which consist of *two metal tracks* and a floating roller of either metal or nylon (Fig. 18-20). These come in sets with manufacturer's instructions as to installation procedures.

Standard Measurements of Kitchen Cupboards

The standard height for bottom cupboard units is 915 mm from the floor to the countertop. For practical purposes, the top of the upper cupboards should be no higher than 2130 mm. The space between upper and lower cupboards should be a minimum of 380 mm, with this space increased to a minimum of 600 mm when cupboards are situated over a sink or range. A toe space 100 mm deep and 89 mm high should be left at the base of lower kitchen cupboards (Fig. 18-21). The space between the top of the upper cupboards and the ceiling may be left open or closed in with a valance.

Custom Built Kitchen Cupboards

With experience, carpenters develop their own style of construction, incorporating the various drawer and door combinations previously mentioned. The most common material used is plywood. Types range from fir to oak, mahogany, cherry, ash,

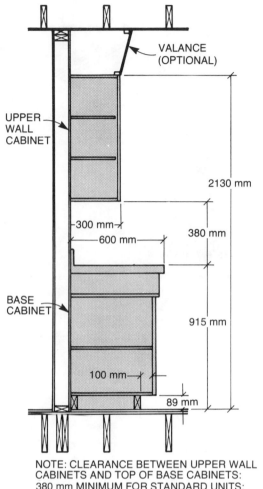

NOTE: CLEARANCE BETWEEN UPPER WALL CABINETS AND TOP OF BASE CABINETS:
380 mm MINIMUM FOR STANDARD UNITS:
600 mm MINIMUM OVER RANGE OR SINK

Figure 18-21   Standard measurements for kitchen cupboards

and include numerous others. The direction of the face grain is an important feature in the appearance of the finished product. As a general rule, the **ends** of the cupboards, **dividers, stiles** and **doors** will have the face grain running *vertically*. **Drawer fronts, shelf rails** and **tops** will have the face grain running *horizontally*. Horizontal members, like the shelves, rails and carcass frames should be set into either a dado or a rabbet to ensure strength and rigidity. All members should be glued and nailed using finishing nails of the correct size. When working with hardwood plywoods, care should be taken when sanding so as not to remove the very thin surface veneer. Do not allow any glue to come in contact with the finish face of any component. The glue will fill the open grain of the wood, preventing the finish (stain) to adhere to that area. This leaves an uneven finish with the glue mark showing. If it is necessary to remove glue, wipe with a

Figure 18-22   Pre-built kitchen units

Courtesy Merit Industries Ltd.

damp cloth immediately and sand carefully when it is dry.

Very often, custom-built cupboards are made from expensive plywood panels. Therefore, pre-planning is important so that the best use can be made of each individual sheet.

Pre-Built Units

Several companies now manufacture kitchen cupboards in units which are simply chosen in various sizes and combinations to suit the space and individual needs of the owner. These units are fastened together by bolts or wood screws. The countertop is fastened to the bottom cupboard after the units are in place (Fig. 18-22).

Wall cabinets are either screwed to the kitchen wall or hung from the ceiling where island cupboards are used. Pre-finished at the company's factory, care must be taken when installing them, to ensure that the exposed face and edges are not damaged. As with all pre-built equipment installed in a home, installation is fast and easy providing that manufacturer's instructions provided are followed closely.

REVIEW QUESTIONS

18—CABINETRY

Answer all questions on a separate sheet

A. Write a full answer

1. List the three basic activity centres of a kitchen.

B. Replace the Xs with the correct word or words

2. The standard height for lower kitchen cupboards is XXX mm.

3. The distance between upper and lower kitchen cupboards should be XXX mm, except for those upper cupboards which are located above a range or sink, then the distance should be increased to XXX mm.

4. The standard depth for upper kitchen cupboards is XXX mm.

5. When upper kitchen cupboards are hung on the wall, usually the top of the units are XXX mm above the floor.

C. Write a full answer

6. List three types of doors that can be used on kitchen cupboards.

D. Mark as either TRUE or FALSE

7. As a general rule, the ends of cupboards, dividers, stiles and doors will have the face grain running horizontally.

8. Overlapping doors use the same type of hinges as rabbeted doors.

9. Flush doors may be installed with or without the use of stiles.

10. A lower countertop 600 mm wide is more than adequate to install a dishwasher beneath.

SUPPLEMENT TO CHAPTER 10
ROOF FRAMING

STANLEY METRIC STEEL SQUARE

Figure 10-71   Stanley metric steel square
Courtesy Stanley Works Limited

All the basic information found in Chapter 10 of this book is applicable no matter what framing square a person may have to use. However, below are described the basic units and calculations particular to the use of the Stanley Metric Steel Square. The Stanley square uses 1 m as the unit of run for the common rafter. The unit of run for the hip/valley rafter is the diagonal measurement of a 1 m² which is 1.414 m (1414 mm).

The Stanley square has a rafter table on the face of the body, which gives in mm the unit length for common rafters per metre of run; the unit of run for hip/valley rafters per metre of common rafter run;

the difference in length for jacks at 400 mm o.c.; the difference in lengths for jacks at 600 mm o.c.; the side cut of jacks and the side cut of hip/valley rafters. This information is given for a unit rise of 250 mm, of 300 mm and then of increments of 100, up to 1500 mm.

When it is necessary to find the length of a rafter by calculation, the same methods as described in Chapter 10 would be used, substituting 1000 mm where 250 mm was used and substituting 1414 mm where 353.5 mm was used.

Plumb cuts on the Stanley square use 200 and 283 on the body for common and hip/valley rafters. These points, referred to as set points, are clearly designated on the square as shown in Figure 10-72.

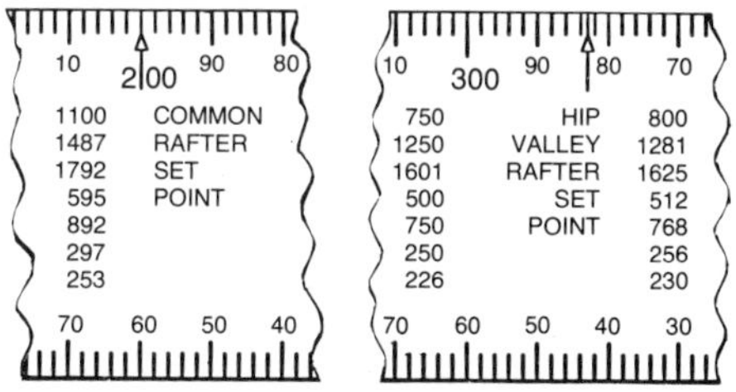

```
┌─────────────────┐ ┌─────────────────┐
│ 10   2|00  90  80│ │ 10   300  90  |80  70│
│ 1100   COMMON    │ │ 750    HIP    800 │
│ 1487   RAFTER    │ │ 1250  VALLEY  1281│
│ 1792   SET       │ │ 1601  RAFTER  1625│
│ 595    POINT     │ │ 500    SET    512 │
│ 892              │ │ 750   POINT   768 │
│ 297              │ │ 250           256 │
│ 253              │ │ 226           230 │
│ 70  60  50   40  │ │ 70  60  50  40  30│
└─────────────────┘ └─────────────────┘
```

Figure 10-72   Set points
Courtesy Stanley Works Limited

To find the number of units of run using the Stanley square, divide the total run by 1000.

Calculating the Line Length of the Common Rafter

Example:
Building Size—7200 mm × 12 000 mm
Roof Slope—1:3
Overhang—750 mm
From the information:
(a) Span = 7200 mm
(b) Total run = 3600 mm
(c) Number of units of run = 3.6
   3600 ÷ 1000 = 3.6
(d) Unit of rise = 333.3 mm
   1:3 = x:1000
   x = 333.3
(e) Unit of Line Length = 1054.08
   $c = \sqrt{a^2 + b^2}$
   $= \sqrt{1000^2 + 333.3^2}$
   $= \sqrt{1\ 000\ 000 + 111\ 088.9}$
   $= \sqrt{1\ 111\ 088.9}$
   $= 1054.08$
(f) Line length of common rafter = 3794.7 mm
   $= 1054.08 \times 3.6$
   $= 3794.7$

Calculating the Line Length of the Common Rafter Tail

From the information on the common rafter, the line length of the rafter tail would be calculated in the following way:

Example:

(a) Run (overhang) = 750 mm
(b) Number of units of run = 0.75
        750 ÷ 1000 = 0.75
(c) Line length of common rafter
        tail = 790.6
        1054.08 × 0.75 = 790.56

Calculating the Line Length of the Hip Rafter

The formula for finding the unit of line length is:

$$\text{Unit of L.L.} = \sqrt{\text{unit of rise}^2 + \text{unit of run for hip}^2}$$

Example:

Slope—1:3
Unit of rise—333.3 mm
Unit of run—1414 mm
Unit of L.L.
$$= \sqrt{333.3^2 + 1414^2}$$
$$= \sqrt{111\,088.9 + 1\,999\,396}$$
$$= \sqrt{2\,110\,484.9}$$
$$= 1452.75$$

After the unit of line length is determined, multiply this figure by the number of units of run for the common rafter. This will give the line length of the hip rafter.

Example:

Building size—7200 × 12 000 mm
Roof slope—1:2.5
Overhang—750 mm
From the information:

(a) Span = 7200 mm
(b) Total run = 3600 mm
(c) Number of units of run = 3.6
(d) Unit of rise = 400 mm
        1:2.5 = x:1000
        2.5x = 1000
        x = 400
(e) Unit of line length = 1470 mm
        found on the 3rd line of the rafter square
(f) Line length of hip rafter = 5292 mm
        1470 × 3.6 = 5292

This measurement is from the plumb cut at the birdsmouth to the centre of the ridge board.

Plumb Cuts and Seat Cuts

The plumb and seat cut line for a common rafter uses 200 on the body and the unit of rise on the tongue of the square. The unit of rise is found in the middle of the tongue and must be transferred to the edge. Figure 10-73 shows the plumb and seat (bottom) cuts for a 500 mm rise on a common rafter.

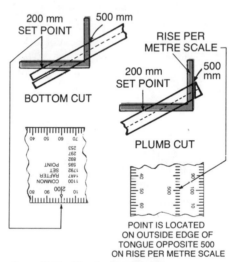

Figure 10-73   Plumb and seat cuts
Courtesy Stanley Works Limited

The plumb and seat cut line for a hip/valley rafter uses 283 on the body and the unit of rise on the tongue. Figure 10-74 shows the plumb and seat cuts for a 500 mm rise on a hip rafter.

Deductions for all rafters will be the same regardless of what framing square is used.

Side Cuts

The side cuts for hip/valley rafters are found on line seven of the rafter table. These are used in conjunction with the "200" common rafter set point.

The side cuts for jack rafters are found on line six of the rafter table. These are used in conjunction with the "200" common rafter set point.

LAYING OUT A HIP RAFTER

"Dropping" the hip rafter will follow the same procedure as described on page 177. In calculating the drop 600 mm is the unit of rise and 1414 mm is the unit of run.

Example:

Using the above information, the following equation is set up:

$$x = \frac{600 \times 19}{1414}$$
$$= 8$$

The amount of drop would be 8 mm.

Rafter Cut Tables

On the back of the body of the Stanley square a table is given showing angles in degrees of the various cuts (Fig. 10-75). This is helpful when using power tools which can be adjusted to cut prescribed angles. "A" is the plumb cut and "B" is the seat cut. Side cut angles for jack, hip or valley rafters are also shown in Figure 10-76.

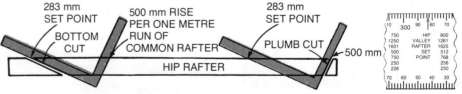

Figure 10-74   Plumb and seat cuts
Courtesy Stanley Works Limited

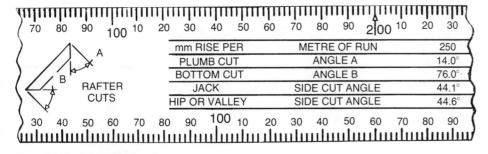

mm RISE PER	METRE OF RUN	250
PLUMB CUT	ANGLE A	14.0°
BOTTOM CUT	ANGLE B	76.0°
JACK	SIDE CUT ANGLE	44.1°
HIP OR VALLEY	SIDE CUT ANGLE	44.6°

Figure 10-75 Table of angle cuts
Courtesy Stanley Works Limited

Framing Squares and Units of Rise and Run

In a shop situation where both the Frederickson and the Stanley framing squares may be used, the following comparison could prove useful. All the information in Chapter 10 is applicable to both squares, substituting these units as shown:

STANLEY	FREDERICKSON	
1000	250	
1414	353.3	
200	250	for layout
283	354	for layout

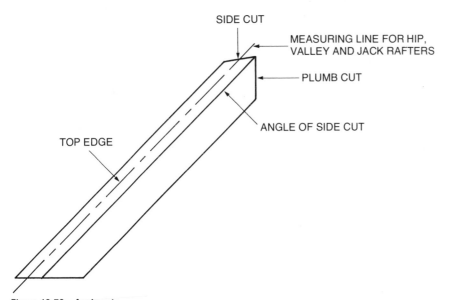

Figure 10-76 Angle cuts
Courtesy Stanley Works Limited

GLOSSARY

A

admixtures–any materials added to the concrete batch before, during or after mixing, other than portland cement, water or aggregates.

aggregate–those inert particles which are mixed with portland cement and water to form concrete.

air-entrained concrete–concrete containing small air bubbles throughout the cement-water paste.

anchor bolt–a steel bolt embedded in concrete foundations to secure structural members in place.

angle bead–a small moulding placed at exterior corners of plastered or gypsum board surfaces to guard against damage.

apex–the uppermost point; two sloped surfaces ending at a point.

apron flashing–a lower flashing where the chimney projects through the roof.

asbestos cement–a fire-resistant weatherproof building material made from portland cement and asbestos, used in the manufacture of siding shingles, pipes etc.

asphalt–a petroleum product used in the manufacture of roof covering, siding, floor tiles etc.

B

backfill–the replacement of earth after excavating for a foundation wall, trench etc.

backing–material used to provide reinforcement or a nailing surface for finish materials.

backsight–a sight taken with a level to a point of known elevation.

balloon framing–a method of wood-frame construction where the studs extend in one piece from the foundation sill to the top plate.

balusters–vertical members that support the handrail on a stair.

balustrade–the complete assembly on a stair, consisting of balusters, newel posts and handrail.

baseboard–a finish board at the base of a wall to conceal the joint between the floor and wall finish.

batterboard–a temporary framework set adjacent to an excavation to assist in locating corners when laying out a foundation.

batt insulation–a blanket type insulating material composed of mineral fibers and manufactured in small units convenient for placing between framing members.

beam–a horizontal structural member usually made of wood, steel or reinforced concrete–used to support overhead loads.

beam pocket–a notch formed at the top of a foundation wall to receive and support the end of a beam.

bearing–the portion of a framing member that rests on a support.

bearing partition–a wall that supports any vertical load in addition to its own weight.

belled pile–a friction pile with the base enlarged for greater support.

bench mark–a relatively permanent object, natural or artificial, used to indicate a definite point from which elevations are set.

bi-fold door–a door consisting of two panels hinged in the middle and sliding on an overhead track.

bi-pass doors–double doors that slide pass one another.

birdsmouth–a notch cut at the bottom of a rafter where it rests on the wall plate, consisting of a plumb cut and a seat cut.

bleed water–excess water in the concrete mixture which surfaces after placing.

blueprint–a series of drawings that give information regarding construction.

bolster–a top piece of either wood or metal placed on a post used to increase the bearing surface at the beam.

brick mould–an exterior moulding for window and door frames. Siding materials such as brick veneer, wood siding, etc. are butted to this mould.

brick veneer–a single row of bricks tied to a wood-frame wall as an exterior finish.

bridging (cross)–small wood or metal members placed in a diagonal position between floor joists to restrain movement.

buck–a wooden frame installed in concrete forms before the concrete is poured to provide an opening for doors or windows.

C

camber–the amount of upward curve given an arch beam girder, etc. to prevent the member from bending downward when loads are imposed on it.

cast-in-place–floor joists that have been set on the forms and when the concrete is poured are embedded in the wall.

chamfer–the corner of a piece of stock beveled at a 45° angle.

chords–the principal members of a roof truss, either top or bottom.

collar tie–a horizontal member usually situated in the middle third of the rafters–used to provide support between opposite rafters.

control joint–a joint made in concrete slabs to control cracking by expansion and contraction.

core tests–samples of the subsoil taken at predetermined intervals.

corner boards–wood members installed at corners of wood-frame structures against which the ends of siding are butted.

counterflashing–a flashing applied above another flashing to shed water over the top of the under flashing–this type of flashing allows some movement without damage to flashing.

cross-section–on a blueprint, a cut at right angles to the longitudinal axis of a drawing.

D

datum point–a reference point from which elevations and measurements are taken.

dead load–the weight of all permanent structural and nonstructural components of a building.

dimension lines–lines drawn to show dimensions of the object drawn.

double headed nails–nails with a second head, making removal convenient even though the nail is securely driven to the first head–used in constructing scaffolds that will be torn down.

double wall–a double frame wall used on low energy houses.

drain tile–tile placed around the perimeter of a building to collect ground water and dispose of it through the sewer line or drywell.

drywall–any exterior finish requiring little or no water for installation.

dwarf wall–a framed wall of less than full normal height.

E

eave–the lower portion of a roof which extends beyond the wall face.

eave soffit–the material used on the under surface of the eave.

elevation–a drawing showing the projection of a building on a vertical plane.

expansion joint–a bituminous fiber strip used to separate blocks or units of concrete to prevent cracking from expansion which occurs because of temperature change.

explosive-actuated tools–uses the energy from a charge of power to drive special steel fasteners into concrete or steel.

exposure–the amount of shingle or siding exposed to the weather.

exterior trim–any trim used on the outside of a building to provide a suitable finish, particularly used on doors and windows.

F

fascia–a finish around the face of eaves and roof projections.

fire cut–joists cut at an angle where they fit into a masonry wall to prevent wall damage in the case of fire.

firestop–blocking within a wall, floor or roof space to prevent the spread of fire or hot gases within the enclosed framing.

flashing–sheet metal or other material used in construction to shed water.

footing–the enlarged base at the bottom of masonry walls and columns.

formwork–an enclosure made from either wood or metal to hold concrete until it sets.

foundation–the lower portion of a building including the footing made of either concrete, masonry or wood.

furring–strips of wood attached to framed walls or ceilings to which finish material will be attached.

G

gable–the upper triangular-shaped portion of an end wall of a building.

gable end roof–a roof composed of rafters meeting at a ridge with the end wall continuous from the bottom plate to the ridge.

gambrel roof–a roof with sloping rafters broken by an obtuse angle so the lower slope is steeper than the upper slope–a common type of roof for a barn.

girder–a large horizontal member used to support overhead loads imposed upon it.

grade–the ground level around a building.

gravel stop–a metal flashing used on the outer edges of flat roofs to prevent the gravel from being washed away.

gusset–a metal or plywood plate used on trusses to stiffen connections.

gypsum–a mineral used in the manufacture of plaster of paris and gypsum board.

gypsum board–a wallboard made of a gypsum plaster core covered on both sides with paper.

H

hardpan–a mixture of sandy gravel soils and clay, producing an extremely hard, compact, soil.

H-clip–a metal device used between sheets of plywood to add rigidity, used instead of wood blocking between framing members.

header–a horizontal structural member that supports the load over door and window openings, also called a lintel.

headroom–the vertical clearance from the open end of the stairwell to the line of flight.

hinge bound–when the two leaves of a hinge come together before the door is properly closed.

hip rafter–the rafter that forms the hip of a roof, extending diagonally from an exterior corner of the building to the ridge.

hip roof–a roof which has four sloping sides that meet at a hip and ridge.

hydration–the chemical reaction which takes place between the cement and water in the curing of concrete.

I

I-beam–a steel beam with a cross-section that resembles the letter I.

insulation–a building material that resists the flow of heat.

interior finish–any materials used to give the inside of a building a finished appearance, this includes wall and ceiling covering as well as any trim required to give a finished effect.

interior trim–a general term for all casing, baseboard or mouldings applied to the interior of buildings to give a finished appearance.

J

jack rafter–any common rafter that has been cut at either the top or bottom to fit into other roof framing members

jamb–the top and side pieces of a door or window frame.

joists–horizontal framing members used to support floor and ceiling loads.

joist hanger–a metal strap used to support and fasten joists that are butted into beams or girders.

K

kerf–a cut made with a saw.

keyway–a groove made in a concrete footing to lock the foundation wall to the footing when it is poured.

kiln dried–wood seasoned in a kiln by means of artificial heat, controlled humidity and air circulation.

L

laminate–layers of wood or paper bonded with glue under heat to produce building materials such as plywood, plastic laminates, etc.

lateral pressure–the sideways pressure exerted by materials such as earth, freshly poured concrete, etc.

ledger strip–a horizontal member attached to framing members to support joists.

lintel–a horizontal member situated over door and window openings to support loads imposed above it, also known as a header.

live load–a load imposed upon structural members resulting from snow, rain, wind, earthquake and loads due to occupancy.

loadbearing–any structural member which must support loads other than its own weight.

lookouts–framing members that support overhanging portions of a roof, also used to support eave soffit between the wall and the rough fascia.

M

mansard roof–a roof with two slopes on all four sides, the lower slope very steep with the upper slope almost flat.

mitre box–a device used to guide a backsaw when cutting various angles used on trimming material.

mitre joint–a cut made at an angle for joining two pieces of board so cut that they will form an angle.

module–a standard or unit of measurement.

moulding–strips of material formed in various curved profiles throughout its length used for finishing and decorative finish.

monolithic–a term used for concrete structures that have been poured and cast as one unit.

mortar–a masonry paste mixture of sand, cement, lime and water, which hardens when exposed to air, used in joining bricks, stone, blocks, etc.

mortar bed–a layer of mortar on which any structural member is laid.

mudsill–a structural member placed directly on the ground as a foundation for a structure.

N

newel post–a post at the beginning or end of a stair which supports the handrail.

non-bearing partition–a wall that supports no load other than its own weight.

nosing–the rounded edge of a stair tread that projects over the riser.

O

object line–a heavy unbroken line outlining the object being drawn.

offset–a term used to describe any recess or variation on a plane surface–also overlapping of panels or doors.

on centre (O.C.)–a term used to define the point from which measurements are taken.

open pit testing–testing of soil during excavation; used where a shallow excavation takes place.

P

parapet–a protective railing or low wall extending above the edge of a roof or balcony.

parging–a thin coat of plaster applied to masonry construction.

partition–an interior wall separating a large area into smaller portions.

pilaster–an enlargement in a portion of wall used where added support may be needed, as in supporting a girder or beam.

pier–a column, usually of concrete–used to support structural members.

plain concrete–unreinforced concrete.

plasticity–a term used to describe the consistency of concrete.

plastic laminate–layers of kraft paper bonded together with resin with a decorative top covering–used as a hard finish on counters and furniture, etc.

plate–a horizontal member at the top and bottom of a wall frame used to support and align wall studs.

platform framing–a type of construction in which the floor platform is framed independently from the wall sections.

plumb–exactly perpendicular or vertical.

plumb cut–a term used in roof framing to indicate vertical cuts on the ends of rafters.

pneumatic power tools–tools dependent on compressed air to operate.

polyethylene membrane–a vapour barrier material made of plastic.

post–a vertical member used as a support for structural members.

pressure-treated–a preservative applied to wood under pressure.

R

rabbet–a groove along the edge of a piece of wood made to receive another piece of wood.

rafter–the sloping member of a roof used to support the roof covering.

rake–an inclination from the perpendicular.

ratio–the relationship in quantity, amount or size between two or more things.

re-bar–deformed steel bars used in the making of reinforced concrete.

reinforced concrete–concrete which has been strengthened by steel bars embedded in it.

release agent–oil applied to forms before pouring so as to facilitate the stripping operation.

reveal–the amount of space the trim is left back from the edge of a door or window jamb.

ribbon–a horizontal member nailed to a structural member to support joists or other building materials–also known as a ledger strip.

ridgeboard–a horizontal member found at the top of a roof where two slopes meet.

ripping–the sawing of a piece of wood lengthwise with the grain.

rise–the vertical distance from one tread to the next in a flight of stairs.

riser–a vertical board under the tread on a flight of stairs.

rough opening–an opening in a frame construction which will receive a door, window, stair, etc.

run–the horizontal distances covered by a flight of stairs; in roofing, the horizontal distance from the wall plate to the centre line of a building.

S

saddle–a flashed roof structure between a chimney and the roof to shed water around the chimney, also known as a cricket.

scab–a short piece of wood used to tie two members together.

screed–a temporary horizontal member placed where concrete is to be poured as an aid in leveling freshly poured concrete.

seat cut–the horizontal cut on the birdsmouth of a rafter.

segration–the separating of the fine and coarse aggregates in a concrete mix.

setback–the horizontal distance between a building and the adjacent street line.

shake–a defect in lumber; a separation of the wood between the annual growth rings; a shingle split from a block of wood used for roofing.

sheathing–material used to cover the framework of buildings on the exterior.

shingle–a small thin piece of building material with one end thicker than the other used in overlapping rows as a roof or wall covering.

shiplap–boards with a rabbet on both sides used to make a lapped flush joint when put together.

shortening–a term used in roofing to describe the amount the rafter has to be shortened to fit into other roof framing members.

sidelap–the amount shingles lap over joints in successive rows; also the amount of lap for battens over boards.

sidelights–window units at the sides of doors.

siding–the finish covering on the outside of frame buildings.

sight in–a term used in surveying meaning to locate a point through a telescope.

sill–the bottom member of a window or exterior door frame.

sill plate–a structural member anchored to the top of a foundation wall, upon which the floor joists rest.

sizing–a term used to describe the method for bringing floor joists to the same height either by shimming or notching where the joists rest on the support.

slump–a term used in concrete batching to describe the consistency of a mix.

soffit–the underside of structures in a building as eave soffit, stair soffit, etc.

sole plate–the bottom horizontal member in a frame wall used to align and fasten the studs.

span–the distance between supports; in roof construction it is the width of the building.

step flashing–overlapping rectangular pieces of flashing used at the junction of shingled roofs and walls.

stop bound–when the edge of a door hits the shoulder of the door frame rabbet or the door stop thus preventing it from closing properly.

storeypole–a strip of wood upon which are marked various measurements to aid in the installation of such building components as exterior siding, cupboards, window and door lintels, etc.

straight edge (concrete)–a wood or metal device of sufficient length to span between screeds or slab forms, used in a sawing action to level freshly placed concrete.

strapping–narrow boards nailed to framing members for the purpose of providing backing for finishes.

stringer–the inclined member which supports the treads and risers of a stair.

strut–a vertical member used to support the ridge and rafters of a roof.

stud–one of a series of vertical structural members used in construction of frame walls.

subfloor–sheathing laid on joists to support the finish floor.

T

tail joist–a regular floor joist that has been cut to allow for a floor opening, butted into a header joist.

template–a gauge used to assist in doing a specific job.

threshold–a strip of wood, metal or other material, usually beveled on each edge and used at the junction of two different floor finishes under doors, or on top of the door sill on exterior doors.

toenailing–driving a nail at a slant through one member to attach to another.

top plate–the horizontal member found at the top of a frame wall structure used to fasten and align the studs.

transit–an instrument used to test for levelness, to transfer points and measure angles on a horizontal plane as well as to measure vertical angles up to 45° from the horizontal plane.

tread–the horizontal member of a set of stairs upon which you walk.

trimmer joists–the joists on either side of a frame floor or ceiling opening.

trimmer studs–the side pieces of a door or window opening in a frame wall that support the lintel or header.

truss rafter–a structural roof frame composed of top and bottom chords and diagonal and vertical members fastened together with gussets, the truss acts as rafter and ceiling joist combined.

U

underlay–a panel material used on top of the subfloor to give a smooth finish to receive the finished floor.

V

valley–the internal angle formed by the two slopes of a roof.

valley rafter–a rafter which forms the intersection of sloping roofs at an internal corner.

vapour barrier–a material used to prevent the passage of vapour or moisture into floor, ceiling or wall cavities.

vibrator–a machine which has a flexible shaft with a vibrating head attached, used in the placement of concrete.

views–those portions of blueprints which show the exterior elevations of a building.

void (concrete)–airlocks causing holes in concrete.

W

walers–horizontal members used to hold concrete forms in place.

wall section–in a blueprint that portion which shows a vertical cut through a building.

water table–an exterior moulding used to divert water away from a wall; the level at which the ground is saturated with water.

weatherstripping–pieces of metal, wood or other materials or a combination of several applied around doors and windows to prevent drafts.

wellhole–the opening in a floor frame where a stair is to be installed.

working drawings–includes all drawings needed by the various tradespeople to complete a building project.

working room–the space needed between the foundation forms and the excavation for erection of forms.

ANSWERS TO REVIEW QUESTIONS

INTRODUCTION

1. Written report
2. Metre (for length), litre (for liquid volume) and gram (for weight).
3. Metre (construction field) expressed as: metre (m), millimetre (mm) for length, square metre (m²) area, cubic metre (m³) for volume.
4. Basic module of 100 mm. Major effect on industry will be standardization.
5. "Hard" conversion is where sizes have been set. "Soft" conversion is where existing sizes have been used and expressed in metric terms.
6. In the future "soft" converted measurements to "hard" may result in slight changes in material size (e.g. 38×89 (soft) $- 35 \times 85$ (hard)).

1—HAND AND POWER TOOLS

A. Hand Tools

1. c
2. b
3. b
4. d
5. a
6. a
7. b
8. b
9. c
10. a
11. with
12. Robertson
13. 25 to 30°
14. sweep
15. nail set
16. True
17. True
18. False
19. False
20. Use the correct tool for any given operation.
    Keep tools properly sharp, adjusted and in good condition.
    Keep all tools dry.
    Lightly lubricate moving parts.

B. Power Tools

1. b
2. d
3. c
4. a
5. chuck
6. front portion of the plane bottom
7. belt or pad size
8. standard (high) and low velocity
9. simultaneously
10. hammer
11. stationary and portable
12. above
13. bits
14. grounded, double insulated
15. See safety of stationary and portable equipment.

2—HOUSE STYLES AND METHODS OF ASSEMBLY

1. Short answer question
2. single-storey, storey and a-half, two-storey and multi-level
3. Short answer question
4. Short answer question (see Chapter 14)
5. Short answer question
6. Short answer question
7. liquid and air
8. water, rock or phase-change material
9. maximize, minimize
10. time lag
11. Short answer question
12. isolated gain (greenhouse)
13. stick framing
14. precision end trimmed
15. fabrication
16. ready-to-move (RTM)
17. packaged
18, 19, 20: Written reports

3—BLUEPRINT READING

1. perspective drawing, working drawing
2. Short answer question
3. b
4. a
5. a
6. c
7. c
8. a
9. c
10. b
11. Principle lines
    Terminates when it meets an extension line
    Postion of permanent accessories
    Extends out from the object
    Indicates a portion of the drawing is left out
    Shows obscure parts of the house
    Indicates a portion of the drawing has been removed for clarity
    Method used for locating parts
12. 1:500, 1:100, 1:50, 1:20, 1:20
13. False
14. True
15. True
16. specifications
17. Canada Mortgage and Housing Corporation
18. owner/builder
19. National Building Code, Associate Committee on the National Building Code
20. Written report

4—SITE PREPARATION AND EXCAVATION

1. Short answer question
2. Pythagorean theorem $a^2 + b^2 = c^2$ or 3, 4, 5 method
3. c
4. a
5. d
6. d
7. c

8. a

9. d

10. c, f, a, e, d, b

11. Gives the vertical distance from a known elevation to the line of sight
Horizontal line through the instrument in any given direction.
Vertical distance from a point of unknown elevation to the line of sight
Vertical distance above or below the bench mark
Obtained by adding the backsight to the point of known elevation
A marked point of known elevation

12. left

13. depth of the excavation

14. batter boards

15. equal

16. Short answer question

17. Short answer question

18. open pit, core

19. water table

20. Short answer question

21. Short answer question

22. mineral salt

23. Short answer question

24. topsoil

25. (a) 305.6 or 306 (b) 153.9 or 154 (c) 212

5—CEMENT AND CONCRETE

1. paste

2. raw materials

3. wet and/or dry

4. clinker

5. 40 kg

6. cement, water, coarse and fine aggregates

7. f, c, a, e, b, and d

8. b

9. b

10. c

11. c

12. a

13. d

14. b

15. b

16. a

17. d

18. c

19. a

20. d

21. calcium chloride, high early

22. tensile

23. smooth and deformed

24. plain

25. construction, isolation and control

6—FOOTINGS AND FOUNDATIONS

1. footing

2. local, frost line

3. continuous (strip), stepped, and independent

4. pilaster, chimney and/or fireplace

5. plumb bob

6. keyways

7. footing forms

8. raised

9. (a) Local code restrictions
(c) Depth below grade
(b) Function of the area
(d) Type of construction/lateral support supplied by the floor framing system.

10. pilings

11. wood, steel or concrete

12. concrete

13. rotting

14. piers

15. Short answer question

16. c

17. a

18. c

19. d

20. a

21. b

22. d

23. c

24. (a) C.L.L. = 45.1 lineal metres
Footing = 6.77 or 7 m^3
Wall = 24.35 or 24.5 m^3
Slab = 11.88 or 12 m^3

(b) C.L.L. = 40.6 lineal metres
Footing = 3.65 or 3.75 m^3
Wall = 14.62 or 14.75 m^3
Slab = 7.27 or 7.5 m^3

25. False

26. True

27. True

28. False

29. True

30. (a) Suspended (b) Sleeper (c) Concrete slab

31. lateral

32. metal galvanized

33. drain tile or gravel bed

34. waterproofed (foundation coating)

35. uniformly

7—FLOOR FRAMING

1. d

2. d

3. a

4. d

5. a

6. b

7. b

8. (a) Platform framing (box sill)—the header joist and joist ends are toe-nailed to a sill plate which is anchored to the foundation wall (b). Balloon framing—there is no header joist, the floor joists are fastened to the studs and toe-nailed to a sill plate anchored to the foundation wall

9. (a) Span (b) Joist spacing (c) Wood type

10. a horizontal member nailed to other framing members to support joists

11. (a) Placing of floor joists with the bow up (b) Shimming or notching of joists to produce a level floor

12. (a) Adjustable steel pipe (b) Solid wood (c) Built-up wood

13. Joists are prevented from twisting at the foundation wall by the header joist, at the beam by toe-nailing to the beam and at intermediate intervals by bridging.

14. (a) Joist spacing   (b) Sizing
    (c) Crowning (d) Floor openings
15. True
16. False
17. False
18. False
19. True
20. True

8—WALL AND CEILING FRAMING

1. d
2. b
3. b
4. d
5. c
6. b
7. c
8. b
9. c
10. d
11. less chance of error
12. to support overhead load
13. Straighten top plate, ensure diagonals are equal.
14. to allow for plumbing and alignment, insulation
15. balloon
16. True
17. True
18. True
19. True
20. False

9—ROOF STYLES AND TERMINOLOGY

1. provides protection against the weather, forms a structural tie between the outside walls, enhances the appearance of the building
2. 45°
3. 90°
4. Yes
5. vertically from the top of the wall plate to the intersecting point of the theory line of a pair of rafters
6. The theoretical length from the top outer edge of the wall plate to the centre of the ridge board.

7. 1100 mm
8. The horizontal distance between the outer faces of the outside wall plates of a building.
9. The horizontal distance as measured from the outer face of the outside wall plate to the centre of the ridge board, one half the span.
10. triangle
11. c
12. b
13. c
14. b
15. c
16. a
17. False
18. True
19. False
20. True

10—ROOF FRAMING

1. c
2. d
3. b
4. c
5. 6038 mm
6. 250
7. both have the same birdsmouth and rafter tail.
8. Find the length by calculation of two adjacent jacks, subtract the lengths and the answer will be the common difference.
9. (i) To the theoretical length add the thickness of the common rafter stock. (ii) To the theoretical length add the thickness of the ridge board plus the 45° thickness of the hip rafter stock (the theoretical length is the length of the building less the width of the building).
10. No, if the span of the main roof is the same as the span of the minor roof, there will be no shortened valley rafter.
11. True
12. True

13. True
14. True
15. False
16. False
17. False
18. True
19. False
20. False

11—CORNICE CONSTRUCTION AND ROOF COVERINGS

1. a
2. c
3. b
4. d
5. b
6. (a) low slope asphalt shingles;
    (b) builtup tar and gravel;
    (c) taper split cedar shakes;
    (d) roll roofing
7. live
8. dead
9. rough fascia, finish fascia, lookout, ribbon, rafter tail, soffit, roof sheathing
10. open, closed, snub
11. wood, metal, asphalt, tile, slate
12. Ensure there is air circulation between soffit and attic space.
13. lookout
14. False
15. False
16. True
17. True
18. False
19. True
20. False

12—EXTERIOR DOORS AND WINDOWS

1. a
2. b
3. b
4. c
5. c
6. b
7. b
8. b
9. d

10. a
11. on the side
12. to provide a drip groove which breaks the flow of water
13. backset
14. When the edge of the door hits the shoulder of the rabbet or doorstop, preventing the door from closing properly
15. Yes
16. (a) Sliding (b) Fixed (c) Hinged
17. False
18. True
19. False
20. True

13—EXTERIOR WALL FINISH

1. bevel, tongue and groove, channel, board and batten
2. single and double coursing
3. storeypole
4. shiplapped
5. modified to support bricks
6. 12 rows, exposure 200 mm
7. metal corners, mitre corners, siding butted to corner boards
8. to prevent rust spots from occurring
9. so the bottom row will be the same angle to the horizontal as the rest of the rows
10. climate, initial cost, maintenance, appearance
11. True
12. False
13. False
14. False
15. True

14—INSULATION AND VAPOUR BARRIER

1. heat loss
2. radiation, conduction and convection
3. sub-divides, air spaces
4. the flow of heat
5, 6, 7, 8: Short answer questions
9. fire hazard
10. loss of air spaces

11. b
12. a
13. d
14. a
15. b
16. True
17. False
18. False
19. air-to-air heat exchanger
20. the air is presently being changed often enough to prevent a buildup of stale air.

15—INTERIOR WALL AND CEILING FINISH

1. plaster
2. 200 mm
3. 1630 mm
4. Plumb wallboard and scribe using metal scribers, plumb wallboard and scribe using wooden block.
5. mitred
6. width—298 mm border with 11 full rows; length—297 mm border with 23 full rows
7. True
8. True
9. True
10. False

16—INTERIOR DOORS AND TRIM

1. d
2. b
3. b
4. c
5. a
6. a
7. c
8. c
9. thickness of wall
10. Door plus clearance will position jamb.

17—STAIR CONSTRUCTION

1. effective depth is sufficient
2. 13
3. distance required to travel
4. primarily to prevent accidents

5. so all the risers will be the same height
6. 18 mm
7. 12.36
8. 2845 mm
9. 2920 mm
10. 176.7 mm, 176.7 mm
11. Decrease the actual run, design the stair with one less riser.
12. line of flight, underside of the upper floor construction
13. mitred
14. 30°, 35°
15. 50°, 20°
16. 140 mm
17. actual rise—182 mm, 15 risers, 12.86 uncovered risers, actual run—225 mm
18. headroom 1961.5 mm
19. (a) Distance from finish floor to finish floor (b) Total rise divided by actual rise (c) Total rise divided by number of risers, vertical distance from the top of one tread to the top of the next (d) Horizontal distance from the face of one riser to the face of the next (e) Total distance the stair travels horizontally (f) Head room plus floor construction divided by actual rise (g) Number of uncovered risers multiplied by actual run (h) Workable opening plus one nosing (i) Finish opening plus allowances for finish (j) Workable opening divided by number of uncovered risers
20. baluster, newel post, tread, riser, stringer, glue block, wedges, tread return, nosing

18—CABINETRY

1. cooking, storage, clean-up
2. 915 mm
3. 380 mm
4. 300 mm
5. 2130 mm
6. flush, rabbeted, overlapping
7. False
8. False
9. True
10. False

INDEX

INDEX

A

Active solar home, 36-38
Adjustable T-Bevel, 2
Aggregates, 76
Air-actuated tools, 28
Air-to-air heat exchanger, 35, 250-252
Air vents, 244, 246, 247
Aluminum roofing, 201
Anchor bolts, 107-109
Architectural symbols, 50-52
Asbestos shingles, 201
Asphalt shingles, 201-204
Asphalt roofing products, 201
Attic access, 35, 245, 246
Auger bits, 14
Automatic drill, 15
Awning windows, 219, 220

B

Backfill, 124
Backing, 150, 151
Backsaw, 8
Balloon framing, 130, 140, 142
Baluster, 271, 272
Band (web) clamps, 16
Bar or pipe clamps, 16
Barge board, 199, 200
Barge rafter, 173
Baseboard, 268, 269
Basement floors, 111, 119-121
Bathroom accessories, 150, 151, 256
Batt insulation, 233, 234
Batter boards, 67, 68
Bay windows, 219, 220

Beams, 127, 128
Beam pocket, 110
Bearing wall, 119, 120, 140
Belled pile, 100
Bench grinder, 21
Bench mark, 65
Bench planes, 11
Beveled siding, 226-229
Bi-fold door, 264
Bi-passing door, 264
Birdsmouth, 168, 169
Bit brace, 14
Bit extension, 15
Bit gauge, 15
Bleeding, concrete, 83
Block plane, 11
Blueprint language, 49, 50
Blueprint reading, 44-52
Bolts, anchor, 107-109
Boring tools, 14, 15
Boston hip, 206, 207
Bow windows, 219, 220
Box cornice, 196-200
Box sill construction, 107, 109
Bracing, 151, 152
Brick and block construction, 112-114
Brick ledge, 111
      mould, 211, 212
      veneer, 111, 226
Bridging, 133-135
Bucks (door and window), 106, 109, 110
Builder's level, 61
      line, 5
Building:
      code, 53-56

layout, 60
line, 59, 60
paper, 224
permit, 55, 56
Built-in cupboards, 292, 293
Built-up beams, 127
      forms, 102
      roofing, 160, 161, 207
Bullfloat, 83
Butt chisel, 10
Butt gauge, 6

C

Cabinet hinges, 289, 290
Cabinet making, 286-293
Canada Mortgage and Housing Corporation, 55
Cant strip, 160, 206, 207
Cantilever, 134, 135
Carpenter's pencil, 5
Casing interior, 267, 268
Casement windows, 219, 220
C-clamps, 16
Ceiling:
      finish, 254, 259
      height of, 142, 143
      joists, 152, 153
      suspended, 259
      tiles, 259, 260
Cement, manufacture of, 73, 74
      types of, 75
Centre line length (C.L.L.), 112
Chalk line, 6
Chisels, 10
Circular saw, 18

Circular saw blades, 19
Collar ties, 165, 171
Columns, steel, 92, 128
        wood, 92, 128
Combination square, 2
Common rafters, 163, 168-171
Compression test, 79, 80
Concrete: 75
        blocks, 113, 114
        calculations, 112
        curing, 85, 86
        finishing, 82, 83, 84
        floors, 111
        mixing, 77, 78, 80
        mixtures, 76, 77
        placing, 81, 82
        reinforcement, 86, 87
        testing, 79, 80
Condensation, 248
Conduction, 232, 233
Construction joints, 87
Continuous footing, 91
Contour lines, 45
Control joints, 84, 87
Convection, 232, 233
Conventions, 50
Coped corners, 269
Coping saw, 8
Corner framing, 144, 145
Cornice construction, 196-201
Countersink bit, 15
Countertop, 286, 288, 292
Crawl space, 98
Cricket, chimney, 209
Cripple rafters, 188, 189
Cripple studs, 147
Cross-bridging, 133, 134
Crosscut saw, 7
Cupboards, types of, 288, 289
Cut-in bracing, 152

D

Dado head, 19
Dampproofing, 123, 124
Darby, 83
Deformed bar, 86
Detail views, 48
Diagonal bracing, 150-152
Dimension lines, 49, 50
Door:
        entrances, 211, 216-218

frame, 211-213
        jamb, 211, 212, 262, 263
        technique in hanging, 214, 215
        type of, 216 262
Double coursing,
        side wall shingling, 224, 225
Double-hung windows, 220
Double plate, 143
Dowelling jig, 15
Drain tile, 122, 123
Drainage, 121-123
Drawer guides, 291, 292
Drawers, 290, 291
Drip groove, 211
Drywall, 254
Dwarf walls, 171, 172

E

Eaves, 172, 173, 196-201
Eave protection, 196, 201, 202
Electric drills, 23
Electric mitre saw, 23
Electrical symbols, 55
Elevations, 65-66
Elevation view, 46, 47
Engineers transit, 61
Excavation, 69, 70
Expansive bit, 15
Exterior doors, 211, 216, 217
Exterior trim, 211
Exterior wall finish, 224-230
Exterior wall sheathing, 155, 157, 158

F

Fabrication of homes, 40-43
Fascia, 172
Fasteners, 27
Fastening tools, 16, 17
Fibreboard tile, 259
Finish floor, 138
Fink truss, 193
Firestop, 120, 140
Firmer chisel, 10
Flashing, 206, 208, 209
Flight, of stairs, 272-275
Floor framing, 127-138
Floor joists, truss, 129, 131
        wood, 129-133
Floor plans, 46
Floor sheathing, 137, 138

Floors, concrete, 111
        wood, 119-121
Flush doors, 262
Folding doors, 264
Footings, concrete, 90-92
        wood, 115, 116
Fore plane, 11
Form, systems, 102-104
        ties, 102, 103
Forms, footings, 93-98
        foundation walls, 101-111
        material, 101, 102
Forstner bit, 15
Foundation:
        concrete, 97-99
        masonry, 112-114
        piles, 99-101
        views, 46
Foundations, wood, 114-122, 136
Framed panel forms, 103, 104
Framing:
        assembly, 149, 150
        balloon, 140, 142
        ceiling, 152, 153
        floor, 127-138
        platform, 140, 141
        roof, 168-194
        rough openings, 147-149
        square, 2, 168
        walls, 140-152
Friction piles, 100

G

Gable ends, 173
Gable end roof, 161
Galvanized sheet metal roofing, 201, 206
Gambrel roof, 162
Gang forming, 103
Gauge, butt, 6
        depth, 6
        marking, 6
Girders and beams, 127, 128
Glass doors, sliding, 218
Grade foundation, 98
Grade, level, 65, 66
Gravel stop, 207, 208
Gravity toggle bolt, 255
Grinder, bench, 22
Grinding, 13
Gusset, 191, 192
Gypsum wallboard, 254, 259

H

Hammers, 8
Hand:
    level, 4
    of a door, 218, 219
    plane, 11
    router, 13, 284
    saw, 6
    tools, 1-17
Hardboard, exterior wall finish, 225
    interior wall finish, 256-258
Hatchet, 9
Header, 129, 132, 147
Headroom, stairs, 272, 277, 281
Heat transmission, 231-233
Hinge bound, 214, 215
Hinges, doors, 214, 267
    cabinet doors, 289, 290
Hip rafters, 174-178
    roof, 161
Hollow wall fastener, 254-256
Horizontal sliding windows, 218-220, 222
House headroom, 142, 143
    styles, 31-33
Howe truss, 193

I

Independent footing, 92
Instrument layout, 61
Insulation 233-237
    application of, 238-247
    types of, 233-237
Interior door frames, 262-268
    doors, 262-264
    trim, 267-269
Interior wall finish, 254, 256-258
Intersecting roof, 182-191
Isolation joints, 87

J

Jack plane, 11
Jack rafters:
    Hip jack, 178-181
    Hip-valley cripple jack, 188, 189
    Valley jack, 186, 187
    Valley cripple jack, 188, 189
Jambs, 211-213, 262, 263
Jointer plane, 11
Jointer (power), 21

Joints, concrete, 87
Joist fill construction, 107, 109
Joist hangers, 122, 131, 153
Joists, 129-133

K

Kitchen cupboards, 288-292
    measurements, 292
    planning, 286-288
Knee brace, 152

L

Laminated beams, 128
Layout tools, 1-4
Ledger, 131
Let-in bracing, 152
Level, hand, 4
    transit, 61
Line level, 4
Line of flight, 271, 272
Lintel, 140, 147
Local building code, 55
Locks, 215, 216, 267
Lookouts, 173
Loose fill insulation, 234-236
Lot layout, 59
Low energy home, 33-35, 155-157, 249-252
Low-slope asphalt shingles, 203, 204
Lumber siding, 224

M

Mansard roof, 162
Marking gauge, 6
Masonry veneer, 111
Masonry walls, 112-114
Material symbols, 51-55
Measuring tools, 1-4
Mechanical vibrator, 82
Mechanical views, 48
Metal roof coverings, 201
Metal siding, 225
Metrication, xi-xv
Metric module, xiii-xv
Metric units, xiii-xv
Mitred siding corner joint, 228, 229
Mitre saw, electric, 23
Mobile home, 41
Mouldings, 268, 269
Mortar, 113, 114

Mortice chisel, 10
Multi-level homes, 32

N

National Building Code, 53, 54
Newel post, 271, 272
Nosing, 272

O

One and a half storey home, 32
Open cornice, 196
Open valley, 206
Overhang, 169
Overlay doors, 288, 290
Outside measurement, 148

P

Packaged homes, 42
Panel doors, 262
Panel forms, 103-109
Panelling, 256-259
Parallel or hand
    screw clamps, 16
Parapet wall, 160
Paring chisel, 10
Partitions, 140
Passive solar home, 38-41
Perspective drawings, 44
Pilaster, 92, 110
Piles, 98-100
Pipe clamps, 16
Placing concrete, 81, 82
Planes, 11-13
Plans, 44-48
Plaster, 254
Plate layout, 143
Plates, 140, 143
Platform framing, 129, 130, 140, 141
Plan view, 45, 46
Plumb bob, 5
Plumbing symbols, 54
Plywood, xiv, 103, 114, 137, 157, 256-258
Pocket door, 263
Pocket tape, 1
Portable:
    electric hand drill, 23
    electric hand saw, 23
    electric jig saw, 23
    electric power plane, 24

electric router, 25
electric sander, 25
Portland cement, 75
Powder actuated tools, 26
Power:
    jointer, 21
    screed, 82, 83
    screwdriver, 23
    tools, 17-28
Prefabricated house, 41-43
Prefabricated panel forms, 103, 104
Preserved wood foundation, 114-122, 136

R

Rabbet plane, 13
Rabbetted doors, 288, 290
Radial arm saw, 20
Radiation, 232, 233
Rafters, 163, 168-193
Rafter table, 3, 176
Ready-to-move home, 41
Reciprocating saw, 24
Reinforced concrete, 86, 87
Residential Standards, 55
Reveal, 212, 213, 267, 268
Ribbon, 140, 142
Ridge board, 165, 171, 181, 190, 191
Rigid insulation, 235-237
Rip saw, 7
Rise, stairs, 272, 278
Risers, stairs, 272, 275, 276
Roof:
    flashing, 203, 206, 208, 209
    intersecting, 182-191
    sheathing, 193, 194
    slope, 164
    terms, 163
    triangle, 165
    trusses, 191-193
    types, 160-162
Roofing materials, 201
Roofing nails, 203, 205
Rotary hammer, 23
Rough buck, 106, 109, 110
Rough openings, 148, 149, 213, 221
Rough sill, 147
Router, 25
Router plane, 13
Run, stairs, 272, 279

S

Scales, 51
Schedules, 49
Screeds, 81
Screw anchor, 256
Screwdrivers, 17
Scribing, 2, 257-259
Sectional house, 41
Sectional views, 47
Shakes, 204-207
Sharpening, 13, 14
Sheathing:
    floor, 137, 138
    roof, 193, 194
    walls, 155, 157, 158
Shingles, 201-207, 224, 225
Siding, 224-230
Single-storey home, 31
Sill plate, 109, 113
Site plan, 45
Site planning, 58
Skylights, 221, 222
Slab on grade, 98
Sliding doors, 264
        windows, 220-222
Sledge hammer, 9
Slope, roof, 164
Slump test, 78, 79
Smooth plane, 11
Snub cornice, 197
Soffit, 196, 199
Soil, study of, 68
        testing, 69
Solar homes, 36-40
Span, 129, 165
Spcciality planes, 11
Specifications, 52, 53
Spiral ratchet screwdriver, 17
Split entries, 33
Spring clamp, 16
Spring toggle bolt, 256
Straight edge, 5, 82
Stairs, building, 282-284
        terms used, 271, 272
        types of, 273, 274
Stairwell, 277, 279-282
Starter strip, 202, 204, 228, 229
Steel I beam, 128, 132
Steel tape, 1
Steel trowel, 84
Stepped footing, 91

Stop bound, 214, 215
Storey and a-half home, 32
Strapping, 260
Stringers, 275, 276
Strip footing, 91
Strip shingles, 201
Struts, 171, 172
Stucco, 225
Studs, 140, 142
Subfloor, 136-138
Sump pit, 115, 123
Sub-surface water, 69
Surface water, 69
Suspended ceilings, 259
Symbols, 50

T

Table saw, 18
Tail joists, 132
Thermal insulation, 233-237
Thresholds, 216, 217
Ties, form, 102
Tile roofing, 201
Total rise, roof, 163
Total rise, stairs, 272, 278, 281
Total run, roof, 163
Total run, stairs, 272, 279, 280
Transit-level, 61
Treads, stair, 272
Trim, exterior, 211
Trim, interior, 267-269
Trimmer joist, 132
Trimmer studs, 147
Trimming plane, 13
Troweled finish, 82-84
Truss joist, 129, 131
Truss rafter, 192-193
Try square, 2
Two-storey home, 32

U

Underlay, 137, 138
Unit of line length, 166
Unit of rise, 166
Unit of run, 166

V

Valley flashing, 204, 206

Valley rafters, 184-190
Vapour barrier, 247, 248
Vapour barrier,
    application of, 249, 251
Veneer, brick, 226
Ventilation, 250-252
Vernier scales, 62
Vertical siding, 224
Vibrator, mechanical, 82
Vinyl siding, 226
Voids (concrete), 82

W

Waler, 101, 103, 104
Wallboard, 254, 256-258
Wall:
    bracing, 151, 152
    fasteners, 254-256
    forms, 101-108
    framing, 140-152
        assembly, 149, 150
    pressure treated, 114-122

    sheathing, 155, 157, 158
    studding, 142
Water control, 69, 121
Waterproofing, foundations, 123, 124
Water table, 69
Whetting, 13
Windows, types of, 218-222
Wood chisel, 10
Wood foundation, 114-122, 136
Wood shingles, 204-207, 224, 225
Working drawings, 44-50